PHYSICS RESEARCH AND TECHNOLOGY

SYMMETRY AND ITS BREAKING IN QUANTUM FIELD THEORY

PHYSICS RESEARCH AND TECHNOLOGY

Additional books in this series can be found on Nova's website
under the Series tab.

Additional E-books in this series can be found on Nova's website
under the E-books tab.

PHYSICS RESEARCH AND TECHNOLOGY

SYMMETRY AND ITS BREAKING IN QUANTUM FIELD THEORY

TAKEHISA FUJITA

Nova Science Publishers, Inc.
New York

For permission to use material from this book please contact us:
Telephone 631-231-7269; Fax 631-231-8175
Web Site: http://www.novapublishers.com

Library of Congress Cataloging-in-Publication Data

Fujita, T.
Symmetry and its breaking in quantum field theory / Takehisa Fujita.
p. cm.
Includes index.
ISBN 978-1-60876-106-7 (softcover)
1. Broken symmetry (Physics) 2. Quantum field theory. I. Title.
QC174.17.S9F85 2010
539.7'25--dc22

2010033090

Published by Nova Science Publishers, Inc. † New York

Contents

Preface

Physics is always difficult, though it is extremely interesting. Many times I thought I understood it sufficiently profoundly, but after some time, it turned out that my understanding of physics was far from satisfactory. In particular, field theory has special complexities which may not be common to other fields of research. The symmetry and its breaking are most exotic and sometimes almost mysterious to even those who can normally understand the basic physics in a clear manner.

In this textbook, I focused on presenting a simple and clear picture of the symmetry and its breaking in quantum field theory. For this purpose, I explained physics of elementary field theory of fermions interacting by gauge fields as well as by four body fermion fields. In this respect, the interpretation of the basic field theory is repeatedly done such that physicists including graduate students may understand the essential points of the symmetry breaking in this textbook.

Also, this book is intended for researchers who look for the basic problems in their investigations. In many fields of research, field theory is used as a computational tool. In this regard, I present some elaborate technical tools which are quite useful and sometimes incentive for new ideas in fundamental researches.

In physics, deeper understanding is more important than quicker understanding. In particular, graduate students should realize that, if someone else can understand the basic physics very quickly, then he is most likely a good interpreter of the textbook knowledge. Slow but deep understanding of physics is most important since it should definitely take much time to understand physics in depth. The shortest path of understanding physics is only one of many paths, and interesting physics may well be found in the paths which are far from the shortest one.

Physics must be simple once we understand it all. For example, I believe that QCD can surely describe the strong interaction physics. However, it may well be difficult to justify the perturbative calculation of the interactions between quarks, unless the gauge independence of the quark-quark interactions is guaranteed. In other words, when the unperturbed as well as interaction Hamiltonians are gauge dependent, we should make it sure that any physical quantities evaluated perturbatively are indeed gauge invariant, which seems to be very difficult.

In this textbook, there are quite a few issues which are still debating. I believe that the present understanding of the basic field theory in this textbook must be reasonably good, and as far as physics of the symmetry and its breaking is concerned, it should be the best of all. The spontaneous symmetry breaking of the global symmetry is by now understood in this textbook in terms of a simple physics terminology, and there is nothing mysterious from the standard way of understanding physics. However, it is still not yet settled whether the local gauge symmetry can be broken in terms of Higgs mechanism or not. At least, the gauge fixing for the non-gauge field is physically not at all easy to understand. For this problem, we need a lot to think over in future what should be physical observables in the Higgs mechanism.

This textbook contains a brief description of the lattice field theory even though it is not directly connected to the symmetry breaking physics. Still it may be interesting for readers to understand the basic point of the lattice field theory. For example, the continuum field theory must be richer than the lattice version, and it is most likely true that the lattice field theory can give only limited information on the continuum field theory, particularly when the latter keeps some symmetry while the former does not.

In Appendix, I explain some elementary physics so that readers may grasp the essence of the symmetry breaking phenomena in fermion field theory with little advanced knowledge. In some sense, Appendix can be read in its own interests since it includes non-relativistic quantum mechanics, Dirac equation and Maxwell equation, in addition to the notations which are often used in field theory. At the same time, Appendix contains some new physics interpretation for bosons, Dirac fields and quantization procedure. In particular, I believe that the first quantization of $[x, p_x] = i\hbar$, etc. may well be the result of the Dirac equation in that the Dirac Lagrangian density can be derived from the gauge principle as well as the Maxwell equations without involving the first quantization procedure. In the final chapter of Appendix, I briefly explain the renormalization in QED which is the most successful theory in quantum field theory. The perturbation theory is not the main issue of this textbook, but nevertheless readers may learn the essence of the renormalization scheme in quantum field theory.

The motive force of writing this textbook is initiated by Frank Columbus who understands the importance of the new picture of spontaneous symmetry breaking physics prior to experts and has encouraged me to write it into a textbook form. Indeed, I started to write this book from intensive discussions and hard works with my collaborators on this subject to achieve deeper but simpler understanding of the symmetry and its breaking in quantum field theory.

I should be grateful to all of my collaborators, in particular, Tomoko Asaga, Makoto Hiramoto, Takashi Homma, Seiji Kanemaki, Sachiko Oshima and Hidenori Takahashi for their great contributions to this book. Quite a few physicists and students also helped me a great deal for their critical reading of this manuscript. However, it is trivial to note that any mistakes in this book are entirely due to my carelessness.

To the Second Edition

The revision of this textbook is made mainly because of the following two reasons. Firstly, the first edition contained the wrong description of the path integral formulation. Even though it is normally found in the field theory textbooks, the path integral description in the field theory textbooks is not a correct one, and therefore I had to rewrite it into a correct formulation which was originally presented by Feynman. Secondly, the revision is concerned with the quantum gravity, and fortunately, the Lagrangian density that includes the gravitational interactions with fermions is properly constructed. Therefore, I included quantum gravity in this textbook, and one can now understand the basic physics of quantum gravity with our standard knowledge of quantum field theory, without referring to the space deformation.

In this occasion, I would like to express my sincere gratitude to late Prof. Kazuhiko Nishijima for his many useful comments and encouragements. His continuous supports for our works encouraged me a great deal, and in particular, the discussions of quantum gravity helped me to improve the description of the graviton propagation.

Finally I should like to thank numerous students and physicists for their interesting comments and suggestions to the first edition as well as the draft of the second edition. In particular, I should be grateful to Atsushi Kusaka, Kazuhiro Tsuda, Naohiro Kanda, Hiroshi Kato, Hiroaki Kubo and Yasunori Munakata for their careful reading of the manuscript.

Takehisa Fujita

Chapter 1

Classical Field Theory of Fermions

The world of elementary particles is basically composed of fermions. Quarks, electrons and neutrinos are all fermions. On the other hand, elementary bosons are all gauge bosons, except Higgs particles though unknown at present. Therefore, if one wishes to understand field theory, then it should be the best to first study fermion field theory models.

In this chapter, we discuss the classical field theory in which "classical field" means that the field is not an operator but a c-number function. First, we treat the Schrödinger field and its equation in terms of the non-relativistic field theory model. In this case, the first quantization of $[x_i, p_j] = i\hbar\delta_{ij}$ is already done since we start from the Lagrangian density. In fact, the Lagrange equation leads to the Schrödinger equation or in other words, the Lagrangian density is constructed such that the Schrödinger equation can be derived from the Lagrange equation. The Dirac field is then discussed in terms of the Lagrangian density and the Lagrange equation. We also discuss the electromagnetic fields which interact with the Dirac field. The gauge invariance will be repeatedly discussed in this textbook, and the first introduction is given here. Finally, the field theory models with self-interacting fields are introduced and their Lagrangian density as well as Hamiltonian are described.

In this textbook, the basic parts of elementary physics can be found in Appendix, and in fact, Appendix is prepared such that it can be read in its own interests independently from the main part of the textbook.

Throughout this book, we employ the natural units

$$c = 1, \quad \hbar = 1.$$

This is, of course, due to its simplicity, and one can easily recover the right dimension of any physical quantities by making use of

$$\hbar c = 197 \text{ MeV} \cdot \text{fm}.$$

1.1. Non-relativistic Fields

If one treats a classical field $\psi(\boldsymbol{r})$, it does not matter whether it is a relativistic field or non-relativistic one. The kinematics becomes important when one solves the equation of motion which is relativistic or non-relativistic. If the kinematics is non-relativistic, then the equation of motion that governs the field $\psi(\boldsymbol{r})$ is the Schrödinger equation. Therefore, we should first study the Schrödinger field from the point of view of the classical field theory.

1.1.1. Schrödinger Equation

Electron in classical mechanics is treated as a point particle whose equation of motion is governed by the Newton equation. When electrons are trapped by atoms, then their motions should be described by quantum mechanics. As long as electrons move much slowly in comparison with the velocity of light c, the equation of their motion is governed by the Schrödinger equation. The Schrödinger equation for electron with its mass m in the external field $U(\boldsymbol{r})$ can be written as [102]

$$\left(i\frac{\partial}{\partial t} + \frac{1}{2m}\nabla^2 - U(\boldsymbol{r}) \right) \psi(\boldsymbol{r},t) = 0, \tag{1.1}$$

where $U(\boldsymbol{r})$ is taken to be a real potential. $\psi(\boldsymbol{r},t)$ corresponds to the electron field in atoms, and $|\psi(\boldsymbol{r},t)|^2$ can be interpreted as a probability density of finding the electron at $(\boldsymbol{r},t)$.

Field $\psi(\boldsymbol{r},t)$ Is Complex

The Schrödinger field $\psi(\boldsymbol{r},t)$ should be a complex function, and the complex field just corresponds to one particle state in the classical field theory. This is a well known fact, but below we will see what may happen when we assume *a priori* that the Schrödinger field $\psi(\boldsymbol{r},t)$ should be a real function.

Real Field Condition Is Unphysical

If one imposes the condition that the field $\psi(\boldsymbol{r},t)$ should be real

$$\psi(\boldsymbol{r},t) = \psi^\dagger(\boldsymbol{r},t)$$

then, one sees immediately that the field $\psi(\boldsymbol{r},t)$ becomes time-independent since eq.(1.1) and its complex conjugate equation give the following constraint for a real field $\psi(\boldsymbol{r},t)$

$$\frac{\partial \psi(\boldsymbol{r},t)}{\partial t} = 0.$$

Also, the field $\psi(\boldsymbol{r})$ should satisfy the following equation

$$\left(-\frac{1}{2m}\nabla^2 + U(r) \right) \psi(\boldsymbol{r}) = 0.$$

Since the general solution of eq.(1.1) can be written as

$$\psi(\boldsymbol{r},t) = e^{-iEt}\phi(\boldsymbol{r})$$

the field $\psi(\boldsymbol{r},t)$ may become a real function only if the energy E of the system vanishes. That is, the energy eigenvalue of E is

$$E = 0.$$

Therefore, the real field cannot propagate and should be unphysical. This means that the real field condition of $\psi(\boldsymbol{r},t)$ is physically too strong as a constraint.

1.1.2. Lagrangian Density for Schrödinger Fields

The Lagrangian density which can produce eq.(1.1) is easily found as

$$\mathcal{L} = i\psi^\dagger \frac{\partial \psi}{\partial t} - \frac{1}{2m}\frac{\partial \psi^\dagger}{\partial x_k}\frac{\partial \psi}{\partial x_k} - \psi^\dagger U \psi, \tag{1.2}$$

where the repeated indices of k mean the summation of $k = 1,2,3$ and, in this text, this notation as well as the vector representation are employed depending on the situations. The repeated indices notation is mostly better for the calculation, but for memorizing the expressions or equations, the vector notation has some advantage.

The Lagrangian density of eq.(1.2) is constructed such that the Lagrange equation can reproduce the Schrödinger equation of eq.(1.1). It may also be important to note that the Lagrangian density of eq.(1.2) has a $U(1)$ symmetry, that is, it is invariant under the change of the field ψ as

$$\psi'(x) = e^{i\theta}\psi(x) \longrightarrow \mathcal{L}' = \mathcal{L},$$

where θ is a real constant. This invariance is clearly satisfied, and it is related to the conservation of vector current in terms of Noether's theorem which will be treated in the later chapters and in Appendix A.

Non-hermiticity of Lagrangian Density

At this point, we should discuss the non-hermiticity of the Lagrangian density. As one notices, the Lagrangian density of eq.(1.2) is not hermitian, and therefore some symmetry will be lost. One can build the Lagrangian density which is hermitian by replacing the first term by

$$i\psi^\dagger \frac{\partial \psi}{\partial t} \longrightarrow \left(\frac{i}{2}\psi^\dagger \frac{\partial \psi}{\partial t} - \frac{i}{2}\frac{\partial \psi^\dagger}{\partial t}\psi\right).$$

However, it is a difficult question whether the Lagrangian density must be hermitian or not since it is not an observable. In addition, when one introduces the conjugate fields

$$\Pi_\psi \equiv \frac{\partial \mathcal{L}}{\partial \dot{\psi}}, \quad \Pi_{\psi^\dagger} \equiv \frac{\partial \mathcal{L}}{\partial \dot{\psi}^\dagger}$$

in accordance with the fields ψ and $\psi^\dagger$, then the symmetry between them is lost. However, the conjugate fields themselves are again not observables, and therefore there is no reason that one should keep this symmetry. In any case, one can, of course, work with the symmetric and hermitian Lagrangian density, but physical observables are just the same as eq.(1.2). In this textbook, we employ eq.(1.2) since it is simpler.

1.1.3. Lagrange Equation for Schrödinger Fields

The Lagrange equation for field theory can be obtained by the variational principle of the action S

$$S = \int \mathcal{L}\, dt\, d^3r$$

and the Lagrange equation is derived in Appendix A. Since the field ψ is a complex field, ψ and $\psi^\dagger$ are treated as independent functional variables. The Lagrange equation for the field ψ is given as

$$\partial_\mu \frac{\partial \mathcal{L}}{\partial(\partial_\mu \psi)} \equiv \frac{\partial}{\partial t}\frac{\partial \mathcal{L}}{\partial \dot{\psi}} + \frac{\partial}{\partial x_k}\frac{\partial \mathcal{L}}{\partial(\frac{\partial \psi}{\partial x_k})} = \frac{\partial \mathcal{L}}{\partial \psi}, \tag{1.3a}$$

where the four dimensional derivative

$$\partial_\mu \equiv \left(\frac{\partial}{\partial x_0}, \frac{\partial}{\partial x_1}, \frac{\partial}{\partial x_2}, \frac{\partial}{\partial x_3}\right) = \left(\frac{\partial}{\partial t}, \frac{\partial}{\partial x}, \frac{\partial}{\partial y}, \frac{\partial}{\partial z}\right)$$

is introduced for convenience. Now, the following equations can be easily evaluated

$$\frac{\partial}{\partial t}\frac{\partial \mathcal{L}}{\partial \dot{\psi}} = i\frac{\partial \psi^\dagger}{\partial t},$$

$$\frac{\partial}{\partial x_k}\frac{\partial \mathcal{L}}{\partial(\frac{\partial \psi}{\partial x_k})} = -\frac{1}{2m}\frac{\partial}{\partial x_k}\frac{\partial \psi^\dagger}{\partial x_k},$$

$$\frac{\partial \mathcal{L}}{\partial \psi} = -\psi^\dagger U$$

and therefore one obtains

$$\left(-i\frac{\partial}{\partial t} + \frac{1}{2m}\boldsymbol{\nabla}^2 - U(r)\right)\psi^\dagger(\boldsymbol{r},t) = 0$$

which is just the Schrödinger equation for $\psi^\dagger$ in eq.(1.1).

It should be interesting to calculate the Lagrange equation for the field $\psi^\dagger$,

$$\frac{\partial}{\partial t}\frac{\partial \mathcal{L}}{\partial \dot{\psi}^\dagger} + \frac{\partial}{\partial x_k}\frac{\partial \mathcal{L}}{\partial(\frac{\partial \psi^\dagger}{\partial x_k})} = \frac{\partial \mathcal{L}}{\partial \psi^\dagger}. \tag{1.3b}$$

In this case, one finds

$$\frac{\partial}{\partial t}\frac{\partial \mathcal{L}}{\partial \dot{\psi}^\dagger} = 0,$$

$$\frac{\partial}{\partial x_k}\frac{\partial \mathcal{L}}{\partial(\frac{\partial \psi^\dagger}{\partial x_k})} = -\frac{1}{2m}\frac{\partial}{\partial x_k}\frac{\partial \psi}{\partial x_k},$$

$$\frac{\partial \mathcal{L}}{\partial \psi^\dagger} = i\frac{\partial \psi}{\partial t} - U\psi$$

and therefore one obtains

$$\left(i\frac{\partial}{\partial t} + \frac{1}{2m}\boldsymbol{\nabla}^2 - U(r)\right)\psi(\boldsymbol{r},t) = 0$$

which is just the same equation as eq.(1.1).

Here, we note that the Lagrangian density is not a physical observable and therefore it does not necessarily have to be determined uniquely. It is by now clear that the Lagrangian density eq.(1.2) reproduces a desired Schrödinger equation and thus can be taken as the right Lagrangian density for Schrödinger fields.

1.1.4. Hamiltonian Density for Schrödinger Fields

From the Lagrangian density, one can build the Hamiltonian density $\mathcal{H}$ which is the energy density of the field $\psi(\boldsymbol{r},t)$. The Hamiltonian density $\mathcal{H}$ is best constructed from the energy momentum tensor $\mathcal{T}^{\mu\nu}$

$$\mathcal{T}^{\mu\nu} \equiv \frac{\partial \mathcal{L}}{\partial(\partial_\mu \psi)}\partial^\nu \psi + \frac{\partial \mathcal{L}}{\partial(\partial_\mu \psi^\dagger)}\partial^\nu \psi^\dagger - \mathcal{L} g^{\mu\nu}$$

which will be derived in eq.(2.32) in Chapter 2. The energy momentum tensor $\mathcal{T}^{\mu\nu}$ satisfies the following equation of conservation law

$$\partial_\mu \mathcal{T}^{\mu\nu} = 0$$

due to the invariance of the Lagrangian density under the translation. Therefore, the conserved charge associated with the $\mathcal{T}^{0\nu}$

$$\mathcal{Q}^\nu = \int \mathcal{T}^{0\nu}\, d^3r$$

should be a conserved quantity. Thus, it is natural that one defines the Hamiltonian in terms of the $\mathcal{Q}^0$.

Hamiltonian Density from Energy Momentum Tensor

The Hamiltonian density $\mathcal{H}$ is defined as

$$\mathcal{H} \equiv \mathcal{T}^{00} = \frac{\partial \mathcal{L}}{\partial \dot{\psi}}\dot{\psi} + \frac{\partial \mathcal{L}}{\partial \dot{\psi}^\dagger}\dot{\psi}^\dagger - \mathcal{L}. \tag{1.4a}$$

Therefore, introducing the conjugate fields Π_ψ and $\Pi_{\psi^\dagger}$ by

$$\Pi_\psi \equiv \frac{\partial \mathcal{L}}{\partial \dot{\psi}} = i\psi^\dagger, \quad \Pi_{\psi^\dagger} \equiv \frac{\partial \mathcal{L}}{\partial \dot{\psi}^\dagger} = 0$$

one can write the Hamiltonian density as

$$\mathcal{H} = \Pi_\psi \dot{\psi} + \Pi_{\psi^\dagger}\dot{\psi}^\dagger - \mathcal{L} = \frac{1}{2m}\boldsymbol{\nabla}\psi^\dagger \cdot \boldsymbol{\nabla}\psi + \psi^\dagger U \psi. \tag{1.4b}$$

1.1.5. Hamiltonian for Schrödinger Fields

The Hamiltonian for the Schrödinger field is obtained by integrating the Hamiltonian density over all space

$$H \equiv \int \mathcal{H}\, d^3r = \int \left[\frac{1}{2m}\boldsymbol{\nabla}\psi^\dagger \cdot \boldsymbol{\nabla}\psi + \psi^\dagger U \psi\right] d^3r. \tag{1.4c}$$

By employing the Gauss theorem

$$\int_V \boldsymbol{\nabla}\cdot(\psi^\dagger \boldsymbol{\nabla}\psi)\, d^3r = \int_S (\psi^\dagger \boldsymbol{\nabla}_n \psi)\, dS_n$$

one can rewrite eq.(1.4c)

$$H = \int \left[-\frac{1}{2m} \psi^\dagger \nabla^2 \psi + \psi^\dagger U \psi \right] d^3r, \tag{1.4d}$$

where the following identity is employed

$$\nabla \cdot (\psi^\dagger \nabla \psi) = \nabla \psi^\dagger \cdot \nabla \psi + \psi^\dagger \nabla^2 \psi.$$

In addition, the surface integral term is neglected since it should vanish at the surface of sphere at infinity.

Now, it may be interesting to note that the Hamiltonian in eq.(1.4d) by itself does not give us much information on the dynamics. As long as we stay in the classical field theory, then the dynamics can be obtained from the equation of motion, that is, the Schrödinger equation. The static Schrödinger equation can be derived from the variational principle of the Hamiltonian with respect to ψ, and this treatment is given in Appendix A.

The Hamiltonian of eq.(1.4c) becomes important when the field ψ is quantized, that is, the field ψ is assumed to be written in terms of the annihilation operator $a_{\boldsymbol{k}}$ as discussed in Chapter 3. In this case, the Schrödinger field becomes an operator and therefore the Hamiltonian as well. This means that one has to prepare the Fock state on which the Hamiltonian can operate, and if one solves the eigenvalue equation for the Hamiltonian, then one can obtain the energy eigenvalue of the Hamiltonian corresponding to the Fock state.

However, the quantization of the Schrödinger field is not needed in the normal circumstances. The field quantization is necessary for the relativistic fields which contain negative energy solutions, and it becomes important when one wishes to treat the quantum fluctuation of the fields which corresponds to the creation and annihilation of particles.

1.1.6. Conservation of Vector Current

From the Schrödinger equation, one can derive the current conservation

$$\frac{\partial \rho}{\partial t} + \nabla \cdot \boldsymbol{j} = 0,$$

where ρ and $\boldsymbol{j}$ are defined as

$$\rho = \psi^\dagger \psi, \quad \boldsymbol{j} = \frac{i}{2m} \left[(\nabla \psi^\dagger) \psi - \psi^\dagger \nabla \psi \right].$$

This continuity equation of the vector current can also be derived as Noether's theorem from the Lagrangian density of eq.(1.2) which is invariant under the global gauge transformation

$$\psi' = e^{i\alpha} \psi.$$

As treated in Appendix A, the Noether current is written as

$$j^\mu \equiv -i \left[\frac{\partial \mathcal{L}}{\partial(\partial_\mu \psi)} \psi - \frac{\partial \mathcal{L}}{\partial(\partial_\mu \psi^\dagger)} \psi^\dagger \right], \quad \text{with } j^\mu = (\rho, \boldsymbol{j})$$

which just gives the above current density ρ and $\boldsymbol{j}$ when one employs the Lagrangian density of eq.(1.2).

It may be interesting to observe that the Lagrange equation, energy momentum tensor and the current conservation are all written in a relativistically covariant fashion when the properties of the Schrödinger field are derived. That is, apart from the shape of the Lagrangian density of the Schrödinger field, all the treatments are just the same as the relativistic description.

1.2. Dirac Fields

Electron in hydrogen atom moves much slowly compared with the velocity of light c. However, if one considers a hydrogen-like $^{209}_{83}$Bi atom where $Z = 83$, for example, then the motion of electron becomes relativistic since its velocity v can be given as

$$\frac{v}{c} \sim (Z\alpha)^2 \sim \left(\frac{83}{137}\right)^2 \sim 0.37$$

which is already comparable with c.

In this case, one should employ the relativistic kinematics, and therefore the Schrödinger equation should be replaced by the Dirac equation which is obtained by a natural extension of the relativistic kinematics. However, the Dirac equation contains new properties which are essentially different from the Schrödinger equation, apart from the kinematics. They have negative energy solutions and spin degrees of freedom. Both properties are very important in physics and will be repeatedly discussed in this textbook.

1.2.1. Dirac Equation for Free Fermion

The Dirac equation for free fermion with its mass m is written as [25, 26]

$$\left(i\frac{\partial}{\partial t} + i\boldsymbol{\nabla}\cdot\boldsymbol{\alpha} - m\beta\right)\psi(\boldsymbol{r},t) = 0, \tag{1.5}$$

where ψ has four components

$$\psi = \begin{pmatrix} \psi_1 \\ \psi_2 \\ \psi_3 \\ \psi_4 \end{pmatrix}.$$

$\boldsymbol{\alpha}$ and β denote the Dirac matrices and can be explicitly written in the Dirac representation as

$$\boldsymbol{\alpha} = \begin{pmatrix} 0 & \boldsymbol{\sigma} \\ \boldsymbol{\sigma} & 0 \end{pmatrix}, \quad \beta = \begin{pmatrix} 1 & 0 \\ 0 & -1 \end{pmatrix},$$

where $\boldsymbol{\sigma}$ denotes the Pauli matrix.

The derivation of the Dirac equation and its application to hydrogen atom are given in Appendix D. One can learn from the procedure of deriving the Dirac equation that the number of components of the electron fields is important, and it is properly obtained in the Dirac equation. That is, among the four components of the field ψ, two degrees of freedom should correspond to the positive and negative energy solutions and another two degrees should correspond to the spin with $s=\frac{1}{2}$. It is also important to note that the factorization procedure indicates that the four component spinor is the minimum number of fields which can take into account the negative energy degree of freedom in a proper way.

Eq.(1.5) can be rewritten in terms of the wave function components by multiplying β from the left hand side

$$(i\partial_\mu\gamma^\mu - m)_{ij}\psi_j = 0 \quad \text{for } i = 1,2,3,4, \tag{1.6}$$

where the repeated indices of j indicate the summation of $j=1,2,3,4$. Here, gamma matrices

$$\gamma_\mu = (\gamma_0, \boldsymbol{\gamma}) \equiv (\beta, \beta\boldsymbol{\alpha})$$

are introduced, and the repeated indices of Greek letters μ indicate the summation of $\mu = 0,1,2,3$ as defined in Appendix A. The expression of eq.(1.6) is called *covariant* since the Lorentz invariance of eq.(1.6) is manifest. It is indeed written in terms of the Lorentz scalars, but, of course there is no deep physical meaning in covariance.

1.2.2. Lagrangian Density for Free Dirac Fields

The Lagrangian density for free Dirac fermions can be constructed as

$$\mathcal{L} = \psi_i^\dagger \left[\gamma_0(i\partial_\mu\gamma^\mu - m)\right]_{ij}\psi_j = \bar{\psi}(i\partial_\mu\gamma^\mu - m)\psi, \tag{1.7}$$

where $\bar{\psi}$ is defined as

$$\bar{\psi} \equiv \psi^\dagger\gamma_0.$$

This Lagrangian density is just constructed so as to reproduce the Dirac equation of (1.6) from the Lagrange equation. It should be important to realize that the Lagrangian density of eq.(1.7) is invariant under the Lorentz transformation since it is a Lorentz scalar. This is clear since the Lagrangian density should not depend on the system one chooses.

Non-hermiticity of Lagrangian Density

This Lagrangian density is not hermitian, and it is easy to construct a hermitian Lagrangian density. However, as we discussed in the context of Schrödinger field, there is no strong reason that one should take the hermitian Lagrangian density since proper physical equations can be obtained from eq.(1.7).

1.2.3. Lagrange Equation for Free Dirac Fields

The Lagrange equation for $\psi_i^\dagger$ is given as

$$\partial_\mu \frac{\partial\mathcal{L}}{\partial(\partial_\mu\psi_i^\dagger)} \equiv \frac{\partial}{\partial t}\frac{\partial\mathcal{L}}{\partial\dot{\psi}_i{}^\dagger} + \frac{\partial}{\partial x_k}\frac{\partial\mathcal{L}}{\partial\left(\frac{\partial\psi_i^\dagger}{\partial x_k}\right)} = \frac{\partial\mathcal{L}}{\partial\psi_i^\dagger} \tag{1.8}$$

and one can easily calculate the following equations

$$\frac{\partial}{\partial t}\frac{\partial\mathcal{L}}{\partial\dot{\psi}_i{}^\dagger} = 0,$$

$$\frac{\partial}{\partial x_k} \frac{\partial \mathcal{L}}{\partial(\frac{\partial \psi_i^\dagger}{\partial x_k})} = 0,$$

$$\frac{\partial \mathcal{L}}{\partial \psi_i^\dagger} = [\gamma_0(i\partial_\mu \gamma^\mu - m)]_{ij} \psi_j$$

and thus, this leads to the following equation

$$[\gamma_0(i\partial_\mu \gamma^\mu - m)]_{ij} \psi_j = 0$$

which is just eq.(1.6). Here, it should be noted that the ψ_i and $\psi_i^\dagger$ are independent functional variables, and the functional derivative with respect to ψ_i or $\psi_i^\dagger$ gives the same equation of motion.

1.2.4. Plane Wave Solutions of Free Dirac Equation

The free Dirac equation of eq.(1.5) can be solved exactly, and it has plane wave solutions. A simple way to solve eq.(1.5) can be shown as follows. First, one writes the wave function ψ in the following shape

$$\psi_s(\boldsymbol{r},t) = \begin{pmatrix} \zeta_1 \\ \zeta_2 \end{pmatrix} \frac{1}{\sqrt{V}} e^{-iEt+i\boldsymbol{p}\cdot\boldsymbol{r}}, \tag{1.9}$$

where ζ_1 and ζ_2 are two component spinors

$$\zeta_1 = \begin{pmatrix} n_1 \\ n_2 \end{pmatrix}, \quad \zeta_2 = \begin{pmatrix} n_3 \\ n_4 \end{pmatrix}.$$

In this case, eq.(1.5) becomes

$$\begin{pmatrix} -m-E & \boldsymbol{\sigma}\cdot\boldsymbol{p} \\ \boldsymbol{\sigma}\cdot\boldsymbol{p} & m-E \end{pmatrix} \begin{pmatrix} \zeta_1 \\ \zeta_2 \end{pmatrix} = 0 \tag{1.10}$$

which leads to

$$E^2 = m^2 + \boldsymbol{p}^2.$$

This equation has the following two solutions.

Positive Energy Solution ($E_p = \sqrt{\boldsymbol{p}^2 + m^2}$)

In this case, the wave function becomes

$$\psi_s^{(+)}(\boldsymbol{r},t) = \frac{1}{\sqrt{V}} u_{\boldsymbol{p}}^{(s)} e^{-iE_p t+i\boldsymbol{p}\cdot\boldsymbol{r}}, \tag{1.11a}$$

$$u_{\boldsymbol{p}}^{(s)} = \sqrt{\frac{E_p+m}{2E_p}} \begin{pmatrix} \chi_s \\ \dfrac{\boldsymbol{\sigma}\cdot\boldsymbol{p}}{E_p+m}\chi_s \end{pmatrix}, \quad \text{with } s = \pm\frac{1}{2}, \tag{1.11b}$$

where χ_s denotes the spin wave function and is written as

$$\chi_{\frac{1}{2}} = \begin{pmatrix} 1 \\ 0 \end{pmatrix}, \quad \chi_{-\frac{1}{2}} = \begin{pmatrix} 0 \\ 1 \end{pmatrix}.$$

Negative Energy Solution ($E_p = -\sqrt{p^2+m^2}$)

In this case, the wave function becomes

$$\psi_s^{(-)}(\boldsymbol{r},t) = \frac{1}{\sqrt{V}} v_{\boldsymbol{p}}^{(s)} e^{-iE_p t + i\boldsymbol{p}\cdot\boldsymbol{r}}, \tag{1.12a}$$

$$v_{\boldsymbol{p}}^{(s)} = \sqrt{\frac{|E_p|+m}{2|E_p|}} \begin{pmatrix} -\dfrac{\boldsymbol{\sigma}\cdot\boldsymbol{p}}{|E_p|+m}\chi_s \\ \chi_s \end{pmatrix}. \tag{1.12b}$$

Some Properties of Spinor

The spinor wave function $u_{\boldsymbol{p}}^{(s)}$ and $v_{\boldsymbol{p}}^{(s)}$ are normalized according to

$$u_{\boldsymbol{p}}^{(s)\dagger} u_{\boldsymbol{p}}^{(s)} = 1,$$

$$v_{\boldsymbol{p}}^{(s)\dagger} v_{\boldsymbol{p}}^{(s)} = 1.$$

Further, they satisfy the following equations when the spin is summed over

$$\sum_{s=1}^{2} u_{\boldsymbol{p}}^{(s)} \bar{u}_{\boldsymbol{p}}^{(s)} = \frac{p_\mu \gamma^\mu + m}{2E_p}, \tag{1.13a}$$

$$\sum_{s=1}^{2} v_{\boldsymbol{p}}^{(s)} \bar{v}_{\boldsymbol{p}}^{(s)} = \frac{p_\mu \gamma^\mu + m}{2E_p}. \tag{1.13b}$$

1.2.5. Quantization in Box with Periodic Boundary Conditions

In field theory, one often puts the theory into the box with its volume $V = L^3$ and requires that the wave function should satisfy the periodic boundary conditions (PBC). This is mainly because the free field solutions are taken as the basis states, and in this case, one can only calculate physical observables if one works in the box. It is clear that the free field can be defined well only if it is confined in the box.

Since the wave function $\psi_s(\boldsymbol{r},t)$ for a free particle in the box should be proportional to

$$\psi_s(\boldsymbol{r},t) \simeq \begin{pmatrix} \zeta_1 \\ \zeta_2 \end{pmatrix} \frac{1}{\sqrt{V}} e^{-iEt + i\boldsymbol{p}\cdot\boldsymbol{r}}$$

the PBC equations become

$$e^{ip_x x} = e^{ip_x(x+L)}, \quad e^{ip_y y} = e^{ip_y(y+L)}, \quad e^{ip_z z} = e^{ip_z(z+L)}. \tag{1.14a}$$

Therefore, one obtains the constraints on the momentum p_k as

$$p_x = \frac{2\pi}{L} n_x, \quad p_y = \frac{2\pi}{L} n_y, \quad p_z = \frac{2\pi}{L} n_z, \quad n_k = 0, \pm 1, \pm 2, \dots. \tag{1.14b}$$

In this case, the number of states N in the large L limit becomes

$$N = \sum_{n_x, n_y, n_z} \sum_{s} = 2 \frac{L^3}{(2\pi)^3} \int d^3 p, \tag{1.15}$$

where a factor of two comes from the spin degree of freedom.

1.2.6. Hamiltonian Density for Free Dirac Fermion

The Hamiltonian density for free fermion can be constructed from the energy momentum tensor $\mathcal{T}^{\mu\nu}$

$$\mathcal{T}^{\mu\nu} \equiv \sum_i \left(\frac{\partial \mathcal{L}}{\partial(\partial_\mu \psi_i)} \partial^\nu \psi_i + \frac{\partial \mathcal{L}}{\partial(\partial_\mu \psi_i^\dagger)} \partial^\nu \psi_i^\dagger \right) - \mathcal{L} g^{\mu\nu}$$

which will be treated in eq.(A.12.3) of Appendix A.

Hamiltonian Density from Energy Momentum Tensor

Now, one defines the Hamiltonian density $\mathcal{H}$ as

$$\mathcal{H} \equiv \mathcal{T}^{00} = \sum_i \left(\frac{\partial \mathcal{L}}{\partial \dot{\psi}_i} \dot{\psi}_i + \frac{\partial \mathcal{L}}{\partial \dot{\psi}_i^\dagger} \dot{\psi}_i^\dagger \right) - \mathcal{L}. \tag{1.16}$$

Since the Lagrangian density of free fermion is given in eq.(1.7) and is rewritten as

$$\mathcal{L} = i\psi_i^\dagger \dot{\psi}_i + \psi_i^\dagger \left[i\gamma_0 \boldsymbol{\gamma} \cdot \boldsymbol{\nabla} - m\gamma_0 \right]_{ij} \psi_j$$

one can introduce the conjugate fields Π_{ψ_i} and $\Pi_{\psi_i^\dagger}$, and calculate them

$$\Pi_{\psi_i} \equiv \frac{\partial \mathcal{L}}{\partial \dot{\psi}_i} = i\psi_i^\dagger, \quad \Pi_{\psi_i^\dagger} = 0. \tag{1.17}$$

In this case, the Hamiltonian density becomes

$$\mathcal{H} = \sum_i \left(\Pi_{\psi_i} \dot{\psi}_i + \Pi_{\psi_i^\dagger} \dot{\psi}_i^\dagger \right) - \mathcal{L} = \bar{\psi}_i \left[-i\boldsymbol{\gamma} \cdot \boldsymbol{\nabla} + m \right]_{ij} \psi_j = \bar{\psi} \left[-i\boldsymbol{\gamma} \cdot \boldsymbol{\nabla} + m \right] \psi. \tag{1.18}$$

1.2.7. Hamiltonian for Free Dirac Fermion

The Hamiltonian for free fermion fields is obtained by integrating the Hamiltonian density over all space

$$H = \int \mathcal{H}\, d^3r = \int \bar{\psi} \left[-i\boldsymbol{\gamma} \cdot \boldsymbol{\nabla} + m \right] \psi\, d^3r. \tag{1.19}$$

As we discussed in the Schrödinger field, the Hamiltonian itself cannot give us much information on the dynamics. One can learn some properties of the system described by the Hamiltonian, but one cannot obtain any dynamical information of the system from the Hamiltonian. In order to calculate the dynamics of the system in the classical field theory model, one has to solve the equation of motions which are obtained from the Lagrange equations for fields.

When one wishes to consider the fluctuations of the fields or, in other words, creations of particles and anti-particles, then one should quantize the fields. In this case, the Hamiltonian becomes an operator. Therefore, one has to prepare the Fock states on which the Hamiltonian can operate. Most of the difficulties of the field theory models should be to find the vacuum of the system.

1.2.8. Conservation of Vector Current

The Lagrangian density of the Dirac field has a global gauge invariance,

$$\psi' = e^{i\alpha}\psi \longrightarrow \mathcal{L}' = \mathcal{L}$$

and therefore there is a Noether current associated with the symmetry. As treated in Appendix A, the Noether current is written as

$$j^\mu \equiv -i\left[\frac{\partial\mathcal{L}}{\partial(\partial_\mu\psi)}\psi - \frac{\partial\mathcal{L}}{\partial(\partial_\mu\psi^\dagger)}\psi^\dagger\right]$$

and therefore the vector current j_μ becomes

$$j^\mu = \bar{\psi}\gamma^\mu\psi.$$

Due to the global gauge invariance of the Lagrangian density, the vector current j_μ satisfies the continuity equation

$$\partial_\mu j^\mu = 0.$$

1.3. Electron and Electromagnetic Fields

The main part of the physical world is governed by the interaction between electrons and electromagnetic fields. Therefore, the Dirac equation, the Maxwell equation and their interactions are most important to understand the basic physics in many fundamental researches.

1.3.1. Lagrangian Density

When electron interacts with electromagnetic fields, the Lagrangian density becomes

$$\mathcal{L} = \bar{\psi}\left(i\partial_\mu\gamma^\mu - gA_\mu\gamma^\mu - m\right)\psi - \frac{1}{4}F_{\mu\nu}F^{\mu\nu}, \tag{1.20}$$

where $F_{\mu\nu}$ denotes the field strength and is given as

$$F_{\mu\nu} = \partial_\mu A_\nu - \partial_\nu A_\mu.$$

A^μ denotes the gauge field with

$$A^\mu = (A_0, \boldsymbol{A}),$$

where A_0 and $\boldsymbol{A}$ are the scalar and vector potentials, respectively. g denotes the gauge coupling constant, and in the classical electromagnetism, it corresponds to the electric charge e.

In the four dimensional field theory of QED, the coupling constant g is dimensionless, and therefore it is renormalizable in the perturbation calculation. In the two dimensional case, the coupling constant g has a mass dimension, and thus it is called *super-renormalizable*. In this case, there appear no infinities from the momentum integral in the perturbative calculations, and therefore one does not have to renormalize the coupling constant.

1.3.2. Gauge Invariance

The Lagrangian density of eq.(1.20) has an interesting feature. The free fermion Lagrangian density part

$$\bar{\psi}(i\partial_\mu\gamma^\mu - m)\psi$$

is just the same as free Dirac Lagrangian density, and the last term in eq.(1.20)

$$-\frac{1}{4}F_{\mu\nu}F^{\mu\nu}$$

corresponds to the field energy term of the electromagnetic fields. The important point is that the shape of the interaction term

$$-g\bar{\psi}A_\mu\gamma^\mu\psi$$

can be determined by the requirement of the invariance under the local gauge transformation.

Local Gauge Transformation

We consider the following local gauge transformation

$$\psi' = e^{-ig\chi}\psi, \quad A'_\mu = A_\mu + \partial_\mu\chi, \tag{1.21}$$

where χ is an arbitrary real function of space and time, that is, $\chi(\boldsymbol{r},t)$ which is therefore called *local*. It is easy to prove that the shape of the field energy term of the electromagnetic fields does not change under the local gauge transformation of eq.(1.21)

$$F'_{\mu\nu} = \partial_\mu A'_\nu - \partial_\nu A'_\mu = \partial_\mu(A_\nu + \partial_\nu\chi) - \partial_\nu(A_\mu + \partial_\mu\chi) = F_{\mu\nu}.$$

In addition, one can easily prove that the Lagrangian density of

$$\bar{\psi}(i\partial_\mu\gamma^\mu - gA_\mu\gamma^\mu - m)\psi$$

does not change its shape under the local gauge transformation of eq.(1.21). That is,

$$\begin{aligned}
&\bar{\psi}'(i\partial_\mu\gamma^\mu - gA'_\mu\gamma^\mu - m)\psi' \\
&= \bar{\psi}e^{-ig\chi}e^{ig\chi}\left(i\partial_\mu\gamma^\mu + g\partial_\mu\chi\gamma^\mu - gA_\mu\gamma^\mu - g\partial_\mu\chi\gamma^\mu - m\right)\psi \\
&= \bar{\psi}\left(i\partial_\mu\gamma^\mu - gA_\mu\gamma^\mu - m\right)\psi.
\end{aligned} \tag{1.22}$$

Therefore, a new Lagrangian density $\mathcal{L}'$ becomes equal to the original one $\mathcal{L}$

$$\mathcal{L}' = \bar{\psi}'\left(i\partial_\mu\gamma^\mu - gA'_\mu\gamma^\mu - m\right)\psi' - \frac{1}{4}F'_{\mu\nu}F'^{\mu\nu} = \mathcal{L}.$$

The invariance of the Lagrangian density under the local gauge transformation determines the shape of the interaction between electron and electromagnetic fields. This is surprising, but it is, in a sense, the same as the classical mechanics as discussed in AppendixE. In this respect, it is interesting to realize that the gauge invariance that arises from the redundancy of the vector potential in solving the Maxwell equations plays an important role for determining the shape of the fundamental interactions.

1.3.3. Lagrange Equation for Dirac Field

The Dirac equation with the electromagnetic interaction can be easily obtained from the Lagrange equation for ψ

$$(i\partial_\mu \gamma^\mu - gA_\mu \gamma^\mu - m)\psi = 0. \tag{1.23}$$

This is the Dirac equation for the hydrogen atom when the potential is static, that is

$$\boldsymbol{A} = 0$$

and

$$gA_0 = -\frac{Ze^2}{r},$$

where we put $g = e$ with e the electric charge.

1.3.4. Lagrange Equation for Gauge Field

The Lagrange equation for the gauge field A_ν is written as

$$\partial_\mu \frac{\partial \mathcal{L}}{\partial(\partial_\mu A_\nu)} = \frac{\partial \mathcal{L}}{\partial A_\nu}.$$

Since one can easily calculate

$$\frac{\partial \mathcal{L}}{\partial A_\nu} = -g\bar{\psi}\gamma^\nu \psi,$$

$$\partial_\mu \frac{\partial \mathcal{L}}{\partial(\partial_\mu A_\nu)} = -\frac{1}{2}\partial_\mu (\partial^\mu A^\nu - \partial^\nu A^\mu) \times 2 = -\partial_\mu F^{\mu\nu}$$

one obtains

$$\partial_\mu F^{\mu\nu} = g\bar{\psi}\gamma^\nu \psi = gj^\nu, \tag{1.24}$$

where the current density j^ν is defined as

$$j^\nu = \bar{\psi}\gamma^\nu \psi = (\bar{\psi}\gamma^0 \psi, \bar{\psi}\boldsymbol{\gamma}\psi). \tag{1.25}$$

Eq.(1.24) is the Maxwell equation, and more explicitly, one can evaluate eq.(1.24)

$$[\nu = 0] \longrightarrow \frac{\partial F^{k0}}{\partial x_k} = \frac{\partial E_k}{\partial x_k} = \nabla \cdot \boldsymbol{E} = gj_0, \tag{1.26a}$$

$$[\nu = k] \longrightarrow \frac{\partial F^{0k}}{\partial t} + \frac{\partial F^{ik}}{\partial x_i} = -\dot{E}_k + \varepsilon_{kij}\frac{\partial B_j}{\partial x_i} = -\dot{E}_k + (\nabla \times \boldsymbol{B})_k = gj_k \tag{1.26b}$$

which are just the Maxwell equations. It is of course easy to see that no magnetic monopole

$$\nabla \cdot \boldsymbol{B} = 0$$

and Faraday's law

$$\nabla \times \boldsymbol{E} = -\frac{\partial \boldsymbol{B}}{\partial t}$$

are automatically satisfied in terms of the vector potential A_μ since

$$\boldsymbol{B} = \nabla \times \boldsymbol{A} \Longrightarrow \nabla \cdot \boldsymbol{B} = \nabla \cdot \nabla \times \boldsymbol{A} = \nabla \times \nabla \cdot \boldsymbol{A} = 0,$$

$$\boldsymbol{E} = -\frac{\partial \boldsymbol{A}}{\partial t} - \nabla A_0 \Longrightarrow \nabla \times \boldsymbol{E} = \nabla \times \left(-\frac{\partial \boldsymbol{A}}{\partial t} - \nabla A_0\right) = -\frac{\partial \boldsymbol{B}}{\partial t}.$$

1.3.5. Hamiltonian Density for Fermions with Electromagnetic Field

Now, one can construct the Hamiltonian density of fermion with electromagnetic field. The Hamiltonian density $\mathcal{H}$ can be defined by the energy momentum tensor $\mathcal{T}^{\mu\nu}$ [eq.(A.12.3)] as

$$\mathcal{H} \equiv \mathcal{T}^{00} = \sum_i \left(\frac{\partial \mathcal{L}}{\partial \dot{\psi}_i} \dot{\psi}_i + \frac{\partial \mathcal{L}}{\partial \dot{\psi}_i^{\dagger}} \dot{\psi}_i^{\dagger} \right) + \sum_k \left(\frac{\partial \mathcal{L}}{\partial \dot{A}_k} \dot{A}_k \right) - \mathcal{L}$$

since $\mathcal{T}^{0\nu}$ is a conserved quantity. By introducing the conjugate fields Π_{ψ_i}, $\Pi_{\psi_i^{\dagger}}$ and Π_{A_k} as

$$\Pi_{\psi_i} \equiv \frac{\partial \mathcal{L}}{\partial \dot{\psi}_i}, \quad \Pi_{\psi_i^{\dagger}} \equiv \frac{\partial \mathcal{L}}{\partial \dot{\psi}_i^{\dagger}}, \quad \Pi_{A_k} = \frac{\partial \mathcal{L}}{\partial \dot{A}_k}$$

one can rewrite the Hamiltonian density as

$$\mathcal{H} = \sum_i \left(\Pi_{\psi_i} \dot{\psi}_i + \Pi_{\psi_i^{\dagger}} \dot{\psi}_i^{\dagger} \right) + \sum_k \Pi_{A_k} \dot{A}_k - \mathcal{L}. \tag{1.27}$$

The conjugate fields Π_{ψ_i}, $\Pi_{\psi_i^{\dagger}}$ and Π_{A_k} can be calculated by employing the Lagrangian density of eq.(1.20)

$$\Pi_{\psi_i} = \frac{\partial \mathcal{L}}{\partial \dot{\psi}_i} = i\psi_i^{\dagger}, \quad \Pi_{\psi_i^{\dagger}} = 0, \quad \Pi_{A_k} = \dot{A}_k + \frac{\partial A_0}{\partial x_k} = -E_k.$$

It should be noted that there is no corresponding conjugate field for A_0 in the Hamiltonian density, and thus there is no kinetic energy term present for A_0. Now, the Hamiltonian density can be calculated as

$$\mathcal{H} = \bar{\psi} \left[-i\gamma_k \frac{\partial}{\partial x_k} + m + gA_\mu \gamma^\mu \right] \psi$$

$$+ \frac{1}{2} \left[\dot{A}_k^{\,2} - \left(\frac{\partial A_0}{\partial x_k} \right)^2 + \left(\frac{\partial A_k}{\partial x_j} \frac{\partial A_k}{\partial x_j} - \frac{\partial A_k}{\partial x_j} \frac{\partial A_j}{\partial x_k} \right) \right]. \tag{1.28a}$$

Eq.(1.28a) can be written in a familiar form

$$\mathcal{H} = \bar{\psi}(-i\boldsymbol{\gamma}\cdot\boldsymbol{\nabla} + m)\psi - g\boldsymbol{j}\cdot\boldsymbol{A} + gj_0 A_0 + \frac{1}{2}\left[\dot{\boldsymbol{A}}^2 - (\boldsymbol{\nabla} A_0)^2 + \boldsymbol{B}^2 \right]. \tag{1.28b}$$

1.3.6. Hamiltonian for Fermions with Electromagnetic Field

The Hamiltonian can be obtained by integrating the Hamiltonian density over all space

$$H = \int \left[\bar{\psi}(-i\boldsymbol{\gamma}\cdot\boldsymbol{\nabla} + m)\psi - g\boldsymbol{j}\cdot\boldsymbol{A} + gj_0 A_0 + \frac{1}{2}\left(\dot{\boldsymbol{A}}^2 - (\boldsymbol{\nabla} A_0)^2 + \boldsymbol{B}^2 \right) \right] d^3r. \tag{1.28c}$$

Now, one makes use of the equation of motion

$$\boldsymbol{\nabla}\cdot\boldsymbol{E} = gj_0$$

in order to rewrite the A_0 in terms of the fermion current density j_0. Since there is a gauge freedom left and one should fix it to avoid the redundancy of the field variables, one may take a Coulomb gauge, for example

$$\nabla \cdot \boldsymbol{A} = 0. \tag{1.29}$$

In this case, the equation of motion for the gauge field A_0 becomes

$$\nabla^2 A_0 = -g j_0 \tag{1.30}$$

which is just a constraint. This is not an equation of motion any more since it does not depend on time. This constraint can be easily solved, and one obtains

$$A_0(r) = \frac{g}{4\pi} \int \frac{j_0(\boldsymbol{r}')\, d^3r'}{|\boldsymbol{r}' - \boldsymbol{r}|}. \tag{1.31}$$

Now, one can make use of the following equation

$$\frac{1}{2} \int (\nabla A_0)^2 \, d^3r = -\frac{1}{2} \int (\nabla^2 A_0) A_0 \, d^3r = \frac{g^2}{8\pi} \int \frac{j_0(\boldsymbol{r}') j_0(\boldsymbol{r})\, d^3r\, d^3r'}{|\boldsymbol{r}' - \boldsymbol{r}|}, \tag{1.32}$$

where the surface integrals are set to zero. Also, $\boldsymbol{E}_T$ is introduced which denotes the transverse electric field

$$\boldsymbol{E}_T = -\dot{\boldsymbol{A}}$$

and it satisfies

$$\nabla \cdot \boldsymbol{E}_T = 0.$$

Therefore, the Hamiltonian of fermions with electromagnetic fields becomes

$$H = \int \{ \bar{\psi} (-i\boldsymbol{\gamma} \cdot \nabla + m) \psi - g \boldsymbol{j} \cdot \boldsymbol{A} \} \, d^3r$$

$$+ \frac{g^2}{8\pi} \int \frac{j_0(\boldsymbol{r}') j_0(\boldsymbol{r})\, d^3r\, d^3r'}{|\boldsymbol{r}' - \boldsymbol{r}|} + \frac{1}{2} \int (\boldsymbol{E}_T^2 + \boldsymbol{B}^2) \, d^3r \tag{1.33}$$

which is a desired form.

1.4. Self-interacting Fermion Fields

Interactions between fermions are mediated by the gauge fields and this is the basic principle for the description of the fundamental field theory models. The reason why the gauge field theory is employed in modern physics is partly because the electromagnetic interaction is described by the gauge field theory but also because the gauge field theory is a renormalizable field theory. This is important since the renormalizable field theory has a predictive power in the perturbative calculations.

On the other hand, the field theory model with current-current interactions is not renormalizable in four dimensions since the coupling constant has the dimension of mass inverse square. Nevertheless, the model proposed by Nambu and Jona-Lasinio has been discussed frequently since it demonstrates, for the first time, the spontaneous symmetry breaking in the vacuum state in fermion field theory models. Therefore, we briefly discuss the Lagrangian density of the Nambu-Jona-Lasinio (NJL) model [93]. In addition, we treat the Thirring model which is the current current interaction model in two dimensions [109]. This model becomes important for the discussion of the spontaneous symmetry breaking which will be discussed in detail in Chapter 4.

1.4.1. Lagrangian and Hamiltonian Densities of NJL Model

The Lagrangian density of the NJL model is given as

$$\mathcal{L} = i\bar{\psi}\gamma_\mu\partial^\mu\psi - m\bar{\psi}\psi + \frac{1}{2}G\left[(\bar{\psi}\psi)^2 + (\bar{\psi}i\gamma_5\psi)^2\right]. \tag{1.34}$$

In this case, the Hamiltonian density of the NJL model can be written as

$$\mathcal{H} = -i\psi^\dagger\boldsymbol{\nabla}\cdot\boldsymbol{\alpha}\psi + m\bar{\psi}\psi - \frac{1}{2}G\left[(\bar{\psi}\psi)^2 + (\bar{\psi}i\gamma_5\psi)^2\right]. \tag{1.35}$$

The coupling constant in this model has a dimension of inverse mass square,

$$G \sim m^{-2}. \tag{1.36}$$

Therefore, the NJL model is not renormalizable in the perturbative sense. Some of physical observables calculated in terms of the first order perturbation theory should have divergences of Λ^2. When the cut-off momentum Λ becomes very large, the physical quantity diverges very quickly, and there is no chance to renormalize this divergence into the coupling constant G.

The NJL model has been discussed often in the context of the spontaneous symmetry breaking physics [83, 84], and therefore we are bound to discuss it here since we will discuss the symmetry and its breaking in the later chapter of this book. Further, it should be fair to mention that, if one solves the field theory model exactly or non-perturbatively, then one may find that the theory has some predictive power. But this problem is too difficult to discuss further.

1.4.2. Lagrangian Density of Thirring Model

There is a popular field theory model in two dimensions with current current interactions. It is called Thirring model which has been extensively studied since it has an exact solution due to the Bethe ansatz technique. This will be treated in detail in the later chapter. Here, we should only introduce the model Lagrangian density. The Thirring model is described by the following Lagrangian density

$$\mathcal{L} = i\bar{\psi}\gamma_\mu\partial^\mu\psi - m_0\bar{\psi}\psi - \frac{1}{2}gj^\mu j_\mu, \tag{1.37}$$

where the fermion current j_μ is given as

$$j_\mu = \bar{\psi}\gamma_\mu\psi. \tag{1.38}$$

The coupling constant g in two dimensional current current interaction model is a dimensionless constant. Therefore, it is renormalizable, and the model has a predictive power in the perturbation calculations.

1.4.3. Hamiltonian Density for Thirring Model

The Hamiltonian density of the Thirring model can be written as

$$\mathcal{H} = -i\bar{\psi}\gamma^1\partial_1\psi + m_0\bar{\psi}\psi + \frac{1}{2}gj^\mu j_\mu. \tag{1.39}$$

Here, the chiral representation for γ matrices in two dimensions is chosen

$$\gamma_0 = \begin{pmatrix} 0 & 1 \\ 1 & 0 \end{pmatrix}, \quad \gamma_1 = \begin{pmatrix} 0 & -1 \\ 1 & 0 \end{pmatrix}, \quad \gamma_5 \equiv \gamma_0\gamma_1 = \begin{pmatrix} 1 & 0 \\ 0 & -1 \end{pmatrix}. \tag{1.40}$$

By introducing the state ψ as

$$\psi = \begin{pmatrix} \psi_a \\ \psi_b \end{pmatrix} \tag{1.41}$$

the Hamiltonian density can be written

$$\mathcal{H} = -i\left(\psi_a^\dagger \frac{\partial}{\partial x}\psi_a - \psi_b^\dagger \frac{\partial}{\partial x}\psi_b\right) + m_0(\psi_a^\dagger\psi_b + \psi_b^\dagger\psi_a) + 2g\psi_a^\dagger\psi_a\psi_b^\dagger\psi_b. \tag{1.42}$$

Therefore, the Hamiltonian of the Thirring model can be written as

$$H = \int dx\left[-i\left(\psi_a^\dagger \frac{\partial}{\partial x}\psi_a - \psi_b^\dagger \frac{\partial}{\partial x}\psi_b\right) + m_0(\psi_a^\dagger\psi_b + \psi_b^\dagger\psi_a) + 2g\psi_a^\dagger\psi_a\psi_b^\dagger\psi_b\right]. \tag{1.43}$$

In Chapter 7, we will discuss the diagonalization procedure of the Thirring model Hamiltonian in terms of the Bethe ansatz technique.

1.5. Quarks with Electromagnetic and Chromomagnetic Interactions

It should be worthwhile writing the total Lagrangian density which is composed of quarks interacting with electromagnetic fields as well as chromomagnetic fields. Normally, one considers either electromagnetic interactions or chromomagnetic interactions separately since they become important at the different physical stages. Here, we write them together since in reality there are always two different types of interactions (QED and QCD) for quarks present in nature. In addition, we include the interaction terms which violate the time reversal invariance as well as parity transformation just for academic interests.

1.5.1. Lagrangian Density

The Lagrangian density of quarks interacting with electromagnetic fields as well as chromomagnetic fields is given as

$$\mathcal{L} = \bar{\psi}_f\left[i\left(\partial_\mu + ig_sA_\mu^aT^a + ie_fA_\mu\right)\gamma^\mu - m_0\right]\psi_f - \frac{1}{4}F_{\mu\nu}F^{\mu\nu} - \frac{1}{4}G_{\mu\nu}^aG^{\mu\nu,a}$$
$$-\frac{i}{2}\tilde{d}_f\bar{\psi}_f\sigma_{\mu\nu}\gamma_5T^a\psi_fG^{\mu\nu,a} - \frac{i}{2}d_f\bar{\psi}_f\sigma_{\mu\nu}\gamma_5\psi_fF^{\mu\nu}, \tag{1.44}$$

where the summation of flavor runs f = up, down, strange, charm, bottom and top quarks. T^a denotes the generator of the $SU(3)$ color group. The last two terms represent the T- and P-violating interactions. $\sigma_{\mu\nu}$ and γ_5 are defined as

$$\sigma_{\mu\nu} = \frac{i}{2}(\gamma_\mu\gamma_\nu - \gamma_\nu\gamma_\mu), \quad \gamma_5 \equiv i\gamma_0\gamma_1\gamma_2\gamma_3.$$

Field Strength of Electromagnetic Field

$F_{\mu\nu}$ denotes the electromagnetic field strength and is written as

$$F_{\mu\nu} = \partial_\mu A_\nu - \partial_\nu A_\mu, \tag{1.45}$$

where A_μ is the gauge field as given in Section 1.3.

Field Strength of Chromomagnetic Field

$G_{\mu\nu}$ denotes the chromomagnetic field strength and is given as

$$G^a_{\mu\nu} = \partial_\mu A^a_\nu - \partial_\nu A^a_\mu - g_s C^{abc} A^b_\mu A^c_\nu, \tag{1.46}$$

where A^a_μ is the color gauge fields. C^{abc} denotes the structure constant in the $SU(3)$ group. The coupling constants g_s and e_f denote the gauge coupling constant of f-flavor quarks interacting with chromomagnetic field and electromagnetic field, respectively.

1.5.2. EDM Interactions

The last two terms in eq.(1.44) represent the interaction terms which violate the time reversal invariance as well as the space reflection at the same time. These terms are given just for references in order to understand the T-violating interactions in future in terms of EDM (Electric Dipole Moments). That is,

$$-\frac{i}{2}\tilde{d}_f \bar{\psi}_f \sigma_{\mu\nu}\gamma_5 T^a \psi_f G^{\mu\nu,a} : \ \text{EDM for chromomagnetic fields,}$$

$$-\frac{i}{2} d_f \bar{\psi}_f \sigma_{\mu\nu}\gamma_5 \psi_f F^{\mu\nu} : \ \text{EDM for electromagnetic fields.}$$

The coupling strengths $\tilde{d}_f$ and d_f denote the strength of the time reversal and parity violating interactions of quark with the chromomagnetic fields and the electromagnetic fields, respectively. The $\tilde{d}_f$ and d_f have the dimension of the mass inverse, and, in fact, they are related to the electric dipole moment.

The existence of the EDM interactions should be determined from experiments. If there is any finite EDM interaction observed in future experiment, it should indicate an existence of a new scale which is different from the quark masses. In this respect, the observation of the EDM interaction must be physically very interesting and important indeed.

Chapter 2

Symmetry and Conservation Law

The Lagrangian density of fermions which is constructed in the previous chapter possesses various symmetries such as Lorentz invariance, time reversal symmetry and so on. These symmetries play a fundamental role for the determination of the vacuum state as well as the spectrum emerged from the model Hamiltonian. Therefore, one should be accustomed to these basic symmetries to understand the field theory.

In this chapter, we explain fundamental symmetry properties of Lorentz invariance, time reversal invariance, parity transformation, charge conjugation, translational invariance, global gauge symmetry, chiral symmetry and $SU(3)$ symmetry in field theory. In particular, the invariance of the Lagrangian density under these symmetries is discussed in detail since it is important to determine the vacuum structure.

If the Lagrangian density has a continuous symmetry, then there is a conserved current associated with this symmetry due to Noether's theorem. From the conservation of the current, one finds a conserved charge which plays an important role for the determination of physical states such as the vacuum state. All of the field theory models discussed here possess the translational invariance of the Lagrangian density, and this leads to the conserved quantity of the energy momentum tensor. From this energy momentum tensor, one can define the Hamiltonian density which is the energy density of the system. Clearly, the Hamiltonian is most important for the quantized field theory models since it can determine all of the physical states as the eigenstate of the Hamiltonian.

2.1. Introduction to Transformation Property

When one considers the transformation of the Lagrangian density or Hamiltonian density under the symmetry operator U, one first evaluates the transformation of the field ψ as

$$\psi' = U\psi.$$

Then, one calculates and sees how the Lagrangian density $\mathcal{L}$ should transform under the symmetry operator U

$$\mathcal{L}' \equiv \mathcal{L}(\psi', \partial_\mu \psi') = \mathcal{L}(U\psi, \partial_\mu(U\psi)).$$

If the Lagrangian density does not change its functional shape,

$$\mathcal{L}' = \mathcal{L}$$

then it is invariant under the transformation of U.

After the field quantization, the transformation procedure becomes somewhat different. The basic physical quantity in quantized field theory becomes the Hamiltonian $\hat{H}$, and the important point is that the Hamiltonian is now an operator and the problem becomes the eigenvalue equation for the field Hamiltonian

$$\hat{H}|\Psi\rangle = E|\Psi\rangle,$$

where $|\Psi\rangle$ is called *Fock state*. Since an operator O transforms under the symmetry operator U as

$$O' = UOU^{-1}$$

the Hamiltonian $\hat{H}$ transforms as

$$\hat{H}' = U\hat{H}U^{-1}.$$

In this case, the Fock state $|\Psi\rangle$ should transform as

$$|\Psi'\rangle = U|\Psi\rangle.$$

The transformation properties of the Lagrangian density should be kept for the quantized Hamiltonian after the field quantization. However, the vacuum state (the Fock state) may break the symmetry and indeed this can happen for the continuous global symmetry like the chiral symmetry. This physical phenomena are called *spontaneous symmetry breaking* which will be treated in Chapter 4.

2.2. Lorentz Invariance

The most important symmetry in physics must be the Lorentz invariance. The Lorentz invariance should hold in the theory of all the fundamental interactions. This is based on the observation that any physical observables should not depend on the systems one chooses if the systems S and S' are related to each other by the Lorentz transformation,

$$x'^{\mu} = \alpha^{\mu}_{\ \nu} x^{\nu}. \tag{2.1}$$

If the S' system is moving with its velocity of v along the x_1-axis, then the matrix $\alpha^{\mu}_{\ \nu}$ can be explicitly written as

$$\{\alpha^{\mu}_{\ \nu}\} = \begin{pmatrix} \frac{1}{\sqrt{1-v^2}} & -\frac{v}{\sqrt{1-v^2}} & 0 & 0 \\ -\frac{v}{\sqrt{1-v^2}} & \frac{1}{\sqrt{1-v^2}} & 0 & 0 \\ 0 & 0 & 1 & 0 \\ 0 & 0 & 0 & 1 \end{pmatrix} = \begin{pmatrix} \cosh\omega & -\sinh\omega & 0 & 0 \\ -\sinh\omega & \cosh\omega & 0 & 0 \\ 0 & 0 & 1 & 0 \\ 0 & 0 & 0 & 1 \end{pmatrix}, \tag{2.2}$$

where

$$\cosh\omega = \frac{1}{\sqrt{1-v^2}}$$

is introduced. In this case, the Dirac wave function ψ should transform by the Lorentz transformation as

$$\psi'(x') = S\psi(x), \tag{2.3}$$

where S denotes a 4×4 matrix. Now, the Lagrangian density for free Dirac field is written both in $\mathcal{S}$ and $\mathcal{S}'$ systems

$$\mathcal{L} = \bar{\psi}(x)(i\partial_\mu\gamma^\mu - m)\psi(x) = \bar{\psi'}(x')(i\partial_\mu{}'\gamma^\mu - m)\psi'(x'). \tag{2.4}$$

From the equivalence between $\mathcal{S}$ and $\mathcal{S}'$ systems, one obtains

$$\bar{\psi'}(x') = \bar{\psi}(x)S^{-1}, \tag{2.5a}$$

$$S\gamma^\mu S^{-1}\alpha_\mu^{\ \nu} = \gamma^\nu. \tag{2.5b}$$

If one solves eq.(2.5b), then one can determine the shape of S explicitly when the $\mathcal{S}'$ system is moving along the x_1-axis

$$S = \exp\left(-\frac{i}{4}\omega\sigma_{\mu\nu}I_n^{\mu\nu}\right),$$

where $\sigma_{\mu\nu}$ and $I_n^{\mu\nu}$ are defined as

$$\sigma_{\mu\nu} = \frac{i}{2}(\gamma_\mu\gamma_\nu - \gamma_\nu\gamma_\mu),$$
$$I_n^{\mu\nu} = \begin{pmatrix} 0 & -1 & 0 & 0 \\ 1 & 0 & 0 & 0 \\ 0 & 0 & 0 & 0 \\ 0 & 0 & 0 & 0 \end{pmatrix}.$$

2.2.1. Lorentz Covariance

If physical quantities like Lagrangian density or equation of motions are written in a manifestly Lorentz invariant fashion, then they are called *Lorentz covariant.* The simplest case is that these equations are written as a Lorentz scalar. In this case, it is trivial to recognize that they are Lorentz invariant. For example, the continuity equation of the vector current reads

$$\frac{\partial\rho}{\partial t} + \nabla\cdot\boldsymbol{j} = 0.$$

This is shown to be Lorentz invariant under the Lorentz transformation. However, it is not manifest, and therefore one defines the four vector current j^μ by

$$j^\mu = (\rho, \boldsymbol{j}).$$

In this case, one can rewrite the current conservation as

$$\partial_\mu j^\mu = 0$$

which is obviously Lorentz invariant since it is written in terms of the Lorentz scalar, and it is a Lorentz covariant expression.

However, one should not stress too much the importance of the Lorentz covariance. The Lorentz invariance is, of course, most important. However, as long as one starts from the Lorentz invariant theory, one does not have to worry about the violation of the Lorentz invariance since it can never be broken unless one makes mistakes in his calculations. In this respect, the Lorentz covariance may play an important role for avoiding careless mistakes if one carries out the perturbative calculation of the S-matrix in a covariant way.

2.3. Time Reversal Invariance

The world we live does not seem to be invariant under the time reversal transformation. Time flows always in the same direction. However, the physical law in the macroscopic world is quite different from the microscopic world, and time arrow defined by the entropy may not necessarily be related to the fundamental interactions.

Almost all of the fundamental interactions are invariant under the time reversal transformation. It is therefore important to understand the time reversal invariance (T-invariance) in field theory models.

2.3.1. T-invariance in Quantum Mechanics

Before going to field theory, we should first understand the definition of the T-invariance in quantum mechanics. When we make $t \to -t$, then the basic operators that appear in physics behave

$$t \to -t: \begin{pmatrix} x_k \to x_k \\ p_k \to -p_k \\ \sigma_k \to -\sigma_k \\ E \to E \end{pmatrix}. \tag{2.6}$$

However, when the momentum p_k and the energy E are replaced by the differential operators as

$$\hat{p}_k = -i\frac{\partial}{\partial x_k}, \quad \hat{E} = i\frac{\partial}{\partial t} \tag{2.7}$$

then, the explicit t-dependences of the p_k and E become just opposite to eq.(2.6), and therefore one should recover them by hand. This can be realized when one makes complex conjugate of the operators $\hat{p}_k$ and $\hat{E}$

$$t \to -t \; : \; \hat{p}_k \to \hat{p}_k{}^* = i\frac{\partial}{\partial x_k} = -\hat{p}_k, \tag{2.8a}$$

$$t \to -t \; : \; \hat{E} \to \left(-i\frac{\partial}{\partial t}\right)^* = i\frac{\partial}{\partial t} = \hat{E} \tag{2.8b}$$

which can reproduce eq.(2.6). Therefore, the time reversal transformation in quantum mechanics means that the operator should be made complex conjugate as $A \to A^*$. This means that if the Hamiltonian contains an imaginary term, then this system violates the T-invariance. As one can see, the complex conjugate operation in accordance with the T-transformation should not be taken for the Pauli matrix σ_k as seen from eq.(2.6). For the Pauli matrix σ_k, one should make just the transformation of $\sigma_k \to -\sigma_k$ for $t \to -t$.

2.3.2. T-invariance in Field Theory

In field theory, momentum operators are all replaced by the differential operators, and therefore T-transformation of the field $\psi(x_k,t)$ means

$$t \to -t \ : \ \psi(x_k,t) \to \psi(x_k,-t)^* \ \text{with} \ \sigma_k \to -\sigma_k. \tag{2.9}$$

As an example, one takes the plane wave solution of eq.(1.11) and makes the T-transformation. Then, one obtains

$$t \to -t \ : \ \psi(x_k,t) \to \left\{ \sqrt{\frac{E_p+m}{2E_p}} \begin{pmatrix} \chi_s \\ \dfrac{\sigma_k p_k}{E_p+m}\chi_s \end{pmatrix} \frac{1}{\sqrt{V}} e^{-iE_pt+ip_kx_k} \right\}^*$$

$$= \psi(x_k,t) \tag{2.10}$$

which is indeed invariant under the T-transformation. The γ_μ matrices transform under the T-transformation

$$t \to -t \ : \ \begin{pmatrix} \gamma_0 \to \gamma_0 \\ \gamma_k \to -\gamma_k \end{pmatrix}. \tag{2.11}$$

Therefore, the free Dirac Lagrangian density of eq.(1.7) transforms

$$t \to -t \ : \ \mathcal{L} \to \left\{ \psi_i^\dagger \left[\gamma_0(-i\partial_0\gamma^0 - i\partial_k\gamma^k - m) \right]_{ij} \psi_j \right\}^* = \mathcal{L} \tag{2.12}$$

which is, of course, T-invariant as expected.

2.3.3. T-violating Interactions (Imaginary Mass Term)

At present, it is most important and fundamental to discover any interactions which violate the T-invariance. The simplest way to introduce the T-violating interaction must be an imaginary mass term,

$$\mathcal{H}_T = i\eta\bar{\psi}\psi, \tag{2.13}$$

where η denotes a real constant. In fact, the CP violating phase is originated from this type of interaction.

2.3.4. T and P-violating Interactions (EDM)

The direct examination of the T-violating interaction is based on the measurement of electric dipole moments (EDM) in isolated systems. The fundamental interaction of the T and P-violations can be written

$$\mathcal{H}_{TP} = \frac{i}{2} d_f \bar{\psi}_f \sigma_{\mu\nu}\gamma_5\psi_f F^{\mu\nu}, \tag{2.14}$$

where d_f denotes the intrinsic EDM of the f-fermion. $F_{\mu\nu}$ and $\sigma_{\mu\nu}$ denote the electromagnetic field strength and the anti-symmetric tensor, respectively and they are defined as

$$F_{\mu\nu} = \partial_\mu A_\nu - \partial_\nu A_\mu,$$

$$\sigma_{\mu\nu} = \frac{i}{2}(\gamma_\mu\gamma_\nu - \gamma_\nu\gamma_\mu).$$

γ_5 is defined as

$$\gamma_5 = i\gamma_0\gamma_1\gamma_2\gamma_3.$$

The Hamiltonian of eq.(2.14) can be obtained by integrating the Hamiltonian density over all space, and in the nonrelativistic limit, the particle Hamiltonian becomes

$$H_{TP} \simeq -d_f \boldsymbol{\sigma}\cdot\boldsymbol{E}. \tag{2.15}$$

The measurements of the neutron EDM have been carried out extensively, and we will see in near future whether the neutron EDM is finite or not. It may be worth quoting the recent experimental measurement on the neutron EDM d_n [64]

$$d_n \simeq (1.9 \pm 5.4) \times 10^{-26} \text{ e} \cdot \text{cm}.$$

2.4. Parity Transformation

The space reflection operation is called *parity* transformation $\hat{P}$, and it is defined as

$$\hat{P}x_k\hat{P}^{-1} = -x_k, \quad \hat{P}t\hat{P}^{-1} = t \tag{2.16a}$$

$$\hat{P}\gamma_k\hat{P}^{-1} = -\gamma_k, \quad \hat{P}\gamma_0\hat{P}^{-1} = \gamma_0. \tag{2.16b}$$

In this case, ψ should also transform into ψ' as

$$\psi'(x_k,t) = \hat{P}\psi(x_k,t) = \gamma_0\psi(x_k,t). \tag{2.16c}$$

The strong and electromagnetic interactions are invariant under the parity transformation. For example, the fermion Lagrangian density with the electromagnetic interaction of eq.(1.20)

$$\mathcal{L} = \bar{\psi}(i\partial_\mu\gamma^\mu - gA_\mu\gamma^\mu - m)\psi - \frac{1}{4}F_{\mu\nu}F^{\mu\nu}$$

can be seen under the parity transformation as follows.

$$\bar{\psi}\hat{P}^{-1}i\partial_0\gamma_0\hat{P}\psi = \bar{\psi}i\partial_0\gamma_0\psi, \quad \bar{\psi}\hat{P}^{-1}i\partial_k\gamma_k\hat{P}\psi = \bar{\psi}i\partial_k\gamma_k\psi,$$
$$\bar{\psi}\hat{P}^{-1}A_0\gamma_0\hat{P}\psi = \bar{\psi}A_0\gamma_0\psi, \quad \bar{\psi}\hat{P}^{-1}A_k\gamma_k\hat{P}\psi = \bar{\psi}A_k\gamma_k\psi,$$

where the following relations are employed

$$\hat{P}A_0\hat{P}^{-1} = A_0, \quad \hat{P}A_k\hat{P}^{-1} = -A_k. \tag{2.17}$$

Therefore, one sees that the fermion Lagrangian density with the electromagnetic interaction is invariant under the parity transformation.

Interaction with Parity Violation

For parity violating interactions, one takes for example

$$\mathcal{L}_I = g'\bar{\psi}\gamma_\mu\gamma_5 A^\mu \psi. \tag{2.18}$$

Under the parity transformation, one finds

$$\bar{\psi}\hat{P}^{-1}\gamma_k\gamma_5\hat{P}\psi = \bar{\psi}\gamma_k\gamma_5\psi$$

which shows that the Lagrangian density of $\mathcal{L}_I$ is odd under the parity transformation.

2.5. Charge Conjugation

The Lagrangian density for electrons interacting with the gauge field is invariant under the charge conjugation operation. The charge conjugate operation starts from the Maxwell equation which is invariant under the sign change of the vector potential.

2.5.1. Charge Conjugation in Maxwell Equation

The Maxwell equation is invariant under the sign change of the vector potential

$$\text{Charge Conjugation} \Longrightarrow A_c^\mu \equiv -A^\mu. \tag{2.19a}$$

This is clear since the Lagrangian density of the gauge field is written as

$$\mathcal{L} = -\frac{1}{4}(\partial_\mu A_\nu - \partial_\nu A_\mu)(\partial^\mu A^\nu - \partial^\nu A^\mu)$$

which is obviously invariant under the operation of eq.(2.19a)

$$\mathcal{L}_c = -\frac{1}{4}(\partial_\mu A_{c\nu} - \partial_\nu A_{c\mu})(\partial^\mu A_c{}^\nu - \partial^\nu A_c{}^\mu) = \mathcal{L}.$$

When the gauge field interacts with the fermion current, then the Lagrangian density becomes

$$\mathcal{L} = -gj_\mu A^\mu - \frac{1}{4}(\partial_\mu A_\nu - \partial_\nu A_\mu)(\partial^\mu A^\nu - \partial^\nu A^\mu).$$

This Lagrangian density should be invariant under the charge conjugation operation and therefore gj_μ should change its sign

$$\text{Charge Conjugation} \Longrightarrow (gj_\mu)_c = -gj_\mu. \tag{2.19b}$$

This is a constraint on the Dirac field and it is indeed realized in the Dirac equation.

2.5.2. Charge Conjugation in Dirac Field

The invariance of the charge conjugation on the Dirac field starts from the Dirac equation with the electromagnetic interaction

$$i(\partial_\mu \gamma^\mu)_{ij}\psi_j - g(A_\mu \gamma^\mu)_{ij}\psi_j - m\psi_i = 0. \tag{2.20}$$

Now, one can make the complex conjugate of the above equation and multiply γ_0 from the left. This can be rewritten with the transposed representation of the gamma matrix γ_μ^T

$$-i(\partial^\mu \gamma_\mu^T)_{ij}\bar{\psi}_j - g(A^\mu \gamma_\mu^T)_{ij}\bar{\psi}_j - m\bar{\psi}_i = 0. \tag{2.21}$$

Now, the $\bar{\psi}$ is transformed as

$$\psi^c \equiv C\gamma^0\psi^* = C\bar{\psi}^T, \tag{2.22}$$

where C is a 4×4 matrix, and ψ^c denotes the state with charge conjugation and corresponds to an anti-particle state. Further, the operator C is assumed to satisfy the following equation

$$\gamma_\mu = -C(\gamma_\mu)^T C^{-1}. \tag{2.23}$$

In this case, one obtains

$$i(\partial_\mu \gamma^\mu)_{ij}(\psi^c)_j + g(A_\mu \gamma^\mu)_{ij}(\psi^c)_j - m(\psi^c)_i = 0. \tag{2.24}$$

This equation is just the same as eq.(2.20) if the sign of g is reversed, and indeed, the sign change of g is the requirement of the charge conjugation of eq.(2.19b). The operator C that satisfies eq.(2.23) is found to be

$$C = i\gamma_2\gamma_0 \tag{2.25}$$

which is the charge conjugation operator in the Dirac field.

2.5.3. Charge Conjugation in Quantum Chromodynamics

The Lagrangian density of QCD

$$\mathcal{L} = \bar{\psi}\left[i(\partial^\mu + ig_s A^{\mu,a}T^a)\gamma_\mu - m_0\right]\psi - \frac{1}{4}G^a_{\mu\nu}G^{\mu\nu,a}$$

with

$$G^{\mu\nu,a} = \partial^\mu A^{\nu,a} - \partial^\nu A^{\mu,a} - g_s C^{abc}A^{\mu,b}A^{\nu,c}$$

is invariant under the charge conjugation operation

$$A_c^{\mu,a} \equiv -A^{\mu,a}, \quad (g_s)_c \equiv -g_s.$$

This can be easily seen since

$$(G^{\mu\nu,a})_c = \partial^\mu A_c^{\nu,a} - \partial^\nu A_c^{\mu,a} - (g_s)_c C^{abc}A_c^{\mu,b}A_c^{\nu,c} = -G^{\mu\nu,a}.$$

In addition, the Dirac field part of the Lagrangian density is invariant in the same way as the QED case, and therefore one sees that the Lagrangian density of QCD is invariant under the charge conjugation operation

$$\mathcal{L}_c = \bar{\psi}_c\left[i(\partial^\mu + i(g_s)_c A_c^{\mu,a}T^a)\gamma_\mu - m_0\right]\psi_c - \frac{1}{4}(G^a_{\mu\nu})_c(G^{\mu\nu,a})_c = \mathcal{L}.$$

2.6. Translational Invariance

When one transforms the coordinate x_k into $x_k + a_k$ with a_k a constant, then the wave function $\psi(x_k)$ becomes

$$\psi(x_k) \longrightarrow \psi(x_k + a_k). \tag{2.26}$$

This translation operation $\hat{R}_{a_k}$ can be written for a very small a as

$$\hat{R}_{a_k}\psi(x_k) = \psi(x_k + a_k) = \left(1 + a_k \frac{\partial}{\partial x_k}\right)\psi(x_k). \tag{2.27}$$

For the finite a_k, one can write

$$\psi(x_k + a_k) = \lim_{n \to \infty}\left(1 + \frac{a_k}{n}\frac{\partial}{\partial x_k}\right)^n \psi(x_k) = e^{ip_k a_k}\psi(x_k). \tag{2.28}$$

Therefore, one finds the translation operation $\hat{R}_{\boldsymbol{a}}$ in three dimensions as

$$\hat{R}_{\boldsymbol{a}} = e^{ip_k a_k} = e^{i\boldsymbol{p}\cdot\boldsymbol{a}}. \tag{2.29}$$

2.6.1. Energy Momentum Tensor

If the Lagrangian density is invariant under the translation, then there is a conserved quantity associated with this symmetry, which is called *energy momentum tensor* as will be defined below. Under the infinitesimal translation of a^ν, the field ψ transforms as

$$\psi' = \psi + \delta\psi, \quad \delta\psi = (\partial_\nu\psi)a^\nu,$$
$$\partial_\mu\psi' = \partial_\mu\psi + \delta(\partial_\mu\psi), \quad \delta(\partial_\mu\psi) = (\partial_\mu\partial_\nu\psi)a^\nu + (\partial_\nu\psi)(\partial_\mu a^\nu).$$

Since the Lagrangian density is invariant under the infinitesimal translation of a^ν, one has

$$\delta\mathcal{L} \equiv \mathcal{L}(\psi', \partial_\mu\psi') - \mathcal{L}(\psi, \partial_\mu\psi) = \frac{\partial\mathcal{L}}{\partial\psi}\delta\psi + \frac{\partial\mathcal{L}}{\partial(\partial_\mu\psi)}\delta(\partial_\mu\psi)$$

$$= \frac{\partial\mathcal{L}}{\partial\psi}(\partial_\nu\psi)a^\nu + \frac{\partial\mathcal{L}}{\partial(\partial_\mu\psi)}(\partial_\mu\partial_\nu\psi)a^\nu + \frac{\partial\mathcal{L}}{\partial(\partial_\mu\psi)}(\partial_\nu\psi)(\partial_\mu a^\nu) = 0. \tag{2.30a}$$

By making use of the following equation

$$\partial_\mu\left(\frac{\partial\mathcal{L}}{\partial(\partial_\mu\psi)}\partial_\nu\psi a^\nu\right) = \partial_\mu\left(\frac{\partial\mathcal{L}}{\partial(\partial_\mu\psi)}\partial_\nu\psi\right)a^\nu + \frac{\partial\mathcal{L}}{\partial(\partial_\mu\psi)}\partial_\nu\psi(\partial_\mu a^\nu)$$

and using the fact that the total divergence does not contribute to the action, one can obtain the following equation

$$\delta\mathcal{L} = \left[\frac{\partial\mathcal{L}}{\partial\psi}(\partial_\nu\psi) + \frac{\partial\mathcal{L}}{\partial(\partial_\mu\psi)}(\partial_\mu\partial_\nu\psi) - \partial_\mu\left(\frac{\partial\mathcal{L}}{\partial(\partial_\mu\psi)}\partial_\nu\psi\right)\right]a^\nu. \tag{2.30b}$$

In addition, the following identity can be employed

$$\partial_\nu\mathcal{L} = \frac{\partial\mathcal{L}}{\partial\psi}(\partial_\nu\psi) + \frac{\partial\mathcal{L}}{\partial(\partial_\mu\psi)}(\partial_\mu\partial_\nu\psi)$$

and one finds

$$\delta\mathcal{L} = \partial_\mu \left[\mathcal{L} g^{\mu\nu} - \frac{\partial \mathcal{L}}{\partial(\partial_\mu \psi)} \partial^\nu \psi \right] a_\nu = 0. \tag{2.31}$$

The same thing should hold as well for the field $\psi^\dagger$ which is an independent functional variable in the Lagrangian density, and therefore eq.(2.31) should be modified

$$\delta\mathcal{L} = \partial_\mu \left[\mathcal{L} g^{\mu\nu} - \frac{\partial \mathcal{L}}{\partial(\partial_\mu \psi)} \partial^\nu \psi - \frac{\partial \mathcal{L}}{\partial(\partial_\mu \psi^\dagger)} \partial^\nu \psi^\dagger \right] a_\nu = 0. \tag{2.31$'$}$$

This means that, if one defines the energy momentum tensor $\mathcal{T}^{\mu\nu}$ as

$$\mathcal{T}^{\mu\nu} \equiv \frac{\partial \mathcal{L}}{\partial(\partial_\mu \psi)} \partial^\nu \psi + \frac{\partial \mathcal{L}}{\partial(\partial_\mu \psi^\dagger)} \partial^\nu \psi^\dagger - \mathcal{L} g^{\mu\nu} \tag{2.32}$$

then $\mathcal{T}^{\mu\nu}$ is a conserved quantity, that is

$$\partial_\mu \mathcal{T}^{\mu\nu} = 0. \tag{2.33}$$

The reason why $\mathcal{T}^{\mu\nu}$ is called *energy momentum tensor* is because $\mathcal{T}^{0\nu}$ is related to the Hamiltonian and momentum densities.

2.6.2. Hamiltonian Density from Energy Momentum Tensor

Since the energy momentum tensor $\mathcal{T}^{0\nu}$ is a conserved quantity, one can define the Hamiltonian density $\mathcal{H}$ by $\mathcal{T}^{00}$

$$\mathcal{H} \equiv \mathcal{T}^{00} = \frac{\partial \mathcal{L}}{\partial \dot{\psi}} \dot{\psi} + \frac{\partial \mathcal{L}}{\partial \dot{\psi}^\dagger} \dot{\psi}^\dagger - \mathcal{L} = \Pi_\psi \dot{\psi} + \Pi_{\psi^\dagger} \dot{\psi}^\dagger - \mathcal{L}, \tag{2.34}$$

where the conjugate fields Π_ψ and $\Pi_{\psi^\dagger}$ are introduced as

$$\Pi_\psi = \frac{\partial \mathcal{L}}{\partial \dot{\psi}}, \quad \Pi_{\psi^\dagger} = \frac{\partial \mathcal{L}}{\partial \dot{\psi}^\dagger}.$$

The Hamiltonian H can be obtained by integrating the Hamiltonian density over all space

$$H = \int \mathcal{H}\, d^3r$$

and it corresponds to the total energy of the field ψ.

2.7. Global Gauge Symmetry

If one transforms the field ψ into ψ' as

$$\psi' = e^{i\alpha}\psi \quad (\alpha \text{ is a real constant}) \tag{2.35}$$

then it is called *global gauge transformation* in which α does not depend on the coordinate x. This is a simple phase transformation which is also found in quantum mechanics since physical observables do not depend on the value of α.

Now, we discuss the invariance of the global gauge symmetry in the Lagrangian density. As examples, we consider the Lagrangian density of QED and the Thirring model

$$\mathcal{L} = \bar{\psi}(i\partial_\mu\gamma^\mu - gA_\mu\gamma^\mu - m)\psi - \frac{1}{4}F_{\mu\nu}F^{\mu\nu}, \tag{2.36a}$$

$$\mathcal{L} = i\bar{\psi}\gamma_\mu\partial^\mu\psi - m\bar{\psi}\psi - \frac{1}{2}gj^\mu j_\mu, \text{ with } j_\mu = \bar{\psi}\gamma_\mu\psi. \tag{2.36b}$$

Obviously, the Lagrangian densities of eqs.(2.36) are invariant under the global gauge transformation

$$\delta\mathcal{L} \equiv \mathcal{L}(\psi', \partial_\mu\psi', \psi'^\dagger, \partial_\mu\psi'^\dagger) - \mathcal{L}(\psi, \partial_\mu\psi, \psi^\dagger, \partial_\mu\psi^\dagger) = 0.$$

In this case, the Noether current associated with the global gauge symmetry is conserved as discussed in Appendix A.11

$$\delta\mathcal{L} = i\alpha\partial_\mu\left[\frac{\partial\mathcal{L}}{\partial(\partial_\mu\psi)}\psi - \frac{\partial\mathcal{L}}{\partial(\partial_\mu\psi^\dagger)}\psi^\dagger\right] = 0 \tag{2.37a}$$

which leads to the conservation of the vector current

$$\partial_\mu j^\mu = 0 \text{ with } j^\mu = -i\left[\frac{\partial\mathcal{L}}{\partial(\partial_\mu\psi)}\psi - \frac{\partial\mathcal{L}}{\partial(\partial_\mu\psi^\dagger)}\psi^\dagger\right]. \tag{2.37b}$$

For the Lagrangian densities of eqs.(2.36), the vector current j_μ in eq.(2.37b) just becomes

$$j_\mu = \bar{\psi}\gamma_\mu\psi.$$

In any field theory models, the conservation of the vector current is known to hold at any level of quantization or regularization, and therefore the charge Q associated with the vector current is always conserved.

2.8. Chiral Symmetry

In eqs.(2.36), if the fermion is massless ($m = 0$), then there is another symmetry which is called *chiral symmetry*. If one transforms the field ψ into ψ' as

$$\psi' = e^{i\alpha\gamma_5}\psi \ (\alpha \text{ is a real constant}), \tag{2.38}$$

then one finds that the Lagrangian densities of the massless QED and the massless Thirring model are invariant under the chiral transformation. This is clear since the γ_5 anti-commutes with γ_μ

$$\{\gamma_5, \gamma_\mu\} = 0$$

and therefore one obtains for $\mu = 0, 1, 2, 3$

$$e^{-i\alpha\gamma_5}\gamma^\mu = \gamma^\mu e^{i\alpha\gamma_5}. \tag{2.39}$$

Thus, one sees that the $\bar{\psi}\gamma^\mu\psi$ transforms by the chiral symmetry as

$$\bar{\psi}'\gamma^\mu\psi' = \psi^\dagger e^{-i\alpha\gamma_5}\gamma_0\gamma^\mu e^{i\alpha\gamma_5}\psi = \bar{\psi}\gamma^\mu\psi.$$

Since the Lagrangian density is invariant under the chiral transformation, the axial vector current

$$j_5^\mu = \bar{\psi}\gamma^\mu\gamma_5\psi$$

is conserved, that is,

$$\partial_\mu j_5^\mu = 0.$$

2.8.1. Expression of Chiral Transformation in Two Dimensions

The chiral transformation of eq.(2.38) can be explicitly written in two dimensions for the field ψ.

Chiral Representation

In the chiral representation of the γ-matrix, the γ_5 and $e^{i\alpha\gamma_5}$ become

$$\gamma_5 = \begin{pmatrix} 1 & 0 \\ 0 & -1 \end{pmatrix}, \quad e^{i\alpha\gamma_5} = \begin{pmatrix} e^{i\alpha} & 0 \\ 0 & e^{-i\alpha} \end{pmatrix}.$$

Therefore, one has

$$\psi' = \begin{pmatrix} \psi'_a \\ \psi'_b \end{pmatrix} = e^{i\alpha\gamma_5} \begin{pmatrix} \psi_a \\ \psi_b \end{pmatrix} = \begin{pmatrix} e^{i\alpha}\psi_a \\ e^{-i\alpha}\psi_b \end{pmatrix}. \tag{2.40}$$

Dirac Representation

In the Dirac representation of the γ-matrix, the γ_5 and $e^{i\alpha\gamma_5}$ become

$$\gamma_5 = \begin{pmatrix} 0 & 1 \\ 1 & 0 \end{pmatrix}, \quad e^{i\alpha\gamma_5} = \begin{pmatrix} \cos\alpha & i\sin\alpha \\ i\sin\alpha & \cos\alpha \end{pmatrix}.$$

Therefore, one has

$$\psi' = \begin{pmatrix} \psi'_a \\ \psi'_b \end{pmatrix} = e^{i\alpha\gamma_5} \begin{pmatrix} \psi_a \\ \psi_b \end{pmatrix} = \begin{pmatrix} \psi_a\cos\alpha + i\psi_b\sin\alpha \\ \psi_b\cos\alpha + i\psi_a\sin\alpha \end{pmatrix}. \tag{2.41}$$

2.8.2. Mass Term

The mass term

$$m\bar{\psi}\psi$$

is not invariant under the chiral transformation

$$m\bar{\psi}'\psi' = m\bar{\psi}e^{2i\alpha\gamma_5}\psi \neq m\bar{\psi}\psi.$$

Therefore, if the system has a finite fermion mass, then the chiral symmetry is not preserved. In this respect, when one takes the massless limit

$$m \to 0$$

then the massless system may not necessarily be connected to the massive one if the chiral symmetry plays an important role for the determination of the vacuum. In fact, the massless limit is the singular point in the Thirring model, and the vacuum structures between the massive and massless Thirring models are completely different from each other. This is reasonable since the massless Thirring model has a vacuum which breaks the chiral symmetry while the massive Thirring model does not possesses the chiral symmetry and therefore its vacuum cannot be connected to the symmetry broken state. In addition, the massless Thirring model has no scaleful parameters, and thus physical observables should be measured in terms of the cutoff Λ, while, in the massive Thirring model, they are described by the mass m which cannot be set to zero after the system is solved.

Transformation of Mass Term in Two Dimensions

The mass term in two dimensions in the chiral representation transforms explicitly by the chiral transformation as

$$m\bar{\psi}'\psi' = m\left(\psi'^{\dagger}_{a}\psi'_{b}+\psi'^{\dagger}_{b}\psi'_{a}\right) = m\left(e^{-2i\alpha}\psi^{\dagger}_{a}\psi_{b}+e^{2i\alpha}\psi^{\dagger}_{b}\psi_{a}\right) \neq m\bar{\psi}\psi$$

which shows again that the mass term is not invariant under the chiral transformation.

2.8.3. Chiral Anomaly

The conservation of the axial vector current is violated if there is a chiral anomaly. The chiral anomaly is closely related to the conflict between the local gauge invariance and the axial vector current conservation when the vacuum is regularized consistently with the local gauge invariance.

Four Dimensional QED

In four dimensional QED, the axial vector current is not conserved due to the anomaly and the conservation of the axial vector current is modified as

$$\partial_\mu j_5^\mu = \frac{g^2}{16\pi^2}\varepsilon^{\rho\sigma\mu\nu}F_{\rho\sigma}F_{\mu\nu}, \tag{2.42}$$

where $\varepsilon^{\rho\sigma\mu\nu}$ denotes the anti-symmetric symbol in four dimensions. $F_{\mu\nu}$ denotes the electromagnetic field strength as given in eq.(1.44).

Two Dimensional QED

The same anomaly equation is found in the two dimensional QED and is written

$$\partial_\mu j_5^\mu = \frac{g}{2\pi}\varepsilon_{\mu\nu}F^{\mu\nu}, \tag{2.43}$$

where $\varepsilon_{\mu\nu}$ denotes the anti-symmetric symbol in two dimensions. The explicit derivation of the anomaly equation eq.(2.43) will be treated in detail in the context of the two dimensional QED in Chapter 5.

Two Dimensional QCD

There is no chiral anomaly in the two dimensional QCD. This can be easily understood when one writes a possible anomaly equation in QCD_2

$$\partial_\mu j_5^\mu \Longleftrightarrow \frac{g}{2\pi}\varepsilon_{\mu\nu}G^{\mu\nu,a}.$$

However, the right hand side has the color index while the left hand side is a color singlet object, and there is no way to construct a color singlet object in the right hand side.

Four Dimensional QCD

Contrary to the two dimensional QCD, there is an anomaly in four dimensional QCD. The axial vector current conservation is modified as

$$\partial_\mu j_5^\mu = \frac{g^2}{32\pi^2}\varepsilon^{\rho\sigma\mu\nu}G^a_{\rho\sigma}G^a_{\mu\nu}, \tag{2.44}$$

where $G^a_{\rho\sigma}$ is the chromomagnetic field strength as given in eq.(1.46).

2.8.4. Chiral Symmetry Breaking in Massless Thirring Model

The massless Thirring model has no local gauge invariance, and therefore there is no anomaly. Thus, the axial vector current is always conserved. Therefore, the axial charge is also a conserved quantity.

In the quantum field theory of the massless Thirring model, one quantizes the fermion fields and therefore the Hamiltonian becomes an operator. Thus, the eigenvalue equation for the Hamiltonian should be solved, and the lowest state is the vacuum where all the negative energy states are occupied by the negative energy particles. The construction of the vacuum state is very difficult since one has to solve infinite many body problems in the negative energy particles. Apart from the exactness of the vacuum state, one can discuss some properties of the vacuum state. One example is the symmetry of the Lagrangian density, and the vacuum can break the symmetry possessed in the Lagrangian density. When the symmetry broken vacuum is realized because it is the lowest energy, then it is called *spontaneous symmetry breaking* phenomenon if the current associated with the symmetry is conserved. We will discuss physics of the symmetry breaking in the vacuum state in detail in chapters 4 and 7.

2.9. $SU(3)$ Symmetry

In quantum mechanics, if the particle Hamiltonian H is invariant under the unitary transformation of $SU(3)$ group,

$$UHU^{-1} = H$$

then the eigenvalues of the particle Hamiltonian H are specified by the eigenvalues of $SU(3)$ group. The same transformation can be applied to the field theory models. Suppose ψ should have 3 degenerate states, and one transforms ψ as

$$\psi' = U\psi.$$

If the Lagrangian density is invariant under the $SU(3)$ transformation, then the Hamiltonian constructed from this Lagrangian density is specified by the eigenvalues of the $SU(3)$ group. In particular, hadron masses predicted by the Hamiltonian should be specified by the eigenvalues of the $SU(3)$ group. In the description of light baryons, one can assume that u, d and s quarks belong to the same multiplet. In this case, one can write ψ as

$$\psi(x) = \begin{pmatrix} \psi_u(x) \\ \psi_d(x) \\ \psi_s(x) \end{pmatrix}. \tag{2.45}$$

By the unitary transformation of 3×3 matrix U, ψ transforms $\psi' = U\psi$ or explicitly

$$\begin{pmatrix} \psi'_u(x) \\ \psi'_d(x) \\ \psi'_s(x) \end{pmatrix} = U \begin{pmatrix} \psi_u(x) \\ \psi_d(x) \\ \psi_s(x) \end{pmatrix} = \begin{pmatrix} u_{11} & u_{12} & u_{13} \\ u_{21} & u_{22} & u_{23} \\ u_{31} & u_{32} & u_{33} \end{pmatrix} \begin{pmatrix} \psi_u(x) \\ \psi_d(x) \\ \psi_s(x) \end{pmatrix}. \tag{2.46}$$

If the Hamiltonian $\hat{H}$ is invariant under the unitary transformation, then the hadron mass $\mathcal{M}$ can be described in terms of some function $G(a)$ as

$$\mathcal{M} = G([\lambda, \mu]), \tag{2.47}$$

where $[\lambda, \mu]$ denotes the quantum number of the symmetric group which specifies the representation of the $SU(3)$ group.

2.9.1. Dimension of Representation $[\lambda, \mu]$

The dimension $D_{[\lambda,\mu]}$ of the state represented by $[\lambda, \mu]$ becomes

$$D_{[\lambda,\mu]} = \frac{1}{2}(\lambda+1)(\mu+1)(\lambda+\mu+2). \tag{2.48}$$

In fact, $[1,0]$ or $[0,1]$ are three dimensional representation which should just correspond to ψ or its anti-particle state. In this way, one sees that the $[1,1]$ representation should have 8 states which are in fact found in nature as octet baryons

$$p,\ n,\ \Lambda,\ \Sigma^{\pm},\ \Sigma^{0},\ \Xi^{\pm}. \tag{2.49}$$

Indeed, their masses are found at around 1 GeV/c^2.

This success of the flavor $SU(3)$ is due to the fact that the interaction Hamiltonian is invariant under the $SU(3)$ transformation. Therefore, the flavor $SU(3)$ invariance of the Hamiltonian is broken by the mass term of the quarks. In particular, the mass of s-quark is assumed to be much larger than the masses of u- and d- quarks by one order of magnitude. However, the quark mass is still smaller than the hadron mass at least by an order of magnitude, and this is probably the main reason why the flavor $SU(3)$ works well. It is interesting to realize that, the fact that hadron masses are much larger than those of quarks indicates that the basic ingredients of generating hadron mass must come from the kinetic energy of quarks inside hadron which should give always positive energy contributions to the mass of hadron. Since the confinement of quarks must be due to the non-abelian character of the gauge fields, it should be most important to understand the properties of the non-abelian gauge field theory.

2.9.2. Useful Reduction Formula

Here, we summarize some examples of useful reduction formula of the $SU(3)$ product representations. First, we show the representation in terms of the dimension of the representation.

$$[1,0] = \mathbf{3}, \quad [0,1] = \mathbf{3}^*, \quad [0,0] = \mathbf{1}, \quad [1,1] = \mathbf{8}, \tag{2.50a}$$

$$[2,0] = \mathbf{6}, \quad [3,0] = \mathbf{10}, \quad [0,2] = \mathbf{6}^*, \quad [0,3] = \mathbf{10}^*. \tag{2.50b}$$

The following reduction formula may be useful

$$\mathbf{3} \otimes \mathbf{3} = \mathbf{3}^* \oplus \mathbf{6}, \quad \mathbf{3} \otimes \mathbf{3}^* = \mathbf{1} \oplus \mathbf{8}, \quad \mathbf{3} \otimes \mathbf{6} = \mathbf{8} \oplus \mathbf{10}. \tag{2.51}$$

For example, we can find the following results

$$\mathbf{3} \otimes \mathbf{3} \otimes \mathbf{3} = (\mathbf{3}^* \oplus \mathbf{6}) \otimes \mathbf{3} = \mathbf{1} \oplus \mathbf{8} \oplus \mathbf{8} \oplus \mathbf{10}, \tag{2.52}$$

$$\mathbf{8} \otimes \mathbf{8} = \mathbf{1} \oplus \mathbf{8} \oplus \mathbf{8} \oplus \mathbf{10} \oplus \mathbf{10}^* \oplus \mathbf{27}. \tag{2.53}$$

Chapter 3

Quantization of Fields

In Chapter 1, we saw that the Lagrange equations for the fermion fields reproduce the Dirac equation which is the relativistic quantum mechanical equation of spin 1/2 fermions. In the Dirac equation, the field ψ would never disappear since the classical fields should always be present since they are c-number functions. The energy spectrum of the hydrogen atom can be described quite well by the Dirac equation.

However, if the hydrogen atom is in the excited state, then it naturally decays into the ground state at the final stage. During the process of transitions, the hydrogen atom emits photons. For example, when the hydrogen atom is in the $2p_{\frac{1}{2}}$-state, then it decays into the ground state of $1s_{\frac{1}{2}}$-state by emitting a photon. In this case, the photon is created during the transition. Therefore, one has to invent some scheme which takes into account the creation or annihilation of electromagnetic fields, and indeed the gauge fields should be quantized in terms of the commutation relations for the creation and annihilation operators.

The Dirac field should be always quantized because of the Pauli principle. The experimental observations in atoms show that one quantum state can be occupied only by one electron (Pauli principle). In order to accomodate the Pauli principle, one has to quantize the Dirac field in terms of anti-commutation relations for the creation and annihilation operators. The quantization of Dirac field with anti-commutation relations is also required from the presence of the negative energy states as the physical observables, and the negative energy states can be well fit into the theoretical framework in terms of the Pauli principle. The field quantization is also consistent with the observation that the pair of electron and positron can be created from virtual photons in the scattering process if some physical conditions are satisfied. In this sense, the field quantization should be made in terms of the creation and annihilation operators of fermion fields, and therefore, the field becomes an operator, and consequently the Hamiltonian becomes an operator.

$$H \Longrightarrow \hat{H} \ \ (\text{operator after field quantization}).$$

In this case, one should solve an eigenvalue equation for the Hamiltonian with a corresponding eigenstate $|\Psi\rangle$

$$\hat{H}|\Psi\rangle = E|\Psi\rangle, \tag{3.1}$$

where E denotes the energy eigenvalue. The state $|\Psi\rangle$ is called *Fock state*, and in quantum field theory, the problem is now how one can solve the eigenvalue equation and determine the energy and eigenstate of the Hamiltonian. In general, it is extremely difficult to solve the eigenvalue equation in quantum field theory. The basic difficulty of the quantized field theory comes mainly from the fact that one has to construct the vacuum state $|\Omega\rangle$ which is composed of infinite numbers of negative energy particles interacting with each other. In addition, $|\Psi\rangle$ should be constructed on this vacuum state by creating particles and anti-particles, and it should satisfy eq.(3.1).

In this chapter, we first treat the quantization of free fermion fields. In most of the field theory models with interactions, the field quantization is done for free fields since one cannot directly quantize the interacting fields. Then, we discuss the quantization of the Thirring model since it can be solved exactly in terms of the Bethe ansatz method. Since the Thirring model gives a non-trivial field theory model, we can learn a lot from this field theory model. We also present the quantization of gauge fields so that we can calculate some scattering processes between electrons in the later chapter.

3.1. Quantization of Free Fermion Field

Classical fermion fields with the Dirac equation can describe the spectrum of the hydrogen atom quite well. However, the representation of the classical field has a limitation since experimental observations indicate that fermion and anti-fermion pairs can be created from the vacuum if the conditions of pair creations are satisfied. This means that the fermion fields cannot be taken as a c-number field. Instead, one should consider the fermion field as an operator. If the field becomes operator, then the value of the field should vary, depending on the state (Fock state) which should be prepared in accordance with the process one wishes to calculate in the perturbation theory.

3.1.1. Creation and Annihilation Operators

We start from the quantization of free Dirac fields. The Lagrangian density of free Dirac field is written as

$$\mathcal{L} = \bar{\psi}(i\partial_\mu \gamma^\mu - m)\psi = i\psi_i^\dagger \dot{\psi}_i + \psi_i^\dagger \left[i\gamma_0 \boldsymbol{\gamma} \cdot \boldsymbol{\nabla} - m\gamma_0 \right]_{ij} \psi_j .$$

In this case, the Hamiltonian can be obtained as given in eq.(1.19)

$$H = \int \mathcal{H}\, d^3r = \int \bar{\psi} \left[-i\boldsymbol{\gamma} \cdot \boldsymbol{\nabla} + m \right] \psi \, d^3r. \tag{1.19}$$

Now, we write the free Dirac field as

$$\psi(\boldsymbol{r},t) = \sum_{\boldsymbol{n},s} \frac{1}{\sqrt{L^3}} \left(a_{\boldsymbol{n}}^{(s)} u_{\boldsymbol{n}}^{(s)} e^{i\boldsymbol{p}_n \cdot \boldsymbol{r} - iE_n t} + b_{\boldsymbol{n}}^{(s)\dagger} v_{\boldsymbol{n}}^{(s)} e^{i\boldsymbol{p}_n \cdot \boldsymbol{r} + iE_n t} \right), \tag{3.2}$$

where $u_{\boldsymbol{n}}^{(s)}$ and $v_{\boldsymbol{n}}^{(s)}$ denote the spinor part of the plane wave solutions as given in Chapter 1, and can be written as

$$u_{\boldsymbol{n}}^{(s)} = \sqrt{\frac{E_n + m}{2E_n}} \begin{pmatrix} \chi_s \\ \dfrac{\boldsymbol{\sigma} \cdot \boldsymbol{p}_n}{E_n + m} \chi_s \end{pmatrix}, \tag{3.3a}$$

$$v_{\boldsymbol{n}}^{(s)} = \sqrt{\frac{E_n+m}{2E_n}} \begin{pmatrix} -\frac{\boldsymbol{\sigma}\cdot\boldsymbol{p_n}}{E_n+m}\chi_s \\ \chi_s \end{pmatrix}, \tag{3.3b}$$

where

$$\boldsymbol{p_n} = \frac{2\pi}{L}\boldsymbol{n}, \quad E_{\boldsymbol{n}} = \sqrt{\boldsymbol{p}_{\boldsymbol{n}}^2 + m^2}$$

and s denotes the spin index with $s = \pm\frac{1}{2}$. Inserting this field into eq.(1.19), one can express the Hamiltonian as

$$H = \sum_{\boldsymbol{n},s} E_n \left(a^{\dagger(s)}_{\boldsymbol{n}} a^{(s)}_{\boldsymbol{n}} - b^{\dagger(s)}_{\boldsymbol{n}} b^{(s)}_{\boldsymbol{n}} \right) + \text{some constants.}$$

The Hamiltonian is a conserved quantity, and therefore we can quantize it. Here, the basic method to quantize the fields is to require that the annihilation and creation operators $a_{\boldsymbol{n}}^{(s)}$ and $a^{\dagger(s')}_{\boldsymbol{n}'}$ for positive energy states and $b_{\boldsymbol{n}}^{(s)}$ and $b^{\dagger(s')}_{\boldsymbol{n}'}$ for negative energy states become operators which satisfy the anti-commutation relations.

Anti-commutation Relations

The creation and annihilation operators for positive and negative energy states should satisfy the following anti-commutation relations,

$$\left\{a_{\boldsymbol{n}}^{(s)}, a^{\dagger(s')}_{\boldsymbol{n}'}\right\} = \delta_{s,s'}\delta_{\boldsymbol{n},\boldsymbol{n}'}, \quad \left\{b_{\boldsymbol{n}}^{(s)}, b^{\dagger(s')}_{\boldsymbol{n}'}\right\} = \delta_{s,s'}\delta_{\boldsymbol{n},\boldsymbol{n}'}. \tag{3.4}$$

All the other cases of the anti-commutations vanish, for examples,

$$\left\{a_{\boldsymbol{n}}^{(s)}, a_{\boldsymbol{n}'}^{(s')}\right\} = 0, \quad \left\{b_{\boldsymbol{n}}^{(s)}, b_{\boldsymbol{n}'}^{(s')}\right\} = 0, \quad \left\{a_{\boldsymbol{n}}^{(s)}, b_{\boldsymbol{n}'}^{(s')}\right\} = 0. \tag{3.5}$$

This corresponds to the field quantization, and the quantization in terms of creation and annihilation operators should be the fundamental quantization procedure.

3.1.2. Equal Time Quantization of Field

The quantization of fields can also be written in terms of equal time anti-commutation relations for fields as,

$$\left\{\psi_i(\boldsymbol{r},t), \pi_j(\boldsymbol{r}',t)\right\} = i\delta_{ij}\delta(\boldsymbol{r}-\boldsymbol{r}'), \tag{3.6}$$

where the conjugate field π_i is given by the Lagrangian density as

$$\pi_i = \frac{\partial \mathcal{L}}{\partial \dot{\psi}_i} = i\psi_i^\dagger.$$

Therefore, the quantization condition of eq.(3.6) becomes

$$\left\{\psi_i(\boldsymbol{r},t), \psi_j^\dagger(\boldsymbol{r}',t)\right\} = \delta_{ij}\delta(\boldsymbol{r}-\boldsymbol{r}'). \tag{3.7}$$

It is important to note that the field quantization must be done always with equal time. Further, one may consider the quantization conditions of fields in terms of eqs.(3.4) and (3.5) as being more fundamental than eq.(3.6), and eq.(3.6) should be taken as the field quantization method which can be derived from eqs.(3.4) and (3.5) together with eq.(3.2).

3.1.3. Quantized Hamiltonian of Free Dirac Field

Now one finds the quantized Hamiltonian which is given as

$$\hat{H} = \sum_{\boldsymbol{n},s} E_n \left(a^{\dagger(s)}_{\boldsymbol{n}} a^{(s)}_{\boldsymbol{n}} - b^{\dagger(s)}_{\boldsymbol{n}} b^{(s)}_{\boldsymbol{n}} \right) + C_0, \tag{3.8}$$

where C_0 is a constant and normally it is discarded since it does not affect on physical observables. This Hamiltonian $\hat{H}$ is written in terms of the creation and annihilation operators.

It may be worthwhile noting that the Hamiltonian is obtained by integrating the Hamiltonian density over all space, and therefore it does not depend on space and also it is a conserved quantity. Instead it is not a c-number but the operator. Therefore, one should find Fock states which must be the eigenstates of the Hamiltonian. For free Dirac fields, one can easily find the eigenstates of the Hamiltonian, and indeed the vacuum state is constructed by filling out all the negative energy states by the negative energy particles as will be treated below.

Anti-particle Representation

The representation of $b^{(s)}_{\boldsymbol{n}}$ corresponds to the negative energy state. The anti-particle representation can be obtained by defining a new operator $\mathrm{b}^{(s)}_{\boldsymbol{n}}$ as

$$\mathrm{b}^{(s)}_{\boldsymbol{n}} = b^{\dagger(s)}_{-\boldsymbol{n}}.$$

In this case, the operator $\mathrm{b}^{(s)}_{\boldsymbol{n}}$ describes the annihilation of an anti-particle. In this representation, the Hamiltonian of free Dirac field becomes

$$\hat{H} = \sum_{\boldsymbol{n},s} E_n \left(a^{\dagger(s)}_{\boldsymbol{n}} a^{(s)}_{\boldsymbol{n}} + \mathrm{b}^{\dagger(s)}_{\boldsymbol{n}} \mathrm{b}^{(s)}_{\boldsymbol{n}} \right) + C_0'. \tag{3.9}$$

Perturbative Vacuum

This expression is employed in most of the field theory textbooks, and it is suitable for describing the processes of fermion creation and annihilation. However, one cannot treat the interacting vacuum state since it is assumed that the vacuum is a simple one which satisfies

$$a^{(s)}_{\boldsymbol{n}}|0\rangle = 0, \quad \mathrm{b}^{(s)}_{\boldsymbol{n}}|0\rangle = 0$$

which defines the vacuum state $|0\rangle$ in the perturbative sense. The construction of the vacuum state in interacting systems is quite difficult and mostly impossible in four dimensional field theory models.

3.1.4. Vacuum of Free Field Theory

When there is no interaction, one can easily construct the exact vacuum state. In this case, one considers the maximum number of freedom to be N, and the particles are put into the box with its length L. The momenta and energies of the negative energy particles can be written as

$$\boldsymbol{p}_n = \frac{2\pi}{L}\boldsymbol{n}, \quad E_{\boldsymbol{n}} = -\sqrt{m^2 + \boldsymbol{p}_{\boldsymbol{n}}^2}, \tag{3.10}$$

where n_k are integers and run

$$n_k = 0, \pm 1, \pm 2, \ldots, \pm N.$$

Further, one defines the cut-off momentum Λ by

$$\Lambda = \frac{2\pi}{L} N \tag{3.11}$$

and one lets N and L as large as required, keeping Λ finite. If the model field theory has no scaleful parameter with massless fermions, then physical observables must be measured by the Λ. But they should not depend on either N nor L.

Fock Space Vacuum

The Fock space vacuum can be written as

$$|0\rangle = \prod_{n_k} b^{\dagger(s)}_{\boldsymbol{n}} |0\rangle\rangle, \quad \text{with } E_{\boldsymbol{n}} = -\sqrt{m^2 + \boldsymbol{p}_{\boldsymbol{n}}^2}, \tag{3.12}$$

where $|0\rangle\rangle$ denotes a null vacuum state which is defined as

$$b^{(s)}_{\boldsymbol{n}} |0\rangle\rangle = 0, \quad a^{(s)}_{\boldsymbol{n}} |0\rangle\rangle = 0. \tag{3.13}$$

This exact vacuum state of the free Dirac field is called *perturbative vacuum state* since it is often employed for the quantum field theory with interactions.

It should be noted that the construction of the exact vacuum state is most important since from this vacuum one can create any physical states by applying creation operators. However, it is, at the same time, clear that the exact vacuum state cannot be normally obtained in the field theory models in four dimensions. There are two fermion field theory models which can be solved exactly. That is, the Schwinger model and Thirring model can be solved exactly, but they are two dimensional field theory models and will be treated in the later chapter.

3.2. Quantization of Thirring Model

Now, we present the quantized Hamiltonian of the Thirring model as an example. In the chiral representation of the γ matrices, the Hamiltonian of the Thirring model becomes

$$\hat{H} = \int dx \left[-i \left(\psi_a^\dagger \frac{\partial}{\partial x} \psi_a - \psi_b^\dagger \frac{\partial}{\partial x} \psi_b \right) + m_0 (\psi_a^\dagger \psi_b + \psi_b^\dagger \psi_a) + 2g \psi_a^\dagger \psi_a \psi_b^\dagger \psi_b \right]. \tag{3.14}$$

Now, the fermion fields are quantized in one space dimension with a box length of L

$$\psi(x) = \begin{pmatrix} \psi_a(x) \\ \psi_b(x) \end{pmatrix} = \frac{1}{\sqrt{L}} \sum_n \begin{pmatrix} a_n \\ b_n \end{pmatrix} e^{ip_n x}, \tag{3.15}$$

where

$$p_n = \frac{2\pi}{L} n, \quad \text{with } n = 0, \pm 1, \ldots.$$

The creation and annihilation operators satisfy the following anti-commutation relations

$$\{a_n, a_m^\dagger\} = \{b_n, b_m^\dagger\} = \delta_{nm}, \quad \{a_n, a_m\} = \{b_n, b_m\} = \{a_n, b_m\} = 0. \tag{3.16}$$

Quantized Hamiltonian in Chiral Representation

In this case, the quantized Hamiltonian can be written

$$\hat{H}=\sum_n\left[p_n\left(a_n^\dagger a_n-b_n^\dagger b_n\right)+m_0\left(a_n^\dagger b_n+b_n^\dagger a_n\right)+\frac{2g}{L}\tilde{j}_a(p_n)\tilde{j}_b(p_n)\right], \tag{3.17}$$

where the currents $\tilde{j}_a(p_n)$ and $\tilde{j}_b(p_n)$ in the momentum representation are given by

$$\tilde{j}_a(p_n)=\sum_l a_l^\dagger a_{l+n}, \tag{3.18a}$$

$$\tilde{j}_b(p_n)=\sum_l b_l^\dagger b_{l+n}. \tag{3.18b}$$

3.2.1. Vacuum of Thirring Model

In general, it is very difficult to construct the vacuum state for interacting field theory models. One has to solve the eigenvalue equation of the Hamiltonian with the infinite number of the negative energy particles, and normally it is impossible.

Fortunately, however, the massless Thirring model can be solved exactly by the Bethe ansatz technique . Furthermore, the solution of the vacuum state is given analytically. Since detailed discussions will be given in Chapter 7, we give only the vacuum state which is constructed by operating creation operators.

Exact Vacuum

The exact vacuum state of the massless Thirring model $|\Omega\rangle$ can be written as

$$|\Omega\rangle=\prod_{k_i^\ell}a_{k_i^\ell}^\dagger\prod_{k_j^r}b_{k_j^r}^\dagger|0\rangle\rangle, \tag{3.19}$$

where $|0\rangle\rangle$ denotes the null vacuum state with

$$a_{k_i^\ell}|0\rangle\rangle=0,\quad b_{k_j^r}|0\rangle\rangle=0. \tag{3.20}$$

The momenta k_i^ℓ and k_j^r should satisfy the periodic boundary condition (PBC) equations which are solved analytically and the momenta k_i^ℓ for left mover and k_j^r for right mover are given as

$$k_1^r=\frac{2N_0}{L}\tan^{-1}\left(\frac{g}{2}\right)\quad \text{for } n_1=0, \tag{3.21a}$$

$$k_j^r=\frac{2\pi n_j}{L}+\frac{2N_0}{L}\tan^{-1}\left(\frac{g}{2}\right)\quad \text{for } n_j=1,2,\ldots,N_0, \tag{3.21b}$$

$$k_i^\ell=\frac{2\pi n_i}{L}-\frac{2(N_0+1)}{L}\tan^{-1}\left(\frac{g}{2}\right)\quad \text{for } n_i=-1,-2,\ldots,-N_0. \tag{3.21c}$$

In this case, the vacuum energy becomes

$$E_v^{\text{true}}=-\Lambda\left\{N_0+1+\frac{2(N_0+1)}{\pi}\tan^{-1}\left(\frac{g}{2}\right)\right\},$$

where Λ denotes the cutoff momentum.

Cut-off momentum Λ

Here, g, L and N_0 denote the coupling constant, the box length and the particle number in the negative energy state, respectively. The cut-off Λ is defined as

$$\Lambda = \frac{2\pi N_0}{L}.$$

Since there is no mass scale in the massless Thirring model, all the observables must be measured in terms of the Λ. The number N_0 and the box length L can be set to any large number as required, and any physical observables should not depend on neither N_0 nor L.

3.3. Quantization of Gauge Fields in QED

In this section, we present the quantization of the gauge fields A_μ in QED. This can be found in any textbooks, and therefore we discuss it briefly so that we can calculate some of the scattering S-matrix in QED processes.

Here, we employ the quantization procedure with the Coulomb gauge fixing condition. The Hamiltonian of the electromagnetic fields is written as

$$\hat{H}_{\text{em}} = \frac{1}{2}\int \left[\Pi_k^2 - \left(\frac{\partial A_0}{\partial x_k}\right)^2 + \left(\frac{\partial A_k}{\partial x_j}\frac{\partial A_k}{\partial x_j} - \frac{\partial A_k}{\partial x_j}\frac{\partial A_j}{\partial x_k}\right)\right] d^3r, \tag{3.22}$$

where Π_k is a conjugate field to A_k and is given as

$$\Pi_k = -\dot{A}_k.$$

It should be noted that there is no term corresponding to the $\dot{A}_0$ since there is no kinetic energy term arising from the A_0 term. In this sense, the A_0 is not a dynamical variable any more.

Coulomb Gauge Fixing

Therefore, one should quantize the gauge field $\boldsymbol{A}$. However, one should be careful for the number of the degree of freedom of the gauge fields since there is a gauge fixing condition. For example, if one takes the Coulomb gauge.

$$\nabla \cdot \boldsymbol{A} = 0 \tag{3.23}$$

then, the gauge field $\boldsymbol{A}$ should have two degrees of freedom. In this case, the gauge field $\boldsymbol{A}$ can be expanded in terms of the free field solutions

$$\boldsymbol{A}(x) = \sum_{\boldsymbol{k}}\sum_{\lambda=1}^{2} \frac{1}{\sqrt{2V\omega_{\boldsymbol{k}}}} \boldsymbol{\varepsilon}(\boldsymbol{k},\lambda)\left[c_{\boldsymbol{k},\lambda}e^{-ikx} + c_{\boldsymbol{k},\lambda}^{\dagger}e^{ikx}\right], \tag{3.24}$$

where

$$\omega_{\boldsymbol{k}} = |\boldsymbol{k}|.$$

The polarization vector $\boldsymbol{\varepsilon}(\boldsymbol{k},\lambda)$ should satisfy the following relations

$$\boldsymbol{\varepsilon}(\boldsymbol{k},\lambda)\cdot\boldsymbol{k} = 0, \quad \boldsymbol{\varepsilon}(\boldsymbol{k},\lambda)\cdot\boldsymbol{\varepsilon}(\boldsymbol{k},\lambda') = \delta_{\lambda,\lambda'} \tag{3.25}$$

since the gauge field $\boldsymbol{A}$ should satisfy eq.(3.23).

Commutation Relations

Since the gauge fields are bosons, the quantization procedure must be done in the commutation relations, instead of anti-commutation relations. Therefore, the quantization can be done by requiring that $c_{\boldsymbol{k},\lambda},\ c^{\dagger}_{\boldsymbol{k},\lambda}$ should satisfy the following commutation relations

$$[c_{\boldsymbol{k},\lambda},\ c^{\dagger}_{\boldsymbol{k}',\lambda'}] = \delta_{\boldsymbol{k},\boldsymbol{k}'}\delta_{\lambda,\lambda'} \tag{3.26}$$

and all other commutation relations vanish.

In this case, the Hamiltonian $\hat{H}_{\mathrm{em}}$ of the electromagnetic fields is written in terms of the creation and annihilation operators as

$$\hat{H}_{\mathrm{em}} = \sum_{\boldsymbol{k}}\sum_{\lambda=1}^{2} \omega_{\boldsymbol{k}}\left(c^{\dagger}_{\boldsymbol{k},\lambda}c_{\boldsymbol{k},\lambda} + \frac{1}{2}\right). \tag{3.27}$$

From eq.(3.27), one sees that there are two degrees of freedom for the quantized gauge fields. Since the gauge field $\boldsymbol{A}$ has always a gauge freedom, it may be the best to quantize the gauge field $\boldsymbol{A}$ in terms of the creation and annihilation operators $c_{\boldsymbol{k},\lambda},\ c^{\dagger}_{\boldsymbol{k},\lambda}$ after the gauge fixing is done.

Zero Point Energy

Eq.(3.27) contains a zero point energy. That is, the vacuum state where there is no electromagnetic field present has an infinite energy

$$E_{\mathrm{vac}} = 2 \times \frac{1}{2}\sum_{\boldsymbol{k}} \omega_{\boldsymbol{k}} = \sum |\boldsymbol{k}| \to \infty.$$

However, there is nothing serious since the vacuum state cannot be observed. Therefore, one should measure the energy of excited states from the vacuum, and thus

$$\Delta E_{\mathrm{em}} = \sum_{\boldsymbol{k}}\sum_{\lambda=1}^{2} \omega_{\boldsymbol{k}}\left(c^{\dagger}_{\boldsymbol{k},\lambda}c_{\boldsymbol{k},\lambda} + \frac{1}{2}\right) - E_{\mathrm{vac}} = \sum_{\boldsymbol{k}}\sum_{\lambda=1}^{2} \omega_{\boldsymbol{k}} c^{\dagger}_{\boldsymbol{k},\lambda}c_{\boldsymbol{k},\lambda}$$

must be physical observables.

3.4. Quantization of Schrödinger Field

As we discussed in the first chapter, the non-relativistic fields do not have to be quantized since there are no creation and annihilation of particles in the non-relativistic kinematics.

Nevertheless, one can quantize the Schrödinger field and work out physical observables in the second quantized representation. It is, of course, the same as the classical field theory calculation, but sometime the second quantized representation is easier than the classical field version.

3.4.1. Creation and Annihilation Operators

We consider the Lagrangian density of eq.(1.2)

$$\mathcal{L} = i\psi^\dagger \frac{\partial \psi}{\partial t} - \frac{1}{2m}\nabla\psi^\dagger \cdot \nabla\psi - \psi^\dagger U \psi.$$

In this case, the Schrödinger field $\psi(\boldsymbol{r},t)$ can be expanded in terms of the free field solutions

$$\psi(\boldsymbol{r},t) = \frac{1}{\sqrt{V}} \sum_{\boldsymbol{k}} a_{\boldsymbol{k}} e^{i\boldsymbol{k}\cdot\boldsymbol{r} - iEt}, \tag{3.28}$$

where $a_{\boldsymbol{k}}$ and $a_{\boldsymbol{k}}^\dagger$ are required to satisfy the following anti-commutation relations

$$\{a_{\boldsymbol{k}}, a_{\boldsymbol{k}'}^\dagger\} = \delta_{\boldsymbol{k},\boldsymbol{k}'}, \quad \{a_{\boldsymbol{k}}, a_{\boldsymbol{k}'}\} = \{a_{\boldsymbol{k}}^\dagger, a_{\boldsymbol{k}'}^\dagger\} = 0. \tag{3.29}$$

Since one assumes that the Schrödinger field corresponds to fermions, one can carry out the field quantization in the anti-commutation relations.

3.4.2. Fermi Gas Model

From eq.(3.28), one sees that there is no anti-particle present in this model. This is clear since one starts from the vacuum which has no particle at all. On the other hand, if one starts from the Fermi surface and identifies the vacuum state in which all the states are occupied up to the Fermi energy ε_F, then one can discuss the particle-hole states which have some similarity with the Dirac hole state. In fact, the formulation is just the same as the Dirac vacuum, but there is of course no anti-particle. The hole state is just a hole in the Fermi sea.

Fermi Momentum

In this picture, particles corresponding to the Schrödinger fields are assumed to obey the Pauli principle. In this respect, the Schrödinger field is considered as the fermion field which should satisfy the anti-commutation relations as shown in eq.(3.29). Therefore, in the Fermi gas model, particle states are occupied up to the Fermi momentum k_F, and when the system has the number of particle N, then one finds

$$N = \sum_{n_x,n_y,n_z} = \frac{L^3}{(2\pi)^3} \int_{|\boldsymbol{k}|\le k_F} d^3k = \frac{L^3}{6\pi^2} k_F^3. \tag{3.30}$$

In this case, the Fermi energy ε_F is written as

$$\varepsilon_F = \frac{1}{2M} k_F^2, \tag{3.31}$$

where M denotes the mass of the Schrödinger particle. Eq.(3.30) indicates that the density ρ of the system becomes

$$\rho = \frac{N}{L^3} = \frac{1}{6\pi^2} k_F^3. \tag{3.32}$$

Spin and Isospin

If the particle is a nucleon, it has spin s and isospin t. In this case, eq.(3.32) becomes

$$\rho = \frac{N}{L^3} = (2s+1)(2t+1)\frac{1}{6\pi^2}k_F^3 = \frac{2}{3\pi^2}k_F^3, \tag{3.33}$$

where

$$s = t = \frac{1}{2}. \tag{3.34}$$

3.5. Quantized Hamiltonian of QED and Eigenstates

The Hamiltonian of fermions with electromagnetic fields is given in eq.(1.33)

$$H = \int \left\{ \bar{\psi}(-i\boldsymbol{\gamma}\cdot\boldsymbol{\nabla} + m)\psi - g\boldsymbol{j}\cdot\boldsymbol{A} \right\} d^3r$$

$$+\frac{g^2}{8\pi}\int \frac{j_0(\boldsymbol{r}')j_0(\boldsymbol{r})}{|\boldsymbol{r}'-\boldsymbol{r}|} d^3r\, d^3r' + \frac{1}{2}\int (\boldsymbol{E}_T^2 + \boldsymbol{B}^2)\, d^3r.$$

Now, the fermion field ψ and the gauge field $\boldsymbol{A}$ are quantized as

$$\psi(\boldsymbol{r}) = \sum_{\boldsymbol{n},s} \frac{1}{\sqrt{L^3}} \left(a_{\boldsymbol{n}}^{(s)} u_{\boldsymbol{n}}^{(s)} e^{i\boldsymbol{p}_n\cdot\boldsymbol{r}} + b_{\boldsymbol{n}}^{(s)} v_{\boldsymbol{n}}^{(s)} e^{i\boldsymbol{p}_n\cdot\boldsymbol{r}} \right), \tag{3.35}$$

$$\boldsymbol{A}(\boldsymbol{r}) = \sum_{\boldsymbol{k}} \sum_{\lambda=1}^{2} \frac{1}{\sqrt{2V\omega_{\boldsymbol{k}}}} \boldsymbol{\varepsilon}(\boldsymbol{k},\lambda) \left[c_{\boldsymbol{k},\lambda} e^{-i\boldsymbol{k}\cdot\boldsymbol{r}} + c_{\boldsymbol{k},\lambda}^{\dagger} e^{i\boldsymbol{k}\cdot\boldsymbol{r}} \right]. \tag{3.36}$$

3.5.1. Quantized Hamiltonian

In this case, the quantized Hamiltonian of eq.(1.33) can be written as

$$\hat{H} = \sum_{\boldsymbol{n},s} E_n \left(a_{\boldsymbol{n}}^{\dagger(s)} a_{\boldsymbol{n}}^{(s)} - b_{\boldsymbol{n}}^{\dagger(s)} b_{\boldsymbol{n}}^{(s)} \right) + \sum_{\boldsymbol{k}} \sum_{\lambda=1}^{2} \omega_{\boldsymbol{k}} \left(c_{\boldsymbol{k},\lambda}^{\dagger} c_{\boldsymbol{k},\lambda} + \frac{1}{2} \right)$$

$$-g\int \bar{\psi}(\boldsymbol{r})\boldsymbol{\gamma}\psi(\boldsymbol{r})\cdot\boldsymbol{A}(\boldsymbol{r})\, d^3r + \frac{g^2}{8\pi}\int \frac{(\bar{\psi}(\boldsymbol{r}')\gamma_0\psi(\boldsymbol{r}'))(\bar{\psi}(\boldsymbol{r})\gamma_0\psi(\boldsymbol{r}))}{|\boldsymbol{r}'-\boldsymbol{r}|} d^3r\, d^3r'. \tag{3.37}$$

The interaction terms are so complicated that we do not write them here in terms of the creation-annihilation operators in an explicit fashion. However, one notices that the interaction terms should induce particle-anti-particle creations or destructions.

3.5.2. Eigenvalue Equation

Now, one should solve eq.(3.1) with the above Hamiltonian eq.(3.37)

$$\hat{H}|\Psi\rangle = E|\Psi\rangle, \tag{3.38}$$

where $|\Psi\rangle$ may be written even for simple bosonic excitation cases

$$|\Psi\rangle = \sum_{(\boldsymbol{p}_1,s_1),\dots,(\boldsymbol{p}_n,s_n),(\boldsymbol{q}_1,t_1),\dots,(\boldsymbol{q}_n,t_n)} f\Big((\boldsymbol{p}_1,s_1),\dots,(\boldsymbol{p}_n,s_n),(\boldsymbol{q}_1,t_1),\dots,(\boldsymbol{q}_n,t_n)\Big)$$

$$\times a^{\dagger(s_1)}_{\boldsymbol{p}_1}\cdots a^{\dagger(s_n)}_{\boldsymbol{p}_n} b^{(t_1)}_{\boldsymbol{q}_1}\cdots b^{(t_n)}_{\boldsymbol{q}_n}|\Omega\rangle, \tag{3.39}$$

where f is the wave function which should be determined so as to satisfy eq.(3.1). The energy eigenvalue E in eq.(3.38) may be calculated by the diagonalization procedure where the space is spanned in terms of $f((\boldsymbol{p}_1,s_1),\dots,(\boldsymbol{p}_n,s_n),(\boldsymbol{q}_1,t_1),\dots,(\boldsymbol{q}_n),t_n))$. For a practical evaluation, one has to truncate the space significantly so as to carry out any numerical calculations. Numerical calculations in two dimensional QED with finite fermion mass will be discussed in Chapter 5.

3.5.3. Vacuum State $|\Omega\rangle$

$|\Omega\rangle$ denotes the vacuum state which should satisfy the eigenvalue equation of eq.(3.1)

$$\hat{H}|\Omega\rangle = E_\Omega|\Omega\rangle,$$

where E_Ω denotes the vacuum energy, and $|\Omega\rangle$ is full of negative energy particles

$$|\Omega\rangle = \prod_{\boldsymbol{p},s} b^{\dagger(s)}_{\boldsymbol{p}}|0\rangle\rangle,$$

where $|0\rangle\rangle$ denotes the null vacuum which satisfies

$$b^{(s)}_{\boldsymbol{p}}|0\rangle\rangle = 0, \quad a^{(s)}_{\boldsymbol{p}}|0\rangle\rangle = 0.$$

One sees clearly that it is practically impossible to solve the eigenvalue equation of the Hamiltonian in eq.(3.1).

Chapter 4

Goldstone Theorem and Spontaneous Symmetry Breaking

The continuous symmetry of the Lagrangian density leads to the conservation of currents and therefore the system should have a conserved charge associated with the symmetry. The best example must be a global gauge symmetry in which the Lagrangian density is invariant under the transformation of the Dirac field ψ as

$$\psi' = e^{i\alpha}\psi \Longrightarrow \mathcal{L}' = \mathcal{L},$$

where α is a real constant. In this case, the vector current

$$j_\mu = \bar{\psi}\gamma_\mu\psi$$

is conserved. That is,

$$\partial_\mu j^\mu = 0. \qquad (4.1)$$

In fermion field theory models, the global gauge symmetry is not broken at any level even though the vacuum state can, in principle, break its symmetry. In particular, the gauge invariance of the local gauge field theory should hold rigorously since the violation of the local gauge invariance should lead to the breakdown of defining physical observables.

On the other hand, the chiral symmetry behaves quite differently from the global gauge symmetry. When the Lagrangian density is invariant under the chiral symmetry transformation

$$\psi' = e^{i\alpha\gamma^5}\psi \Longrightarrow \mathcal{L}' = \mathcal{L}$$

the axial vector current

$$j_5^\mu = \bar{\psi}\gamma^\mu\gamma_5\psi$$

is also conserved, that is,

$$\partial_\mu j_5^\mu = 0. \qquad (4.2)$$

However, the chiral symmetry is broken not only because of the chiral anomaly but also in terms of the spontaneous symmetry breaking mechanism. In the former case, the conservation of the axial vector current is violated by the anomaly term while, in the spontaneous chiral symmetry breaking, the vacuum loses its symmetry since the vacuum state prefers the lowest energy state which does not have to keep its symmetry. The discussion of the anomaly term is given in Chapter 5 in the context of the Schwinger model where the basic mechanism of the chiral anomaly and the violation of the axial vector current are described in a transparent fashion.

In this chapter, we discuss the symmetry breaking phenomena in fermion field theory models. In particular, we clarify what is physics of the spontaneous symmetry breaking and the Goldstone theorem [60, 61]. Normally, the mathematics in the theorem can be understood in a straightforward way, but its physics in connection with the theorem is always difficult since one has to examine all the possible conditions in nature when the symmetry is broken spontaneously.

First, we explain the general feature of the symmetry and its preservation in quantum many body theory, before going to the discussion of the Goldstone theorem. Then, we explain the Goldstone theorem and the problem related to the proof of existence of the Goldstone boson. Also, we present a new interpretation of the Goldstone theorem. In particular, we give a good example of the chiral symmetry breaking in the Thirring model together with the exact solution of the vacuum state.

Further, we comment on the symmetry breaking of boson field theory models together with the Higgs mechanism. However, this part is presented with reservation since there are still some problems which are not yet clarified completely in this textbook. The difficulty of the symmetry breaking physics in boson field theory is partly because the scalar boson field itself has some serious problems as will be discussed in Appendix C and partly because there is no model which can present exact solutions of the boson field theory models.

4.1. Symmetry and Its Breaking in Vacuum

Various symmetries of Hamiltonian play an important role for determining the energy eigenvalues, and in quantum mechanics, one often sees that the lowest state (ground state) preserves the symmetry of the Hamiltonian. In fact, those states which break the symmetry are, in general, higher than the symmetry preserving state for the same quantum numbers or configurations of the wave function. This can be naturally realized in quantum many body theory as our experiences tell us.

However, the physics of the symmetry breaking in quantum field theory is quite different, and phenomena which seem to be in an apparent contradiction with the picture of quantum many body theory can indeed occur. This is called *spontaneous symmetry breaking* and it has been discussed extensively in many field theory textbooks. In the spontaneous symmetry breaking, the vacuum state of the field theory models is realized with the symmetry broken state. That is, the true vacuum prefers the symmetry broken state, contrary to the naive expectation in quantum many body theory.

A question may arise as to what should be an intuitive explanation why the vacuum of the field theory models has the symmetry broken state as the most favorable state, in contradiction with the experiences of quantum many body theory. Here, we show that the symmetry broken state of the vacuum in the field theory models is naturally realized in the context of quantum many body theory. In short, the vacuum of quantum field theory is constructed by the negative energy particles, and therefore, the symmetry preserving state should have the absolute magnitude of its total energy which is smaller than the symmetry broken state. This is just consistent with the prediction of quantum many body theory. However, energies of the vacuum are all negative, and thus the lower state is, of course, the one that breaks the symmetry since the absolute magnitude of its energy is larger than that of the symmetry preserving state. This is exactly what one observes in the Thirring model as will be discussed in Chapter 7.

4.1.1. Symmetry in Quantum Many Body Theory

In quantum mechanics, the ground state energy of the particle Hamiltonian H can be written

$$E_{\rm tot} = \langle 0|H|0\rangle, \tag{4.3}$$

where we denote the ground state by $|0\rangle$ which preserves the symmetry of the Hamiltonian H. In this case, the lowest state is normally the one that keeps the symmetry. The total energy of the states which do not keep the symmetry should be found in higher energies than the symmetry preserving state. This energy of the N particle state may be written more explicitly as

$$E_{\rm tot} = \sum_{i=1}^{N} \mathcal{E}_i(\boldsymbol{k}_i), \tag{4.4}$$

where we denote the energy of the i-th particle by $\mathcal{E}_i(\boldsymbol{k}_i)$. The momentum $\boldsymbol{k}_i$ should be determined by solving the many body equations of motion

$$F_i(\boldsymbol{k}_1,\ldots,\boldsymbol{k}_N) = 0, \;\; i = 1,\ldots,N. \tag{4.5}$$

Now, suppose there is a symmetry in the Hamiltonian H. The solution of the above equations should be specified by the symmetric and the symmetry broken solutions. In quantum many body system, it is often the case that the symmetric solution $E_{\rm tot}^{sym}$ is lower than the symmetry broken solution $E_{\rm tot}^{sym.br}$.

$$E_{\rm tot}^{sym} < E_{\rm tot}^{sym.br}. \tag{4.6}$$

This, of course, depends on the interactions between particles in the system, and one can only claim that there should be some systems in which eq.(4.6) can hold.

4.1.2. Symmetry in Field Theory

In quantum field theory, the symmetry breaking phenomena occur in a completely different fashion. The lowest state which is called *vacuum* sometimes breaks the continuous symmetry since it is found to be lower than the symmetry preserving state. Since the continuous symmetry has the Noether current associated with its symmetry, the current conservation should hold true in the process of determining the lowest state of the model field theory as long as the model has no anomaly.

Why is it possible that the symmetry broken state becomes lower than the symmetry preserving state in an obvious contradiction with the picture of quantum many body theory? Here, we present a simple intuitive picture why it may occur. The basic point is that the vacuum in field theory models is constructed by particles with the negative energies which are solved in the many body Dirac equations. To be more specific, the vacuum energy $E_{\rm vac}$ can be written as

$$E_{\rm vac} = -\lim_{N\to\infty} \sum_{i=1}^{N} \mathcal{E}_i(\boldsymbol{k}_i), \tag{4.7}$$

where the energy of the i-th particle is denoted by $\mathcal{E}_i$. Since the system is infinite, one should make the number N infinity at the end of the calculation. It should be noted that one cannot make the system infinity from the beginning since in this case one cannot define the total energy of the system. This construction of the vacuum in terms of the finite N-particle system and then making the number N infinity must be well justified since the deep negative energy states in the vacuum should not have any effects on the physical properties of the vacuum state.

In order to determine the momenta of the negative energy particles, one should solve the equations of motion which may be similar to eq.(4.5)

$$G_i(\boldsymbol{k}_1,\ldots,\boldsymbol{k}_N) = 0, \;\; i = 1,\ldots,N. \tag{4.8}$$

Again, one can assume that there is a symmetry in the Hamiltonian H. In this case, the solution of the above equations should be specified by the symmetric solution and the symmetry broken solution. Just in the same way as the positive energy case, if one defines the total energy $E_{\rm tot}$ by

$$E_{\rm tot} = \sum_{i=1}^{N} \mathcal{E}_i(\boldsymbol{k}_i) \tag{4.9}$$

then the symmetric solution $E_{\rm tot}^{sym}$ must be lower than the symmetry broken solution $E_{\rm tot}^{sym.br}$

$$E_{\rm tot}^{sym} < E_{\rm tot}^{sym.br}. \tag{4.10}$$

However, the energy of the vacuum is all negative, and therefore one sees that the symmetry broken vacuum state must be the lowest, that is

$$E_{\rm vac}^{sym.br} < E_{\rm vac}^{sym} \tag{4.11}$$

which is just opposite to the prediction of the quantum many body theory.

In this way, the vacuum in field theory models prefers the symmetry broken state. The symmetry preserving state has the lowest energy in magnitude, but due to the negative sign in front (eq.(4.7)), the lowest energy must be the one that breaks the symmetry. This is exactly what happens in the spontaneous symmetry breaking in fermion field theory models. This appearance of the two states, one that preserves the symmetry and the other that breaks the symmetry must depend on the dynamical properties of the models one considers. Up to now, the Thirring model exhibits the two states in the vacuum, and therefore it presents a good example of the spontaneous symmetry breaking physics. In the next section, we discuss the Goldstone theorem, keeping this fact in mind.

4.2. Goldstone Theorem

The physics of the spontaneous symmetry breaking started from the Goldstone theorem. The theorem states that there should appear a massless boson when the symmetry of the vacuum state is spontaneously broken. In this process of the spontaneous symmetry breaking, the current conservation should hold. This is important since the Goldstone theorem is entirely based on the current conservation, and without the current conservation the theorem cannot be proved.

Here, without loss of generality, we can restrict our discussion to the chiral symmetry breaking of the fermion field theory models. In this case, the chiral charge Q_5 must be a conserved quantity.

4.2.1. Conservation of Chiral Charge

When the Lagrangian density has the chiral symmetry which can be represented by the unitary operator $U(\alpha)$, there is a conserved current associated with the symmetry, which is eq.(4.2). In this case, there is a conserved chiral charge Q_5

$$Q_5 = \int j_5^0(x)\, d^3r. \tag{4.12}$$

The quantized Hamiltonian $\hat{H}$ of this system is invariant under the unitary transformation $U(\alpha)$,

$$U(\alpha)\hat{H}U(\alpha)^{-1} = \hat{H}. \tag{4.13}$$

Therefore, the chiral charge operator $\hat{Q}_5$ commutes with the Hamiltonian $\hat{H}$

$$\hat{Q}_5\hat{H} = \hat{H}\hat{Q}_5. \tag{4.14}$$

4.2.2. Symmetry of Vacuum

The symmetry of the vacuum is determined in terms of its energy, and when the lowest energy state is realized, the vacuum may break the symmetry which is possessed in the Hamiltonian. Here, the symmetry of the vacuum can be defined in the following way. The symmetric vacuum is denoted by $|0\rangle$ while the symmetry broken vacuum is denoted by $|\Omega\rangle$. They satisfy the following equations,

$$U(\alpha)|0\rangle = |0\rangle, \tag{4.15a}$$

$$U(\alpha)|\Omega\rangle \neq |\Omega\rangle. \tag{4.15b}$$

These equations can be written in terms of the chiral charge operator $\hat{Q}_5$ as

$$\hat{Q}_5|0\rangle = 0, \tag{4.16a}$$

$$\hat{Q}_5|\Omega\rangle \neq 0. \tag{4.16b}$$

4.2.3. Commutation Relation

In the Goldstone theorem, one starts from the following commutation relation which is an identity equation,

$$\left[\hat{Q}_5, \int \bar{\psi}(x)\gamma_5\psi(x)\, d^3r\right] = -2\int \bar{\psi}(x)\psi(x)\, d^3r. \tag{4.17}$$

Now, one takes the expectation value of the above equation with the vacuum state $|\Omega\rangle$, and obtains

$$\langle\Omega|\left[\hat{Q}_5, \int \bar{\psi}(x)\gamma_5\psi(x)\,d^3r\right]|\Omega\rangle = -2\langle\Omega|\int \bar{\psi}(x)\psi(x)\,d^3r|\Omega\rangle. \tag{4.18}$$

If the right hand side (fermion condensate) has a finite value, then the vacuum state $|\Omega\rangle$ must be a symmetry broken state since it should at least satisfy eq.(4.16b) because of the finite value of the left hand side.

Now, one can rewrite eq.(4.18) and assume that the right hand side is nonzero because of the finite fermion condensate

$$\sum_n (2\pi)^3\delta(\boldsymbol{p}_n)\left[\langle\Omega|j_5^0|n\rangle\langle n|\bar{\psi}\gamma_5\psi|\Omega\rangle e^{-iE_nt} - \langle\Omega|\bar{\psi}\gamma_5\psi|n\rangle\langle n|j_5^0|\Omega\rangle e^{iE_nt}\right] \neq 0, \tag{4.19}$$

where $|n\rangle$ denotes the complete set of the fermion number zero states of the field theory model one considers. Therefore, bosonic states as well as the pair of massless free fermion and anti-fermion states should be included in the intermediate states. From eq.(4.19), one sees that the right hand side is nonzero and time-independent while the left hand side is time dependent unless there is a state $|n\rangle$ that satisfies

$$E_n = 0 \;\text{ for }\; \boldsymbol{p}_n = 0. \tag{4.20}$$

Eq.(4.20) is just consistent with the dispersion relation of a massless boson.

4.2.4. Momentum Zero State

However, one easily notices that the free massless fermion and anti-fermion pair can also satisfy eq.(4.20). To be more specific, one can write the energy and momentum of the state $|n\rangle$ as

$$E_n = E_f + E_{\bar{f}}, \tag{4.21a}$$

$$\boldsymbol{p}_n = \boldsymbol{p}_f + \boldsymbol{p}_{\bar{f}}, \tag{4.21b}$$

where $\boldsymbol{p}_f$ ($\boldsymbol{p}_{\bar{f}}$) and E_f ($E_{\bar{f}}$) denote the momentum and energy of the fermion (anti-fermion), respectively. For the free massless fermion and anti-fermion pair with

$$\boldsymbol{p}_f = 0 \;\text{ and }\; \boldsymbol{p}_{\bar{f}} = 0 \tag{4.22a}$$

one obtains [41, 42, 43, 69]

$$E_f = 0 \;\text{ and }\; E_{\bar{f}} = 0 \tag{4.22b}$$

and therefore eq.(4.20) is indeed satisfied

$$\boldsymbol{p}_n = \boldsymbol{p}_f + \boldsymbol{p}_{\bar{f}} = 0 \Longrightarrow E_n = E_f + E_{\bar{f}} = 0.$$

Dispersion Relation of Massless Boson

From one information of eq.(4.20), one could derive the dispersion relation of a massless boson if the state must be covariant

$$E_n^2 - \boldsymbol{p}_n^2 = (p_n)_\mu p_n^\mu = 0.$$

The requirement of the covariance for the state $|n\rangle$ may be justified when the state is an isolated system. However, it is difficult to show the covariance from only one information on the zero momentum state which is just eq.(4.20) since the vacuum is always in the momentum zero state.

On the other hand, the dispersion relation of a massless boson

$$E = |\boldsymbol{p}|$$

contains information which should be valid for arbitrary momentum $\boldsymbol{p}$. Intuitively, it is clear that one cannot obtain the dispersion relation of a massless boson from only one information which is at $\boldsymbol{p} = 0$. Therefore, one sees that the Goldstone theorem proves the existence of a free massless fermion and anti-fermion pair for the fermion field theory models, and this is, of course, a natural statement. But there exists no massless boson.

4.2.5. Pole in S-matrix

In the spontaneous symmetry breaking, a massless pole in the S-matrix calculations is sometimes found, and they claim that the pole should be related to a massless boson (Goldstone boson). However, the S-matrix in these calculations is evaluated in the trivial (perturbative) vacuum, and it has nothing to do with a physical massless boson. Furthermore, one has to be careful that a pole in the S-matrix may not have to correspond to a bound state, and if one wishes to find a bound state pole, then one should calculate poles in exact Green's function in which the evaluation should be based on the symmetry broken vacuum state. But this is just the same as solving the system exactly.

4.3. New Interpretation of Goldstone Theorem

Here, we present a new interpretation of the Goldstone theorem and clarify what is indeed the physics of the spontaneous symmetry breaking in fermion field theory models. Since the spontaneous symmetry breaking is connected with the structure of the vacuum, we should understand the physical feature of the vacuum. However, most of the difficulties of the field theory models are concentrated in the dynamical evaluation of the vacuum state, and therefore, we should first treat the spontaneous symmetry breaking physics in terms of finite number of freedoms. After that, we should examine whether the procedure can be justified when the number of the freedom is set to infinity.

4.3.1. Eigenstate of Hamiltonian and $\hat{Q}_5$

Since the operator $\hat{Q}_5$ commutes with the Hamiltonian $\hat{H}$ as discussed in eq.(4.14),

$$\hat{Q}_5 \hat{H} = \hat{H} \hat{Q}_5$$

the $\hat{Q}_5$ has the same eigenstate as the Hamiltonian. If one defines the symmetry broken vacuum state $|\Omega\rangle$ by the eigenstate of the Hamiltonian $\hat{H}$ with its energy eigenvalue E_Ω, then one can write

$$\hat{H}|\Omega\rangle = E_\Omega|\Omega\rangle. \tag{4.23}$$

In this case, one can also write the eigenvalue equation for the $\hat{Q}_5$

$$\hat{Q}_5|\Omega\rangle = q_5|\Omega\rangle \tag{4.24}$$

with its eigenvalue q_5. These equations should hold for the exact eigenstates of the Hamiltonian.

Now, if one takes the expectation value of eq.(4.17) with the symmetry broken vacuum $|\Omega\rangle$ which is the eigenstate of the Hamiltonian as well as $\hat{Q}_5$, then one obtains for the left hand side as

$$\langle\Omega|\left[\hat{Q}_5, \int \bar{\psi}(x)\gamma_5\psi(x)\,d^3r\right]|\Omega\rangle$$

$$= \langle\Omega|q_5 \int \bar{\psi}(x)\gamma_5\psi(x)\,d^3r - \left(\int \bar{\psi}(x)\gamma_5\psi(x)\,d^3r\right) q_5|\Omega\rangle = 0 \tag{4.25}$$

with the help of eq.(4.24). This means that the right hand side of eq.(4.18) must vanish, that is,

$$\langle\Omega|\int \bar{\psi}(x)\psi(x)\,d^3r|\Omega\rangle = 0. \tag{4.26}$$

Therefore, the exact eigenstate of the vacuum has no fermion condensate even in the symmetry broken vacuum. The relation of eq.(4.18) has repeatedly been used, and if there is a finite fermion condensate, then the symmetry of the vacuum must be broken since the left hand side of eq.(4.18) vanishes due to eq.(4.16a) for the symmetric vacuum state. However, as seen above, the condensate must vanish even for the symmetry broken vacuum state if the vacuum is the eigenstate of the Hamiltonian, which is a natural consequence.

4.3.2. Index of Symmetry Breaking

The way out of this dilemma is simple. One should not take the expectation value of the vacuum state. Instead, the index of the symmetry breaking in connection with the condensate operator $\int \bar{\psi}(x)\psi(x)\,dx$ should be the following operator equation

$$\left(\int \bar{\psi}(x)\psi(x)\,d^3r\right)|\Omega\rangle = |\Omega'\rangle + C_1|\Omega\rangle, \tag{4.27}$$

where $|\Omega'\rangle$ denotes an operator-induced state which is orthogonal to the $|\Omega\rangle$

$$\langle\Omega|\Omega'\rangle = 0.$$

C_1 is related to the condensate value. For the exact eigenstate which breaks the chiral symmetry, one finds

$$C_1 = 0. \tag{4.28}$$

In this case, the identity equation of (4.17) can be applied to the state $|\Omega\rangle$ and one obtains

$$(\hat{Q}_5 - q_5) \int \bar{\psi}(x)\gamma_5\psi(x)\, d^3r|\Omega\rangle = -2|\Omega'\rangle \tag{4.29}$$

with the help of eq.(4.24). Indeed, eq.(4.29) holds true for the exact eigenstate. It is now clear that one should not take the expectation value of eq.(4.17) by the vacuum state. It just gives a trivial equation of "0" = "0".

4.4. Chiral Symmetry in Quantized Thirring Model

In this section, we show explicitly how the chiral symmetry in the quantized Thirring model Hamiltonian behaves in terms of the creation and annihilation operators.

4.4.1. Lagrangian Density

The Lagrangian density of the Thirring model is written as

$$\mathcal{L} = i\bar{\psi}\gamma_\mu\partial^\mu\psi - \frac{1}{2}g\bar{\psi}\gamma_\mu\psi\,\bar{\psi}\gamma^\mu\psi \tag{4.30}$$

which is invariant under the chiral transformation

$$\psi' = e^{i\alpha\gamma_5}\psi. \tag{4.31}$$

Therefore, the chiral charge

$$\hat{Q}_5 = \int \bar{\psi}(x)\gamma^0\gamma_5\psi(x)\, d^3r$$

is a conserved quantity. In fact, it commutes with the Hamiltonian $\hat{H}$

$$[\hat{H}, \hat{Q}_5] = 0.$$

4.4.2. Quantized Hamiltonian

The field ψ is quantized as

$$\psi(x) = \begin{pmatrix} \psi_a(x) \\ \psi_b(x) \end{pmatrix} = \frac{1}{\sqrt{L}}\sum_n \begin{pmatrix} a_n \\ b_n \end{pmatrix} e^{ip_n x}, \quad \text{with } p_n = \frac{2\pi}{L}n, \tag{4.32}$$

where the operators a_n and b_n should satisfy the anti-commutation relations

$$\{a_n, a_m^\dagger\} = \{b_n, b_m^\dagger\} = \delta_{nm}, \quad \{a_n, a_m\} = \{b_n, b_m\} = \{a_n, b_m\} = 0.$$

In this case, the quantized Hamiltonian of the Thirring model becomes

$$\hat{H} = \sum_n \left[p_n \left(a_n^\dagger a_n - b_n^\dagger b_n\right) + \frac{2g}{L}\left(\sum_l a_l^\dagger a_{l+n}\right)\left(\sum_m b_m^\dagger b_{m+n}\right)\right]. \tag{4.33}$$

4.4.3. Chiral Transformation for Operators

The chiral transformation of eq.(4.31) is written in terms of a_n and b_n as

$$U(\alpha)\begin{pmatrix} a_n \\ b_n \end{pmatrix} U^{-1}(\alpha) = \begin{pmatrix} e^{i\alpha} a_n \\ e^{-i\alpha} b_n \end{pmatrix}, \quad U(\alpha)\begin{pmatrix} a_n^\dagger \\ b_n^\dagger \end{pmatrix} U^{-1}(\alpha) = \begin{pmatrix} e^{-i\alpha} a_n^\dagger \\ e^{i\alpha} b_n^\dagger \end{pmatrix}. \tag{4.34}$$

In this case, one easily sees that the Hamiltonian is also invariant under the transformation of eq.(4.34)

$$U(\alpha)\hat{H}U^{-1}(\alpha) = \sum_n \Bigg[p_n U(\alpha)\left(a_n^\dagger a_n - b_n^\dagger b_n\right) U^{-1}(\alpha)$$

$$+\frac{2g}{L} U(\alpha)\left(\sum_l a_l^\dagger a_{l+n}\right) U^{-1}(\alpha) U(\alpha)\left(\sum_m b_m^\dagger b_{m+n}\right) U^{-1}(\alpha)\Bigg] = \hat{H}. \tag{4.35}$$

4.4.4. Unitary Operator with Chiral Charge $\hat{Q}_5$

Now, the unitary operator $U(\alpha)$ can be explicitly written as

$$U(\alpha) = e^{-i\alpha\hat{Q}_5}, \tag{4.36}$$

where the chiral charge operator $\hat{Q}_5$ is expressed in terms of the creation and annihilation operators as

$$\hat{Q}_5 = \int \bar{\psi}(x)\gamma^0\gamma_5\psi(x)\, d^3r = \sum_n \left[a_n^\dagger a_n - b_n^\dagger b_n\right].$$

In this case, one can confirm the following identities

$$U(\alpha)\begin{pmatrix} a_n \\ b_n \end{pmatrix} U^{-1}(\alpha) = e^{-i\alpha\hat{Q}_5}\begin{pmatrix} a_n \\ b_n \end{pmatrix} e^{i\alpha\hat{Q}_5} = \begin{pmatrix} e^{i\alpha} a_n \\ e^{-i\alpha} b_n \end{pmatrix}, \tag{4.37a}$$

$$U(\alpha)\begin{pmatrix} a_n^\dagger \\ b_n^\dagger \end{pmatrix} U^{-1}(\alpha) = e^{-i\alpha\hat{Q}_5}\begin{pmatrix} a_n^\dagger \\ b_n^\dagger \end{pmatrix} e^{i\alpha\hat{Q}_5} = \begin{pmatrix} e^{-i\alpha} a_n^\dagger \\ e^{i\alpha} b_n^\dagger \end{pmatrix} \tag{4.37b}$$

which are just the same as eq.(4.34).

4.4.5. Symmetric and Symmetry Broken Vacuum

In the Thirring model, the Bethe ansatz solutions show that there are symmetric vacuum $|0\rangle$ and symmetry broken vacuum $|\Omega\rangle$ states, and the energy of the symmetry broken vacuum is found to be lower than the symmetric vacuum energy. They are the eigenstate of the chiral charge $\hat{Q}_5$ and one finds

$$\hat{Q}_5|0\rangle = 0, \tag{4.38a}$$

$$\hat{Q}_5|\Omega\rangle = \pm|\Omega\rangle. \tag{4.38b}$$

Therefore, for the unitary operator $U(\alpha) = e^{-i\alpha\hat{Q}_5}$, the symmetric vacuum does not change

$$U(\alpha)|0\rangle = e^{-i\alpha\hat{Q}_5}|0\rangle = |0\rangle \tag{4.39a}$$

while the symmetry broken vacuum becomes

$$U(\alpha)|\Omega\rangle = e^{-i\alpha\hat{Q}_5}|\Omega\rangle = e^{\pm i\alpha}|\Omega\rangle \neq |\Omega\rangle \tag{4.39b}$$

which indeed satisfies the criteria of the symmetry broken vacuum state in eqs.(4.15).

4.5. Spontaneous Chiral Symmetry Breaking

There is one good example which perfectly satisfies the above requirements of the spontaneous chiral symmetry breaking and zero fermion condensate. That is the Bethe ansatz vacuum of the massless Thirring model which will be discussed in detail in Chapter 7. Here, we employ the results of the Bethe ansatz vacuum of the Thirring model and discuss the vacuum and its properties in the context of the spontaneous symmetry breaking.

4.5.1. Exact Vacuum of Thirring Model

Now, the left and right mover fermion creation operators can be denoted by $a_k^\dagger$, $b_k^\dagger$, respectively, and thus the vacuum state $|\Omega\rangle$ can be written as

$$|\Omega\rangle = \prod_{k_i^\ell} a_{k_i^\ell}^\dagger \prod_{k_j^r} b_{k_j^r}^\dagger |0\rangle\rangle, \tag{4.40}$$

where $|0\rangle\rangle$ denotes the null vacuum state with

$$a_{k_i^\ell}|0\rangle\rangle = 0, \quad b_{k_j^r}|0\rangle\rangle = 0. \tag{4.41}$$

The momenta k_j^ℓ for left mover and k_i^r for right mover should satisfy the periodic boundary condition (PBC) equations which are solved analytically, and therefore one can determine the momenta k_j^ℓ and k_i^r, as will be given in Chapter 7.

4.5.2. Condensate Operator

Now, the condensate operator $\int \bar{\psi}(x)\psi(x)\,dx$ can be written as

$$\int \bar{\psi}(x)\psi(x)\,dx = \sum_n (b_n^\dagger a_n + a_n^\dagger b_n). \tag{4.42}$$

Therefore, eq.(4.27) becomes

$$\begin{aligned} \int \bar{\psi}(x)\psi(x)\,dx|\Omega\rangle &= \sum_n (b_n^\dagger a_n + a_n^\dagger b_n)|\Omega\rangle \\ &= \sum_n \left\{ \prod_{k_i^\ell, k_i^\ell \neq n} a_{k_i^\ell}^\dagger \prod_{k_j^r} b_{k_j^r}^\dagger b_n^\dagger |0\rangle\rangle + \prod_{k_i^\ell} a_{k_i^\ell}^\dagger \prod_{k_j^r, k_j^r \neq n} b_{k_j^r}^\dagger a_n^\dagger |0\rangle\rangle \right\}. \end{aligned} \tag{4.43}$$

Clearly, the right hand side of eq.(4.43) is different from the vacuum state of the Bethe ansatz solution of eq.(4.40), and therefore denoting the right hand side of eq.(4.43) by $|\Omega'\rangle$, one obtains

$$\int \bar{\psi}(x)\psi(x)\,dx|\Omega\rangle = \sum_n (b_n^\dagger a_n + a_n^\dagger b_n)|\Omega\rangle = |\Omega'\rangle. \tag{4.44}$$

Obviously, the value of C_1 in eq.(4.27) is zero in the massless Thirring model, and indeed this confirms eq.(4.29).

It is now clear and most important to note that one cannot learn the basic dynamics of the symmetry breaking phenomena from the identity equation. If one wishes to study the symmetry breaking physics in depth, then one has to solve the dynamics of the vacuum in the field theory model properly even though it is extremely difficult to solve it exactly.

4.6. Symmetry Breaking in Two Dimensions

In two dimensional field theory models, it is well known that there should not exist any physical massless bosons because of the infra-red singularity of the propagator of the massless boson. Therefore, if one assumes the Goldstone theorem, then one finds that there should not occur any spontaneous symmetry breaking in two dimensional field theory models, which is known as Coleman's theorem [22, 31, 91].

4.6.1. Fermion Field Theory in Two Dimensions

However, as we will see in the later chapters, the vacuum states of the massless Thirring model as well as QCD in two dimensions prefer the symmetry broken vacuum states together with the current conservation. Therefore, the spontaneous symmetry breaking of the vacuum indeed takes place in two dimensional field theory models of Thirring and QCD_2. By now, this is not surprising since the Goldstone theorem does not hold in fermion field theory models. On the contrary, the spontaneous symmetry breaking in these models are consistent with the new picture of the symmetry breaking physics. As far as the spontaneous symmetry breaking physics is concerned, the two dimensional field theory is not at all special since there appears no massless boson after the symmetry is spontaneously broken in the vacuum.

4.6.2. Boson Field Theory in Two Dimensions

The spontaneous symmetry breaking should not occur in boson field theory models in two dimensions. This may be reasonable since the Goldstone theorem may hold for the boson field theory models where a massless boson should appear. However, there should not exist any physical massless boson in two dimensions, and therefore, the spontaneous symmetry breaking should be forbidden in two dimensional boson field theory models.

4.7. Symmetry Breaking in Boson Fields

In the subsequent two sections, we stray from the main stream of the spontaneous symmetry breaking physics in fermion field theory models, and come to discussions of the spontaneous symmetry breaking in boson field theory in four dimensions. In most of the field theory textbooks, the discussion of this subject can be found, and therefore, we discuss it briefly in this section.

4.7.1. Double Well Potential

Now, we discuss the spontaneous symmetry breaking in boson field theory models. This can be found in any field theory textbooks, and therefore we only sketch a simple picture why the massless boson appears in the spontaneous symmetry breaking. But it should be noted that the treatment here is approximate, and there is still some unsolved problem left when one wishes to understand the spontaneous symmetry breaking in boson field theory models in an exact fashion. The Hamiltonian density for complex boson fields can be written as

$$\mathcal{H} = \frac{1}{2}(\nabla\phi^{\dagger})(\nabla\phi) + U(|\phi|). \tag{4.45}$$

This has a $U(1)$ symmetry. However, the Hamiltonian density must be real, and therefore the $U(1)$ symmetry of the Hamiltonian density is a trivial constraint. Now, when one takes the potential as a double well type

$$U(|\phi|) = u_0\left(|\phi|^2 - \lambda^2\right)^2, \tag{4.46}$$

where u_0 and λ are constant, then the minimum of the potential $U(|\phi|)$ can be found at

$$|\phi(x)| = \lambda.$$

However, one must notice that this is a minimum of the potential, but not the minimum of the total energy.

4.7.2. Change of Field Variables

The minimum of the total energy must be found together with the kinetic energy term. Now, one rewrites the complex field as

$$\phi(x) = (\lambda + \eta(x))e^{i\frac{\xi(x)}{\lambda}}, \tag{4.47}$$

where η is assumed to be much smaller than the λ,

$$|\eta(x)| \ll \lambda.$$

In this case, one can rewrite eq.(4.45) as

$$\mathcal{H} = \frac{1}{2}\left[(\nabla\xi)(\nabla\xi) + (\nabla\eta)(\nabla\eta)\right] + U(|\lambda + \eta(x)|) + \cdots. \tag{4.48}$$

Here, one finds the massless boson ξ which is associated with the degeneracy of the vacuum energy. The important point is that this infinite degeneracy of the potential vacuum is converted into the massless boson degrees of freedom when the degeneracy of the potential vacuum is resolved by the kinetic energy term.

This is the spontaneous symmetry breaking which is indeed found by Goldstone, and he pointed out that there should appear a massless boson associated with the symmetry breaking. The degeneracy of the potential vacuum is converted into a massless boson degree of freedom. This looks plausible, and at least approximately there is nothing wrong with this treatment of the spontaneous symmetry breaking phenomena in contrast to the fermion field theory model. However, the treatment is still approximate, and one should confirm that the terms neglected in eq.(4.48) may not cause any troubles. At least, one cannot claim that the massless boson which appears after the spontaneous symmetry breaking is an isolated particle of the system. Also, the Goldstone theorem shows that there should be a boson state as given in eq.(4.20)

$$E_n = 0 \quad \text{for} \quad \boldsymbol{p}_n = 0$$

which is consistent with a massless boson. However, to be rigorous, one may still have to prove that the state with the above constraint is an isolated system.

In this respect, it should be most important to solve the boson field theory model with the double well potential in an exact fashion, and then one may learn the essence of the symmetry breaking physics in the boson field theory in depth, and this is indeed a future problem.

4.7.3. Current Density of Fields

In addition, the boson fields ξ and η are real fields, and therefore the current density $j_0(x)$ of the boson fields ξ and η must vanish as the classical field

$$j_0(x) = i\left(\xi^\dagger(x)\frac{\partial\xi(x)}{\partial t} - \frac{\partial\xi^\dagger(x)}{\partial t}\xi(x)\right) = 0 \ \text{ since } \ \xi^\dagger(x) = \xi(x)$$

and the same equation holds for the η field as well. However, they diverge when they are quantized as will be discussed in the Appendix C. Therefore, both of the fields cannot propagate as a physical particle. In this sense, one may say that the model field theory of eq.(4.45) is not realistic.

4.8. Breaking of Local Gauge Symmetry?

At present, all of the realistic field theory models have the local gauge symmetry. Quantum electrodynamics (QED), quantum chromodynamics (QCD) and Weinberg-Salam model have the local gauge invariance. The local gauge symmetry in QED and QCD should hold rigorously, and this is just what we observe from experiments.

However, this local gauge invariance seems to be broken in the Higgs mechanism, and therefore we should discuss the essence of the spontaneous symmetry breaking of the local gauge invariance [67]. This concept is employed in the electro-weak theory by Weinberg-Salam, and the $SU(2)\otimes U(1)$ gauge field model is quite successful in describing many experimental observations.

4.8.1. Higgs Mechanism

The Lagrangian density of the complex scalar field $\phi(x)$ which interacts with the $U(1)$ gauge field can be written as

$$\mathcal{L} = \frac{1}{2}(D_\mu\phi)^\dagger(D^\mu\phi) - u_0\left(|\phi|^2 - \lambda^2\right)^2 - \frac{1}{4}F_{\mu\nu}F^{\mu\nu}, \tag{4.49}$$

where

$$D_\mu = \partial_\mu + igA_\mu, \quad F_{\mu\nu} = \partial_\mu A_\nu - \partial_\nu A_\mu.$$

Here, the same double well type potential as in eq.(4.46) is assumed for the complex scalar field. Now, one rewrites the complex scalar field as

$$\phi(x) = (\lambda + \eta(x))e^{i\frac{\xi(x)}{\lambda}}, \tag{4.50}$$

where $\eta(x)$ and $\xi(x)$ denote new fields, and therefore one can obtain a new Lagrangian density

$$\mathcal{L} = \frac{1}{2}(\partial_\mu \eta)(\partial^\mu \eta) - u_0\left(|\lambda+\eta|^2 - \lambda^2\right)^2$$
$$+\frac{1}{2}g^2\lambda^2\left(A_\mu + \frac{1}{g\lambda}\partial_\mu \xi\right)\left(A^\mu + \frac{1}{g\lambda}\partial^\mu \xi\right) - \frac{1}{4}F_{\mu\nu}F^{\mu\nu} + \cdots . \tag{4.51}$$

This Lagrangian density is still invariant under the following gauge transformation

$$A'_\mu(x) = A_\mu(x) + \partial_\mu \chi(x), \quad \xi'(x) = \xi(x) - g\lambda\chi(x),$$

where $\chi(x)$ is an arbitrary function of space and time.

4.8.2. Gauge Fixing

Now, one fixes the gauge such that

$$\xi(x) = 0$$

which is called *unitary gauge*. This means that one takes the following gauge fixing

$$\chi(x) = \frac{1}{g\lambda}\xi(x).$$

In this case, the Lagrangian density of eq.(4.51) becomes

$$\mathcal{L} = \frac{1}{2}(\partial_\mu \eta)(\partial^\mu \eta) - u_0\left(|\lambda+\eta|^2 - \lambda^2\right)^2 + \frac{1}{2}g^2\lambda^2 A_\mu A^\mu - \frac{1}{4}F_{\mu\nu}F^{\mu\nu} + \cdots . \tag{4.52}$$

This new Lagrangian density shows that the gauge field becomes massive and the complex scalar field has lost one degree of freedom and becomes a real scalar field $\eta(x)$. Since the new Lagrangian density is obtained by fixing the gauge, it does not have a gauge freedom any more. The peculiarity of this gauge fixing is that the $\xi(x) = 0$ has nothing to do with the redundancy of the gauge field A_μ itself. In fact, this gauge fixing does not reduce the number of freedoms of the gauge field. If all the physical observables can be reproduced by this gauge fixing, then this gauge fixing can be justified [53].

4.8.3. What Is Physics Behind Higgs Mechanism?

In the Higgs mechanism, the gauge field acquires the mass by the spontaneous symmetry breaking of the Higgs fields. This is mainly because one takes the double well type potential for a scalar field and the potential has a minimum at

$$|\phi(x)| = \lambda.$$

Therefore, the kinetic energy part of the Higgs field which couples with the gauge field A_μ becomes a constant λ. Therefore, this part has lost a coupling between the Higgs and the gauge fields. However, this mass-term-like interactions should be still gauge invariant and it cannot be considered as a mass term of the gauge field. The mass term of the gauge field is obtained by fixing the gauge at the Lagrangian density level. Therefore, after fixing the gauge, there is no gauge freedom any more in the Lagrangian density.

Normally, one fixes the gauge at the point where one calculates the physical observables. In terms of physics, the gauge fixing becomes necessary when one wishes to determine the gauge field solutions from the equations of motion. This is clear since the gauge field has redundant variables at the level of solving the equations of motion. By fixing the gauge, one can determine the gauge field, but of course physical observables should not depend on the choice of the gauge fixing.

At the same time, one should be careful for the choice of the gauge fixing. There is no guarantee that any kind of the gauge fixing can give the same physical observables. At least, one should examine whether the gauge choice one takes can indeed reproduce the right physical observables or not.

Here, the $U(1)$ gauge field is discussed, but it is straightforward to extend it to the non-abelian case. In the non-abelian gauge field theory, gauge fields themselves are not gauge invariant, and therefore they cannot be a physical observable. However, after the spontaneous symmetry breaking of the complex scalar fields, the non-abelian gauge fields become physical observables after the gauge fixing as Weinberg-Salam model shows. The physics behind this statement is difficult to understand, and in this respect, the Higgs mechanism should be understood more in depth in future.

Chapter 5

Quantum Electrodynamics

The field theory model that can describe electrons interacting with electromagnetic fields is quantum electrodynamics (QED) in four dimensions. The basic field equation is the Maxwell equation, and QED is a quantized field theory which can describe many experimental observations.

However, it is still difficult to solve QED in an exact fashion in four dimensions, and therefore one has to make some kind of approximations. Basically, there are two types of the approximations which are interesting and tractable. The first one is a perturbation theory which is very powerful in evaluating scattering processes and many other physical quantities. In most of the field theory textbooks, the perturbation theory and Feynman rules are well explained, and therefore we only describe some important points in the evaluation of the S-matrix in this text. Also, we describe the renormalization scheme in QED in Appendix I since the concept of the renormalization is quite important and the renormalized QED is most successful in describing many physical observables.

The second approximation is to treat QED in one space and one time dimensions. Even though the two dimensional field theory is not directly related to real world, one can learn a lot about non-perturbative aspects of the field theory models. In particular, QED with massless fermions in two dimensions can be solved exactly, and the exact solution can lead us to the understanding of many interesting physical phenomena which are otherwise difficult to understand. In addition, one can learn some technical developments in quantum field theory even though most of them are rather special to two dimensions and cannot be applied to four dimensions.

In this chapter, we first describe the general properties of QED and discuss the basic ingredients to understand QED. Then, we explain briefly the S-matrix calculation in QED. Further, we discuss the Schwinger model which is a massless QED in two dimensions. Since the Schwinger model can be solved exactly, we can learn a lot about the basic points of the non-perturbative aspect in quantum field theory.

5.1. General Properties of QED

Some of the basic properties of the Lagrangian density of QED are discussed in Chapter 1. Here, we repeat the discussion of the fundamental QED properties which are important and helpful for the understanding of physical phenomena in QED.

5.1.1. QED Lagrangian Density

The Lagrangian density of QED with massive fermions can be written as

$$\mathcal{L} = \bar{\psi}(i\gamma_\mu D^\mu - m)\psi - \frac{1}{4}F_{\mu\nu}F^{\mu\nu}, \tag{5.1}$$

where

$$D_\mu = \partial_\mu + igA_\mu, \quad F_{\mu\nu} = \partial_\mu A_\nu - \partial_\nu A_\mu.$$

Mass Scale or Cut-off Λ

It should be noted that the coupling constant g in four dimensional QED is dimensionless. Therefore, the Lagrangian density of QED in four dimensions has no scale parameter with dimensions if the mass of fermion is set to zero, $m = 0$. Therefore, in this four dimensional QED, one has to introduce some scale by hand. Normally, one takes the momentum cut-off Λ for the scale by which one measures physical observables. For example, if one calculates the spectrum of QED with massless fermions, then all of the excitation energy E_n should be described as

$$E_n = c_n\Lambda, \tag{5.2}$$

where c_n denotes a numerical constant which depends on the quantum number n.

In real nature, electron has a finite mass, and therefore physical observables in QED are measured by the electron mass m_e. For examples, the energy eigenvalue E_n and Bohr radius a_B of the hydrogen-like atom with its potential

$$V(r) = -\frac{Ze^2}{r}$$

in non-relativistic quantum mechanics are given as

$$E_n = -\frac{m_e Z^2 e^4}{2n^2}, \quad a_B = \frac{1}{m_e Z e^2}.$$

Here, the charge e is a dimensionless constant and n runs as $n = 1, 2, \ldots$.

5.1.2. Local Gauge Invariance

The Lagrangian density of QED is invariant under the local gauge transformation

$$A'_\mu = A_\mu + \partial_\mu \chi, \tag{5.3a}$$

$$\psi' = e^{-ig\chi}\psi. \tag{5.3b}$$

Since the following equations can be easily proved

$$D'_\mu \psi' = e^{-ig\chi} D_\mu \psi, \quad F'_{\mu\nu} = F_{\mu\nu} \tag{5.3c}$$

the gauge invariance of the Lagrangian density of eq.(5.1) is guaranteed.

5.1.3. Equation of Motion

The equation of motion for the gauge field can be obtained from the Lagrange equation for the vector field A^ν

$$\partial_\mu \frac{\partial \mathcal{L}}{\partial(\partial_\mu A^\nu)} = \frac{\partial \mathcal{L}}{\partial A^\nu}. \tag{5.4}$$

This leads to the following Maxwell equation

$$\partial_\mu F^{\mu\nu} = g j^\nu, \tag{5.5}$$

where the fermion vector current j_μ is given as

$$j_\mu = \bar{\psi}\gamma_\mu\psi.$$

From the Lagrange equation for ψ,

$$\partial_\mu \frac{\partial \mathcal{L}}{\partial(\partial_\mu \psi)} = \frac{\partial \mathcal{L}}{\partial \psi}$$

one obtains the Dirac equation

$$(i\partial_\mu\gamma^\mu - gA_\mu\gamma^\mu - m)\psi = 0. \tag{5.6}$$

5.1.4. Noether Current and Conservation Law

The Lagrangian density is obviously invariant under the global gauge transformation

$$\psi' = e^{i\theta}\psi, \tag{5.7}$$

where θ is a real constant. In this case, there is a conserved current associated with the global gauge invariance.

FirstCone makes the infinitesimal global gauge transformation as

$$\psi' = (1 + i\theta)\psi = \psi + \delta\psi, \tag{5.8}$$

where θ is assumed to be infinitesimally small. Also, $\delta\psi$ is introduced as

$$\delta\psi = i\theta\psi. \tag{5.9}$$

In this case, one obtains

$$\delta\mathcal{L} \equiv \mathcal{L}(\psi', \partial_\mu\psi') - \mathcal{L}(\psi, \partial_\mu\psi) = \frac{\partial \mathcal{L}}{\partial \psi}\delta\psi + \frac{\partial \mathcal{L}}{\partial(\partial_\mu \psi)}\delta(\partial_\mu\psi) = 0. \tag{5.10}$$

By making use of the equation of motion for ψ, one obtains the conservation law

$$\delta\mathcal{L} = i\theta\left[\left(\partial_\mu \frac{\partial \mathcal{L}}{\partial(\partial_\mu \psi)}\right)\psi + \frac{\partial \mathcal{L}}{\partial(\partial_\mu \psi)}(\partial_\mu\psi)\right] = i\theta\partial_\mu\left[\frac{\partial \mathcal{L}}{\partial(\partial_\mu \psi)}\psi\right] = 0. \tag{5.11}$$

Since one can calculate

$$\frac{\partial \mathcal{L}}{\partial(\partial_\mu \psi)}\psi = -g\bar{\psi}\gamma^\mu\psi = -gj^\mu \tag{5.12}$$

this leads to the conservation of the fermion vector current j^μ

$$\partial_\mu j^\mu = 0. \tag{5.13}$$

From eq.(5.11), one notices that the Noether current does not depend on the interaction terms. This is because the interaction terms should not depend on the field derivative of $\partial_\mu\psi$.

5.1.5. Gauge Invariance of Interaction Lagrangian

The interaction Lagrangian density itself

$$\mathcal{L}_I = -g j_\mu A^\mu \tag{5.14}$$

is not gauge invariant, and therefore if one wishes to make any perturbation calculations, one should check it in advance that the gauge dependent part should not cause any troubles in the perturbative estimation with the interaction Lagrangian density of $\mathcal{L}_I$. Now, the interaction Lagrangian density can be rewritten with the gauge transformation

$$\mathcal{L}_I = -g j_\mu (A^\mu + \partial^\mu \chi) = -g j_\mu A^\mu - g\partial^\mu (j_\mu \chi) + g(\partial^\mu j_\mu)\chi. \tag{5.15}$$

The second term of the last equation is a total divergence and hence does not contribute to any physical observables in perturbation theory, and the last term vanishes as long as the vector current conservation ($\partial_\mu j^\mu = 0$) holds. Therefore, one sees that the gauge dependent parts do not cause any contributions to the perturbative calculation under the condition that the vector current conservation of fermions should be respected.

5.1.6. Gauge Fixing

The total Hamiltonian is gauge invariant, and therefore one should fix the gauge since the gauge field A_μ has a redundancy as variables. There are many ways to fix the gauge, and of course there should be any differences for the observables one calculates from different gauge fixings. The most popular gauge fixing must be a Coulomb gauge

$$\nabla \cdot \boldsymbol{A} = 0. \tag{5.16}$$

This has some advantage in that the time component of the gauge field A_0 is not a dynamical variable any more and becomes just a simple constraint which can be easily solved by employing equations of motion. Another gauge fixing is Lorentz gauge

$$\partial_\mu A^\mu = 0. \tag{5.17}$$

This has some advantage since everything can be done in a covariant fashion. In this case, one may avoid mistakes in calculating physical observables as long as one keeps its covariance. Also, one may take the temporal gauge

$$A_0 = 0. \tag{5.18}$$

In this case, one can recover the Coulomb interactions if one calculates the interactions properly. Any physical observables like the energy spectrum should not depend on the choice of the gauge fixing.

5.1.7. Gauge Choices

The physical observables must be independent from gauge choices. In particular, the Coulomb interaction should be derived also by the temporal gauge $A_0 = 0$. Here, it is shown how one can obtain the Coulomb interaction when one takes the temporal gauge.

In particular, it is also shown that the conservation of the fermion vector current ($\partial_\mu j^\mu = 0$) plays an important role and indeed without the current conservation, the different choices of the gauge fixing, the Coulomb gauge, the temporal gauge and the axial gauge, give different results on the Coulomb interactions.

Temporal Gauge ($A_0 = 0$)

We start from the Hamiltonian of eq.(1.28) in Chapter 1

$$H = \int \left\{ \bar{\psi}(-i\boldsymbol{\gamma}\cdot\boldsymbol{\nabla}+m)\psi - g\boldsymbol{j}\cdot\boldsymbol{A} + gj_0A_0 + \frac{1}{2}\left[\dot{\boldsymbol{A}}^2 - (\boldsymbol{\nabla}A_0)^2 + \boldsymbol{B}^2\right]\right\} d^3r.$$

Now, the $A_0 = 0$ gauge is taken, and therefore the equation of motion for the gauge field becomes

$$\boldsymbol{\nabla}\cdot\frac{\partial \boldsymbol{A}}{\partial t} = -gj_0. \tag{5.19}$$

Here, there is still a gauge freedom left. Namely, the Hamiltonian is invariant under the following transformation

$$\boldsymbol{A} \to \boldsymbol{A} + \boldsymbol{\nabla}\chi(\boldsymbol{r}), \quad \psi \to e^{ig\chi(\boldsymbol{r})}\psi, \tag{5.20}$$

where $\chi(\boldsymbol{r})$ depends only on the coordinate $\boldsymbol{r}$. Therefore, we can write the vector field $\boldsymbol{A}$ as

$$\boldsymbol{A} = \boldsymbol{A}_T + \boldsymbol{\nabla}\xi, \text{ with } \boldsymbol{\nabla}\cdot\boldsymbol{A}_T = 0. \tag{5.21}$$

In this case, the equation of motion for the gauge field becomes

$$\boldsymbol{\nabla}^2\phi_0 = -gj_0, \text{ with } \phi_0 \equiv \dot{\xi}. \tag{5.22}$$

Therefore, $\frac{1}{2}\dot{\boldsymbol{A}}^2$ term in the Hamiltonian can be written with the transverse electric field $\boldsymbol{E}_T$ as defined by

$$\boldsymbol{E}_T = -\frac{\partial \boldsymbol{A}_T}{\partial t}.$$

Therefore, we have

$$\frac{1}{2}\int \dot{\boldsymbol{A}}^2 d^3r = \frac{1}{2}\int \boldsymbol{E}_T^2 d^3r + \int \boldsymbol{E}_T\cdot\boldsymbol{\nabla}\phi_0\, d^3r + \frac{1}{2}\int (\boldsymbol{\nabla}\phi_0)^2 d^3r. \tag{5.23}$$

The second term in the right hand side vanishes since

$$\boldsymbol{\nabla}\cdot\boldsymbol{E}_T = 0$$

holds, and the third term is just the same as the Coulomb interaction as shown in Chapter 1. Therefore, the Hamiltonian with the temporal gauge becomes just the same as eq.(1.33) as obtained by the Coulomb gauge fixing

$$H = \int \left\{ \bar{\psi}(-i\boldsymbol{\gamma}\cdot\boldsymbol{\nabla}+m)\psi - g\boldsymbol{j}\cdot\boldsymbol{A}\right\} d^3r$$

$$+\frac{g^2}{8\pi}\int \frac{j_0(\boldsymbol{r}')j_0(\boldsymbol{r})\, d^3r\, d^3r'}{|\boldsymbol{r}'-\boldsymbol{r}|} + \frac{1}{2}\int \left[\boldsymbol{E}_T^2 + \boldsymbol{B}^2\right] d^3r.$$

Axial Gauge ($A_3 = 0$)

The axial gauge fixing is also employed where one has the gauge condition

$$A_3 = 0.$$

In this case, the vector potential has only the transverse component and therefore one can define the transverse electric field

$$\boldsymbol{E}_T = -\frac{\partial \boldsymbol{A}}{\partial t},$$

where

$$\boldsymbol{\nabla} \cdot \boldsymbol{E}_T = 0$$

holds. In this case, the equation of motion for the gauge field becomes

$$\boldsymbol{\nabla}^2 A_0 = -g j_0$$

and therefore A_0 can be solved just in the same way as the Coulomb gauge case. In addition, the $\frac{1}{2}\dot{\boldsymbol{A}}^2$ term in eq.(1.28c) can be written in terms of $\boldsymbol{E}_T$ as

$$\frac{1}{2}\dot{\boldsymbol{A}}^2 = \frac{1}{2}\boldsymbol{E}_T{}^2.$$

Therefore, the Hamiltonian becomes just the same as eq.(1.33).

5.1.8. Gauge Dependence without $\partial_\mu j^\mu = 0$

The gauge invariance of the interaction Lagrangian density

$$\mathcal{L}_I = -g j_\mu A^\mu$$

does not cause any problem when the vector current conservation holds. However, this suggests that, when there is no conservation of fermion vector current, then there may be some troubles [49]. Here, we should examine explicitly which shape of the Coulomb interactions H_C should emerge from the three different gauge choices when the vector current is not conserved

$$\begin{aligned} \text{Coulomb gauge } \boldsymbol{\nabla} \cdot \boldsymbol{A} &= 0, \\ \text{Temporal gauge } A_0 &= 0, \\ \text{Axial gauge } A_3 &= 0. \end{aligned}$$

Before going to the calculations, one should first specify the current non-conservation. This is closely connected with the equation of motion, and one should naturally modify the equation of motion as

$$\partial_\mu F^{\mu\nu} = g(j^\nu + X^\nu), \tag{5.24}$$

where the extra current X^ν is introduced by hand here, but, in QCD, this corresponds to the gluon currents as we will see in Chapter 6. Therefore, one has

$$\partial_\mu (j^\mu + X^\mu) = 0, \ \text{ but } \partial_\mu j^\mu \neq 0. \tag{5.25}$$

In this case, one can easily calculate the Coulomb interaction for both of the gauge fixing in a perturbative manner

(1) Coulomb gauge ($\nabla \cdot \boldsymbol{A} = 0$)

$$H_C = \frac{g^2}{8\pi} \int \frac{(j_0(\boldsymbol{r}') - X_0(\boldsymbol{r}'))(j_0(\boldsymbol{r}) + X_0(\boldsymbol{r}'))}{|\boldsymbol{r}' - \boldsymbol{r}|} d^3r\, d^3r'. \tag{5.26a}$$

(2) Temporal gauge ($A_0 = 0$)

$$H_C = \frac{g^2}{8\pi} \int \frac{(j_0(\boldsymbol{r}') + X_0(\boldsymbol{r}'))(j_0(\boldsymbol{r}) + X_0(\boldsymbol{r}'))}{|\boldsymbol{r}' - \boldsymbol{r}|} d^3r\, d^3r'. \tag{5.26b}$$

(3) Axial gauge ($A_3 = 0$)

$$H_C = \frac{g^2}{8\pi} \int \frac{(j_0(\boldsymbol{r}') - X_0(\boldsymbol{r}'))(j_0(\boldsymbol{r}) + X_0(\boldsymbol{r}'))}{|\boldsymbol{r}' - \boldsymbol{r}|} d^3r\, d^3r'. \tag{5.26c}$$

Therefore, one sees that the different choices of the gauge fixing give different Coulomb interactions.

In this way, one can prove that the gauge dependence of the interaction term is crucially dependent on the fermion vector current conservation. If the additional current X^μ is absent, then it is clear that the interaction term is indeed gauge independent which is just the case for the electron-electron potential. However, in the case of the potential between quarks, the vector current of quarks is not conserved, and the equation of motion is just similar to eq.(5.24). Therefore, it is difficult to define the potential between quarks in the perturbation theory, starting from the interaction Lagrangian density. This is closely related to the fact that the color charges of quarks are time dependent and therefore it should be difficult to define any interactions between quarks whose color charges change their values as the function of time.

5.2. *S*-matrix in QED

In QED, one can carry out the perturbative calculations of many physical quantities, such as $g-2$ or lepton-proton scattering cross section. Here, we present some basic ingredients to make the S-matrix expansion and carry out the perturbative calculation of the S-matrix in the scattering process. In the evaluation of the S-matrix in quantum field theory, unperturbed states are free fields. This is important since one assumes $a - priori$ the existence of free fields if the coupling constant g is sufficiently small. This seems to be a trivial statement, but this condition is satisfied only when the vector current of fermion is conserved since, in this case, the interaction term can keep the gauge invariance. In this respect, one should always keep in mind that the vector current of fermions must be conserved.

5.2.1. Definition of *S*-matrix

The interaction term $gj_\mu A^\mu$ in QED is gauge independent since the vector current is conserved, and therefore one can carry out the S-matrix evaluation starting from the interaction term. First, one starts from the Hamiltonian which is described as

$$H = H_0 + H_I, \tag{5.27}$$

where H_0 and H_I denote the free field Hamiltonian and the interaction Hamiltonian

$$H_0 = \int d^3r \left\{ \bar{\psi}(-i\boldsymbol{\gamma}\cdot\boldsymbol{\nabla}+m)\,\psi + \frac{1}{2}\left[\dot{A}_k^{\,2} - \left(\frac{\partial A_0}{\partial x_k}\right)^2 + \left(\frac{\partial A_k}{\partial x_j}\frac{\partial A_k}{\partial x_j} - \frac{\partial A_k}{\partial x_j}\frac{\partial A_j}{\partial x_k}\right)\right]\right\}, \tag{5.28a}$$

$$H_I = g\int j_\mu A^\mu \, d^3r, \tag{5.28b}$$

respectively.

Interaction Picture

Next, one employs the interaction picture for convenience and transforms the state vector $|\Psi(t)\rangle_S$ of the normal Schrödinger picture to the state vector $|\Psi(t)\rangle$ of the interaction picture by

$$|\Psi(t)\rangle = e^{iH_0t}|\Psi(t)\rangle_S.$$

In this case, the state $|\Psi(t)\rangle$ satisfies the following Schrödinger-like equation

$$i\frac{d}{dt}|\Psi(t)\rangle = H_I(t)|\Psi(t)\rangle, \tag{5.29}$$

where $H_I(t)$ denotes the interaction Hamiltonian in the interaction picture and is related to the original one by

$$H_I(t) = e^{iH_0t}H_Ie^{-iH_0t}. \tag{5.30}$$

Since the evaluation of the perturbation theory is based on the eigenstates of H_0, the time dependence of e^{iH_0t} becomes a simple constant and should not affect on the final result of the S-matrix if one treats the time dependence properly.

S-matrix

The S-matrix is defined in terms of $|\Psi(t)\rangle$ as

$$|\Psi(\infty)\rangle = S|\Psi(-\infty)\rangle. \tag{5.31}$$

Therefore, one should first solve eq.(5.29) by the iteration method as

$$|\Psi(t)\rangle = |\Psi(-\infty)\rangle + (-i)\int_{-\infty}^{t} dt_1 H_I(t_1)|\Psi(-\infty)\rangle$$

$$+(-i)^2\int_{-\infty}^{t} dt_1 \int_{-\infty}^{t_1} dt_2 H_I(t_1)H_I(t_2)|\Psi(-\infty)\rangle + \cdots. \tag{5.32}$$

However, this expression is not very convenient to handle since the integration is not from $-\infty$ to ∞ for all the variables of t_i. Therefore, one first seeks for the symmetrized expression of the integrand since, in this case, one can easily extend the integration region to from $-\infty$ to ∞.

T-product

Thus, one defines the T-product for a symmetrized expression for fermion field operators

$$T\{A(t_1)B(t_2)\} = A(t_1)B(t_2)\theta(t_1-t_2) - B(t_2)A(t_1)\theta(t_2-t_1), \tag{5.33}$$

where $\theta(t)$ denotes a step function which is defined as

$$\theta(t) = \begin{pmatrix} 0 & t<0 \\ 1 & t>0 \end{pmatrix}.$$

Therefore, it is easy to obtain

$$\int_{-\infty}^{t} dt_1 \int_{-\infty}^{t_1} dt_2 H_I(t_1)H_I(t_2) = \int_{-\infty}^{t} dt_1 \int_{-\infty}^{t_1} dt_2 T\{H_I(t_1)H_I(t_2)\}$$

$$= \frac{1}{2}\int_{-\infty}^{t} dt_1 \int_{-\infty}^{t} dt_2 T\{H_I(t_1)H_I(t_2)\}. \tag{5.34}$$

This leads to the following S-matrix expression

$$S = \sum_{n=0}^{\infty} \frac{(-i)^n}{n!} \int_{-\infty}^{\infty} dt_1 \cdots \int_{-\infty}^{\infty} dt_n T\{H_I(t_1)\cdots H_I(t_n)\}. \tag{5.35}$$

This can be rewritten in terms of the Hamiltonian density as

$$S = \sum_{n=0}^{\infty} \frac{(-i)^n}{n!} \int d^4x_1 \cdots \int d^4x_n T\{\mathcal{H}_I(x_1)\cdots \mathcal{H}_I(x_n)\}. \tag{5.36}$$

It is important to note that S-matrix of eq.(5.36) is still written in terms of the field operators, and therefore one should make the expectation value of the S-matrix between the initial and final Fock states. Here, one must prepare these initial and final Fock states which are composed of free field solutions. These Fock states should be determined by the physical processes one wishes to consider.

5.2.2. Fock Space of Free Fields

The free field Fock space is based on the unperturbed Hamiltonian H_0 which is constructed from the unperturbed Lagrangian density $\mathcal{L}_0$

$$\mathcal{L}_0 = \bar{\psi}(i\partial_\mu \gamma^\mu - m)\psi - \frac{1}{4}F_{\mu\nu}F^{\mu\nu}.$$

Clearly, the gauge field term is gauge invariant. The free fermion part changes its shape under the local gauge transformation of $\psi' = e^{-ig\chi}\psi$ as

$$\begin{aligned} \bar{\psi}'(i\partial_\mu \gamma^\mu - m)\psi' &= \bar{\psi}(i\partial_\mu \gamma^\mu - m)\psi + g(\partial^\mu \chi) j_\mu \\ &= \bar{\psi}(i\partial_\mu \gamma^\mu - m)\psi + g\partial^\mu(\chi j_\mu) - g(\partial^\mu j_\mu)\chi. \end{aligned}$$

However, the second term $g\partial^\mu(\chi j_\mu)$ in the last equation is a total divergence and thus does not contribute to the dynamics while the last term $g(\partial^\mu j_\mu)\chi$ vanishes due to the fermion current conservation ($\partial^\mu j_\mu = 0$). Therefore, the unperturbed Lagrangian density is gauge invariant and the Fock space of the free fields is constructed from the gauge invariant Hamiltonian as long as the fermion current j_μ is conserved. In other words, one has to consider the current conservation properly when one carries out the perturbative calculations, and this is the important point of the S-matrix evaluation in QED.

5.2.3. Electron-Electron Interactions

Now, one can calculate the electron-electron interaction in QED. The basic interaction is given as

$$\mathcal{H}_I(x) = -\mathcal{L}_I = g j_\mu(x) A^\mu(x). \tag{5.37}$$

For the evaluation of the S-matrix, one should prepare initial $|\mathrm{i}\rangle$ and final $|\mathrm{f}\rangle$ two electron states whose momenta and spins are specified as $(\boldsymbol{p}_1, s_1)$, $(\boldsymbol{p}_2, s_2)$ for initial states and $(\boldsymbol{k}_1, s_1)$, $(\boldsymbol{k}_2, s_2)$ for final states, respectively

$$|\mathrm{i}\rangle = a_{\boldsymbol{p}_1}^{\dagger(s_1)} a_{\boldsymbol{p}_2}^{\dagger(s_2)} |0\rangle, \tag{5.38a}$$

$$|\mathrm{f}\rangle = a_{\boldsymbol{k}_1}^{\dagger(s_1')} a_{\boldsymbol{k}_2}^{\dagger(s_2')} |0\rangle. \tag{5.38b}$$

In this case, one can easily calculate the S-matrix of the electron-electron scattering process in the second order perturbation theory as

$$\langle \mathrm{f}|S|\mathrm{i}\rangle = \frac{i^2}{2} \int dx_1^4 dx_2^4 \langle \mathrm{f}| T\{\mathcal{H}_I(x_1)\mathcal{H}_I(x_2)\}|\mathrm{i}\rangle. \tag{5.39}$$

The evaluation of eq.(5.39) can be carried out in a straightforward way.

Quantization of Fermion Field in Anti-particle Representation

Here, one employs the anti-particle representation and therefore ψ is expanded as

$$\psi(x) = \sum_{\boldsymbol{n},s} \frac{1}{\sqrt{L^3}} \left(a_{\boldsymbol{n}}^{(s)} u_{\boldsymbol{n}}^{(s)} e^{i\boldsymbol{p}_n \cdot \boldsymbol{r} - iE_n t} + \mathrm{b}_{\boldsymbol{n}}^{\dagger(s)} v_{\boldsymbol{n}}^{(s)} e^{-i\boldsymbol{p}_n \cdot \boldsymbol{r} + iE_n t} \right), \tag{5.40}$$

where $\mathrm{b}_{\boldsymbol{n}}^{\dagger(s)}$ denotes the creation operator of the positron. The creation and annihilation operators $a_{\boldsymbol{n}}^{(s)}$, $a_{\boldsymbol{n}}^{(s)\dagger}$, $\mathrm{b}_{\boldsymbol{n}}^{(s)}$, $\mathrm{b}_{\boldsymbol{n}}^{(s)\dagger}$ satisfy the anti-commutation relations.

$$\{a_{\boldsymbol{n}}^{(s)}, a^{\dagger}{}_{\boldsymbol{n}'}^{(s')}\} = \delta_{s,s'}\delta_{\boldsymbol{n},\boldsymbol{n}'}, \quad \{\mathrm{b}_{\boldsymbol{n}}^{(s)}, \mathrm{b}^{\dagger}{}_{\boldsymbol{n}'}^{(s')}\} = \delta_{s,s'}\delta_{\boldsymbol{n},\boldsymbol{n}'}$$

and all the other anti-commutations vanish.

Quantization of Gauge Field

The gauge field A_μ is also quantized as

$$\boldsymbol{A}(x) = \sum_{\boldsymbol{k}} \sum_{\lambda=1}^{2} \frac{1}{\sqrt{2V\omega_{\boldsymbol{k}}}} \boldsymbol{\varepsilon}(\boldsymbol{k},\lambda) \left[c_{\boldsymbol{k},\lambda} e^{-ikx} + c_{\boldsymbol{k},\lambda}^{\dagger} e^{ikx} \right], \tag{5.41}$$

where $c_{\boldsymbol{k},\lambda}$ and $c_{\boldsymbol{k},\lambda}^{\dagger}$ satisfies commutation relations

$$[c_{\boldsymbol{k},\lambda}, c_{\boldsymbol{k}',\lambda'}^{\dagger}] = \delta_{\boldsymbol{k},\boldsymbol{k}'} \delta_{\lambda,\lambda'}$$

and all the other commutation relations vanish.

In the calculation, one has to make use of the anti-commutation relations between the fermion and anti-fermion creation and annihilation operators and commutation relations between photon operators. It should be noted that fermion operators and photon operators commute with each other, that is,

$$[c_{\boldsymbol{k},\lambda}, a_{\boldsymbol{n}}^{(s)}] = 0, \quad [c_{\boldsymbol{k},\lambda}, \mathrm{b}_{\boldsymbol{n}}^{(s)}] = 0, \quad \text{etc.} \tag{5.42}$$

When one calculates eq.(5.39), one first rewrites it as

$$\mathcal{H}_I(x) = gN\left\{ (\bar{\psi}^+ + \bar{\psi}^-)(\gamma^\mu A_\mu^+ + \gamma^\mu A_\mu^-)(\psi^+ + \psi^-) \right\}, \tag{5.43}$$

where N denotes a normal ordering of the operators where N is defined as

$$N\{a_{\boldsymbol{n}}^{(s)} a_{\boldsymbol{n}'}^{(s')\dagger}\} \equiv -a_{\boldsymbol{n}'}^{(s')\dagger} a_{\boldsymbol{n}}^{(s)}, \quad N\{c_{\boldsymbol{k},\lambda} c_{\boldsymbol{k}',\lambda'}^{\dagger}\} \equiv c_{\boldsymbol{k}',\lambda'}^{\dagger} c_{\boldsymbol{k},\lambda}.$$

Here, ψ^+ ($\bar{\psi}^-$), $\bar{\psi}^+$ (ψ^-) and A_μ^+ (A_μ^-) are proportional to the creation (annihilation) operators of positron, electron and photon, respectively. For example, ψ^+ and ψ^- are written as

$$\psi^- = \sum_{n_1,n_2,n_3,s} \frac{1}{\sqrt{L^3}} a_{\boldsymbol{n}}^{(s)} u_{\boldsymbol{n}}^{(s)} e^{i\boldsymbol{p}_{\boldsymbol{n}}\cdot\boldsymbol{r} - iE_{\boldsymbol{n}}t}, \tag{5.44a}$$

$$\psi^+ = \sum_{n_1,n_2,n_3,s} \frac{1}{\sqrt{L^3}} \mathrm{b}_{\boldsymbol{n}}^{\dagger(s)} v_{\boldsymbol{n}}^{(s)} e^{-i\boldsymbol{p}_{\boldsymbol{n}}\cdot\boldsymbol{r} + iE_{\boldsymbol{n}}t}. \tag{5.44b}$$

Vacuum State

Further, the vacuum state is defined such that the following relations should hold

$$a_{\boldsymbol{p}}^{(s)}|0\rangle = 0, \quad \mathrm{b}_{\boldsymbol{p}}^{(s)}|0\rangle = 0, \quad c_{\boldsymbol{p}}^{(\lambda)}|0\rangle = 0. \tag{5.45}$$

S-matrix and Invariant Amplitude

Therefore, the S-matrix in eq.(5.39) is evaluated to be

$$\langle \mathrm{f}|S|\mathrm{i}\rangle = \left\{ (2\pi)^4 \delta^{(4)}(p_1 + p_2 - k_1 - k_2) \sqrt{\frac{1}{L^6}} \right\} (M_a + M_b), \tag{5.46}$$

where the invariant amplitudes M_a and M_b are given as

$$M_a = -g^2 \left(\bar{u}_{\boldsymbol{k}_1}^{(s_1')}\gamma^\alpha u_{\boldsymbol{p}_1}^{(s_1)}\right) D_{\alpha\beta}(k_1 - p_1) \left(\bar{u}_{\boldsymbol{k}_2}^{(s_2')}\gamma^\alpha u_{\boldsymbol{p}_2}^{(s_2)}\right), \tag{5.47a}$$

$$M_b = g^2 \left(\bar{u}_{\boldsymbol{k}_2}^{(s_2')}\gamma^\alpha u_{\boldsymbol{p}_1}^{(s_1)}\right) D_{\alpha\beta}(k_2 - p_1) \left(\bar{u}_{\boldsymbol{k}_1}^{(s_1')}\gamma^\alpha u_{\boldsymbol{p}_2}^{(s_2)}\right). \tag{5.47b}$$

Here, $D_{\alpha\beta}(q)$ denotes the photon propagator and is written as

$$D_{\alpha\beta}(q) = \frac{-g_{\alpha\beta}}{q^2 + i\varepsilon}. \tag{5.48}$$

5.2.4. Feynman Rules for QED

Eq.(5.39) can be evaluated also by employing Feynman rules with the help of Feynman diagrams. They are simple and useful, but the Feynman rules can be found in most of the field theory textbooks. Therefore, we should ask readers to refer to these textbooks. There is no special point to note, and one should just follow the Feynman rules when one wishes to calculate any corresponding diagrams. The S-matrix evaluation employing the Feynman rules is carried out in the momentum space. Below, we give some of the ingredients which are important in the evaluation of the Feynman diagrams.

Vertex and propagator in Feynman diagrams

1. For each photon-fermion vertex, a factor

$$ig\gamma_\mu$$

is written.

2. For each internal photon line which is labeled by the momentum k,

$$D_{\alpha\beta} = -\frac{g_{\alpha\beta}}{k^2 + i\varepsilon}$$

is written. $D_{\alpha\beta}$ denotes the photon propagator.

3. For each internal fermion line which is labeled by the momentum p,

$$S_F = \frac{1}{p_\mu\gamma^\mu - m + i\varepsilon}$$

is written. S_F denotes the fermion propagator.

These are just simplest examples for the S-matrix evaluations. However, if one wishes to avoid any mistakes, it should be better to evaluate the S-matrix from eq.(5.39). The only thing one has to be careful is to prepare the right initial and final states which are eigenstates of the free field Hamiltonian.

5.3. Schwinger Model (Massless QED$_2$)

QED in four dimensions cannot be solved in an exact fashion. The simplest system of QED with electrons must be a positronium which is an electron-positron bound state. However, this bound state energy cannot be calculated exactly. Fortunately, the positronium can be described almost in the non-relativistic kinematics, and therefore a major part of the binding energy of positronium can be reproduced by the non-relativistic QED calculation.

Still, if there is any system which can be solved exactly, then one may learn a lot about the properties of QED from the structure of the exact solution. Unfortunately, no system of QED in four dimensions is solved exactly. However, the situation becomes very different when the space-time dimension is reduced. In particular, QED in one space and one time dimension is solved exactly if fermions are massless, and it is called Schwinger model [104]. This exact solution in this context means that the Schwinger model can be rewritten into a new free field theory model. Indeed, the Schwinger model can be bosonized and becomes a free massive boson field theory model. This is interesting, but we believe that the Schwinger model is very special in that the interaction between fermions and anti-fermions is attractive, but it is a confining potential. Therefore, there exists no free fermion state, and that should be a good reason why the Schwinger model becomes identical to the free bosonic fields.

5.3.1. QED with Massless Fermions in Two Dimensions

The Schwinger model Lagrangian density can be written as

$$\mathcal{L} = \bar{\psi} i\gamma_\mu D^\mu \psi - \frac{1}{4} F_{\mu\nu}F^{\mu\nu}, \tag{5.49}$$

where

$$D_\mu = \partial_\mu + igA_\mu, \quad F_{\mu\nu} = \partial_\mu A_\nu - \partial_\nu A_\mu.$$

Hamiltonian of Massless QED$_2$

In this case, the Hamiltonian of massless QED$_2$ becomes

$$H = \int \bar{\psi}(x)\left[-i\gamma^1\partial_1 + g\gamma^1 A_1\right]\psi(x)\,dx + \frac{1}{2}\int \dot{A}_1^2\,dx$$
$$-\frac{1}{2}\int (\partial_1 A_0)^2\,dx + g\int j_0(x)A_0(x)\,dx. \tag{5.50}$$

It should be noted that the massless QED in two dimensions has a mass scale, that is, the coupling constant g has a mass dimension. Therefore, all the physical observables such as a boson mass are measured by the coupling constant g. In this respect, the massless limit in QED$_2$ corresponds to the strong coupling limit in this field theory model.

Also, it should be important to note that the two dimensional QED is not singular at the massless limit of $m \to 0$ since the mass scale is already given by the coupling constant g. That is, all the physical observables are measured by the coupling constant g, and therefore physical observables are continuous functions of the mass m for $m \to 0$. On the other hand, QED in four dimensions is quite different from the two dimensional case. The coupling constant g in four dimensional QED is dimensionless and therefore the only mass scale in this model is the fermion mass m and all the observables must be described in terms of the fermion mass m. Thus, the massless limit must be a singular point since there is no other scale present. When one treats QED with massless fermions, then all of the physical observables must be measured by the cut-off momentum Λ.

5.3.2. Gauge Fixing

In the treatment of the bosonization procedure in the Schwinger model, it is most convenient to take the Coulomb gauge

$$\partial^1 A_1 = 0 \tag{5.51}$$

instead of the axial gauge fixing ($A_1 = 0$). In this case, A_1 depends only on time. The equation of motion for A_0 is

$$\partial_1^2 A_0(x) = -g j_0$$

and the solution is given as

$$A_0(x) = -\frac{1}{2} g \int |x-x'| j_0(x')\, dx'.$$

Therefore, the Hamiltonian of eq.(5.50) can be rewritten as

$$H = \int \bar{\psi}(x) \left[-i\gamma^1\partial_1 + g\gamma^1 A_1\right] \psi(x)\, dx + \frac{1}{2}\int \dot{A}_1^2\, dx$$

$$-\frac{g^2}{4}\int j_0(x)|x-x'| j_0(x')\, dx\, dx'. \tag{5.50'}$$

The energy spectrum does not depend on the choice of the gauge. However, it is normally difficult to take into account the zero mode of boson in the axial gauge fixing. Indeed, A_1 corresponds to the zero mode as we will see it below.

5.3.3. Quantized Hamiltonian of Schwinger Model

Now, the fermion field should be quantized in a box with the length L

$$\psi(x) = \frac{1}{\sqrt{L}} \sum_n \begin{pmatrix} a_n \\ b_n \end{pmatrix} e^{i\frac{2\pi}{L}nx}, \tag{5.52}$$

where the creation and annihilation operators a_n and b_n should satisfy the following anti-commutation relations

$$\{a_n, a_m^\dagger\} = \delta_{nm}, \quad \{b_n, b_m^\dagger\} = \delta_{nm} \tag{5.53}$$

and all other anti-commutation relations vanish. Further, the chiral representation of the γ matrices is employed here

$$\gamma_0 = \begin{pmatrix} 0 & 1 \\ 1 & 0 \end{pmatrix}, \quad \gamma_1 = \begin{pmatrix} 0 & -1 \\ 1 & 0 \end{pmatrix}, \quad \gamma_5 = \gamma_0\gamma_1 = \begin{pmatrix} 1 & 0 \\ 0 & -1 \end{pmatrix}. \tag{5.54}$$

This is simply because of the convenience.

Schwinger Model Hamiltonian

Therefore, the quantized Hamiltonian of the Schwinger model can be written as

$$H = \frac{L}{2}\dot{A}_1^{\,2} + \sum_n \left(\frac{2\pi}{L}n + gA_1\right) a_n^\dagger a_n + \sum_n \left(-\frac{2\pi}{L}n - gA_1\right) b_n^\dagger b_n$$

$$+\frac{g^2 L}{8\pi^2}\sum_{p\neq 0}\frac{1}{p^2}\tilde{j}_0(p)\tilde{j}_0(-p), \tag{5.55}$$

where the currents $\tilde{j}_0(p)$ and $\tilde{j}_1(p)$ denote the momentum representation of the fermion currents $j_0(x)$ and $j_1(x)$, and can be written

$$\tilde{j}_0(p) = \sum_k \left[a_{k+p}^\dagger a_k + b_{k+p}^\dagger b_k\right], \tag{5.56}$$

$$\tilde{j}_1(p) = \sum_k \left[a_{k+p}^\dagger a_k - b_{k+p}^\dagger b_k\right]. \tag{5.57}$$

The Schwinger model is solved by several methods. The bosonization is one of them and will be treated below [90]. Also, the Schwinger model has been solved by the Bogoliubov transformation method . The Bogoliubov transformation method is an approximate scheme for the four fermion interaction models. However, the correct mass of the Schwinger boson is obtained by the Bogoliubov transformation method. Further, the Bogoliubov transformation method reproduces the right condensate value of the Schwinger model which is obtained analytically. This suggests that the Bogoliubov vacuum state may well be a good vacuum state since the condensate value should exhibit some information of the vacuum structure.

5.3.4. Bosonization of Schwinger Model

In the Schwinger model, the Coulomb gauge is taken, and in this case, the vector potential A_1 depends only on time and corresponds to the zero mode of the boson field. The momentum representation $\tilde{j}_\mu(p)$ of the current can be written in terms of $\rho_a(p)$ and $\rho_b(p)$

$$\tilde{j}_0(p) = \rho_a(p) + \rho_b(p), \tag{5.58a}$$

$$\tilde{j}_1(p) = \rho_a(p) - \rho_b(p), \tag{5.58b}$$

where $\rho_a(p)$ and $\rho_b(p)$ are defined as

$$\rho_a(p) = \sum_k a_{k+p}^\dagger a_k, \tag{5.59a}$$

$$\rho_b(p) = \sum_k b_{k+p}^\dagger b_k. \tag{5.59b}$$

Commutation Relations of Currents

Now, one can easily prove that $\rho_a(p)$ and $\rho_b(p)$ satisfy the following commutation relations,

$$[\rho_a(p),\rho_a(q)]|\text{phys}\rangle = -p\delta_{p,-q}|\text{phys}\rangle, \tag{5.60a}$$

$$[\rho_b(p),\rho_b(q)]|\text{phys}\rangle = p\delta_{p,-q}|\text{phys}\rangle. \tag{5.60b}$$

The above commutation relations can only be valid when these equations are operated on the physical states $|\text{phys}\rangle$. Here, the physical states mean that the negative energy states must be completely occupied if the negative energy levels are sufficiently deep. Further, in the physical states, there should be no particles present in the positive energy states if the particle energy is sufficiently high. Under these conditions, one can prove that eqs.(5.60) hold true as operator equations.

Boson Fields $\Phi(p)$ and $\Pi(p)$

In this case, $\tilde{j}_0(p)$ and $\tilde{j}_1(p)$ are related to the boson field and its conjugate field as

$$\tilde{j}_0(p) = ip\sqrt{\frac{4\pi}{L}}\Phi(p) \ \text{ for } \ p \neq 0, \tag{5.61a}$$

$$\tilde{j}_1(p) = \sqrt{\frac{L}{\pi}}\Pi(p) \ \text{ for } \ p \neq 0, \tag{5.61b}$$

where $\Phi(p)$ and $\Pi(p)$ denote the boson field and its conjugate field which satisfy the bosonic commutation relation

$$[\Phi(p),\Pi^\dagger(p')] = i\delta_{p,p'}.$$

5.3.5. Chiral Anomaly

It is very important to note that $\Phi(0)$ and $\Pi(0)$ are not defined in eqs.(5.61). In the Schwinger model, they are related to the chiral charge and its time derivative as

$$\Phi(0) = \frac{\pi}{g\sqrt{L}}Q_5, \tag{5.62a}$$

$$\Pi(0) = \frac{\pi}{g\sqrt{L}}\dot{Q}_5, \tag{5.62b}$$

where $\dot{Q}_5$ is described by the vector field A_1 due to the anomaly equation

$$\dot{Q}_5 = \frac{Lg}{\pi}\dot{A}_1. \tag{5.63}$$

Here, we briefly discuss how one obtains the chiral anomaly when one regularizes the charge and the energy of the vacuum.

Large Gauge Transformation

It is important to note that one should regularize the charge in the gauge invariant way since the Hamiltonian has still the invariance of a large gauge transformation

$$A_1 \to A_1 + \frac{2\pi}{Lg}N, \quad N \text{ integer.}$$

Regularized Charges

First, the regularized charges of the left and the right movers are defined by

$$Q_L^{(\lambda)} \equiv \sum_{n=-\infty}^{N_L} e^{\lambda(n+\frac{LgA_1}{2\pi})}, \tag{5.64a}$$

$$Q_R^{(\lambda)} \equiv \sum_{n=N_R}^{\infty} e^{\lambda(-n-\frac{LgA_1}{2\pi})}, \tag{5.64b}$$

where the charges are regularized in terms of the ζ function regularization by taking into account the gauge invariance, and one should make $\lambda \to 0$. For the infinitesimally small λ, one obtains

$$Q_L^{(\lambda)} = \frac{1}{\lambda} + \left(N_L + \frac{LgA_1}{2\pi} + \frac{1}{2}\right) + \frac{\lambda}{2}\left(N_L + \frac{LgA_1}{2\pi} + \frac{1}{2}\right)^2 - \frac{\lambda}{24}, \tag{5.64a'}$$

$$Q_R^{(\lambda)} = \frac{1}{\lambda} + \left(-N_R - \frac{LgA_1}{2\pi} + \frac{1}{2}\right) + \frac{\lambda}{2}\left(-N_R - \frac{LgA_1}{2\pi} + \frac{1}{2}\right)^2 - \frac{\lambda}{24}. \tag{5.64b'}$$

In this case, the charge and the chiral charge of the vacuum state are defined as

$$Q = Q_L^{(\lambda)} + Q_R^{(\lambda)}, \tag{5.65a}$$

$$Q_5 = Q_L^{(\lambda)} - Q_R^{(\lambda)}. \tag{5.65b}$$

The regularized charge and chiral charge become

$$Q = \frac{2}{\lambda} + N_L + 1 - N_R + O(\lambda), \tag{5.66a}$$

$$Q_5 = N_L + N_R + \frac{LgA_1}{\pi}. \tag{5.66b}$$

Vacuum Charge

Since the charge of the vacuum must be zero, one can set

$$Q = 0. \tag{5.67}$$

Therefore,

$$N_L + 1 - N_R = 0$$

should hold. In addition, the $\frac{2}{\lambda}$ term should be ignored. From eq.(5.66b), one obtains the chiral anomaly equation of eq.(5.63) by making the time derivative of Q_5. Namely, the axial vector current is not conserved any more due to the chiral anomaly.

5.3.6. Regularization of Vacuum Energy

Further, one should regularize the vacuum energy in the same way as the charge. Denoting the left and right movers of the regularized vacuum energy by E_L^{vac} and E_R^{vac}, one can calculate them as

$$E_L^{vac} \equiv \frac{2\pi}{L} \sum_{n=-\infty}^{N_L} \left(n + \frac{LgA_1}{2\pi}\right) e^{\lambda(n+\frac{LgA_1}{2\pi})} = \frac{2\pi}{L}\frac{\partial Q_L^{(\lambda)}}{\partial\lambda}, \qquad (5.68a)$$

$$E_R^{vac} \equiv \frac{2\pi}{L} \sum_{n=N_R}^{\infty} \left(-n - \frac{LgA_1}{2\pi}\right) e^{\lambda(-n-\frac{LgA_1}{2\pi})} = \frac{2\pi}{L}\frac{\partial Q_R^{(\lambda)}}{\partial\lambda}. \qquad (5.68b)$$

Making use of eqs.(5.64) and leaving out irrelevant constant terms, one obtains

$$E_L^{vac} = \frac{\pi}{L}\left(Q_L^{(\lambda)}\right)^2, \qquad (5.69a)$$

$$E_R^{vac} = \frac{\pi}{L}\left(Q_R^{(\lambda)}\right)^2. \qquad (5.69b)$$

Therefore, the total vacuum energy can be written with $\lambda \to 0$

$$E^{vac} = \frac{\pi}{L}\left[\left(Q_L^{(\lambda)}\right)^2 + \left(Q_R^{(\lambda)}\right)^2\right] = \frac{\pi}{2L}\left[Q^2 + Q_5^2\right] = \frac{\pi}{2L}Q_5^2 \qquad (5.70)$$

since $Q = 0$. Thus, the vacuum energy part of the Hamiltonian eq.(5.55) can be written as

$$H^{vac} = \frac{\pi^2}{2g^2L}\dot{Q}_5^2 + \frac{\pi}{2L}Q_5^2. \qquad (5.71)$$

Zero Mode

If one identifies the boson fields $\Phi(0)$, $\Pi(0)$ as given in eqs.(5.62)

$$\Phi(0) = \frac{\pi}{g\sqrt{L}}Q_5, \quad \Pi(0) = \frac{\pi}{g\sqrt{L}}\dot{Q}_5$$

then one can write the vacuum part of the Hamiltonian

$$H^{vac} = \frac{1}{2}\Pi^\dagger(0)\Pi(0) + \frac{g^2}{2\pi}\Phi^\dagger(0)\Phi(0). \qquad (5.72)$$

5.3.7. Bosonized Hamiltonian of Schwinger Model

Further, the positive energy part of the kinetic energy Hamiltonian can be rewritten in terms of the kinetic energy of the boson Hamiltonian

$$\sum_p \left(\frac{2\pi p}{L}\right)\left(a_p^\dagger a_p - b_p^\dagger b_p\right) = \frac{1}{2}\sum_{p\neq 0}\left\{\Pi^\dagger(p)\Pi(p) + \left(\frac{2\pi p}{L}\right)^2 \Phi^\dagger(p)\Phi(p)\right\}. \qquad (5.73)$$

Here, it should be noted that the identification of the kinetic energies between the fermion and boson fields can be considered as operator equations under the condition that all the operations should be done onto the physical states in fermion Fock space as explained above. In this case, eq.(5.73) holds true as operator equations.

The total Hamiltonian of the Schwinger model must be the sum of the zero mode Hamiltonian of eq.(5.72) and the kinetic energy terms of the boson Hamiltonian of eq.(5.73) together with the Coulomb interaction part in eq.(5.55) noting eq.(5.61a) $[\tilde{j}_0(p) = ip\sqrt{\frac{4\pi}{L}}\Phi(p)]$

$$\frac{g^2 L}{8\pi^2}\sum_{p\neq 0}\frac{1}{p^2}\tilde{j}_0(p)\tilde{j}_0(-p) = \frac{g^2}{2\pi}\sum_p \Phi^\dagger(p)\Phi(p).$$

Therefore, one can write the total Hamiltonian for the Schwinger model as

$$H = \sum_p \left\{ \frac{1}{2}\Pi^\dagger(p)\Pi(p) + \frac{1}{2}\left(\frac{2\pi p}{L}\right)^2 \Phi^\dagger(p)\Phi(p) + \frac{g^2}{2\pi}\Phi^\dagger(p)\Phi(p) \right\}. \tag{5.74}$$

This is just the free massive boson Hamiltonian, and the mass of the boson (Schwinger boson $\mathcal{M}_0$) is given as

$$\mathcal{M}_0 = \frac{g}{\sqrt{\pi}}.$$

It should be important to note that the Schwinger model has the right zero mode in the Hamiltonian of the boson field. However, as we will see it in Section 7, there is no corresponding zero mode in the massless Thirring model, and this leads to the finite gap of the spectrum in the massless Thirring model, which is indeed consistent with the fact that there should exist no physical massless boson in two dimensions.

In deriving the boson Hamiltonian from the Schwinger model, there is no mathematical problem as we see above. However, physically speaking, it is not at all simple. This boson should be made out of an infinite sum of fermion and anti-fermion pairs, but we cannot explicitly show how it is composed of the fermion and anti-fermion pairs. In the subsequent section, we see that the mass of the Schwinger boson can be obtained by the Bogoliubov transformation method which takes the special type of the sum of fermion and anti-fermion pairs into account in a non-perturbative fashion. However, the Bogoliubov vacuum state is not an eigenstate of the total Hamiltonian and therefore the Fock state that can reproduce the right Schwinger boson is not necessarily the exact eigenstate.

In the next section, we will see the boson mass calculation in the two dimensional QED with the finite fermion mass. A naive calculation of the boson mass calculation with the perturbative vacuum state shows some difficulty when the fermion mass is much smaller than the Schwinger boson mass. This difficulty can be overcome by introducing the Bogoliubov transformation which is the variational method.

5.4. Quantized QED$_2$ Hamiltonian in Trivial Vacuum

In this section, we discuss the boson spectrum of QED$_2$ in the simplest Fock space calculation which is based on the perturbative vacuum state, before going to the discussion of the Bogoliubov transformation method. This has a good connection to the nonrelativistic calculation when the fermion mass is sufficiently large. Here, the coupling constant g has the mass dimension, and if the fermion mass m_0 is larger than the Schwinger boson mass $\mathcal{M}_0 = \frac{g}{\sqrt{\pi}}$, then one may say that the fermion motion can be treated mostly in the non-relativistic kinematics.

However, the calculated result of the boson mass in QED_2 with the trivial vacuum has the basic difficulty when the fermion mass is sufficiently small, because the calculated boson mass becomes a negative value! This difficulty can be overcome by the Bogoliubov transformation method [15].

5.4.1. Hamiltonian and Gauge Fixing

The Lagrangian density is just the same as eq.(5.1)

$$\mathcal{L} = \bar{\psi}(i\gamma_\mu \partial^\mu - g\gamma_\mu A^\mu - m_0)\psi - \frac{1}{4}F_{\mu\nu}F^{\mu\nu}.$$

In order to solve the equation of motion for A_0, one can fix the gauge by

$$A_1 = 0.$$

In this case, the Hamiltonian of QED_2 can be written as

$$H = \int \left[-i\bar{\psi}\gamma^1 \partial_1 \psi + m_0 \bar{\psi}\psi \right] dx - \frac{g^2}{4} \int j_0(x)|x-x'|j_0(x')\, dx\, dx'. \quad (5.75)$$

5.4.2. Field Quantization in Anti-particle Representation

Now, one can quantize the fields. For the massive case in QED_2, one employs the anti-particle representation because of its convenience.

$$\psi(x) = \frac{1}{\sqrt{L}} \sum_n \left[a_n u_n e^{-i\frac{2\pi n x}{L}} + \mathrm{b}_n^\dagger v_n e^{i\frac{2\pi n x}{L}} \right], \quad (5.76)$$

where a_n and b_n denote the annihilation operators for particles and anti-particles, respectively, and they satisfy the anticommutation relations of eqs.(5.53). The periodic boundary condition is required on the wave function

$$\psi(x) = \psi(x+L). \quad (5.77)$$

5.4.3. Dirac Representation of γ-matrices

The Dirac spinors u_n and v_n are explicitly written as

$$u_n = \frac{1}{\sqrt{2E_n(E_n+m_0)}} \begin{pmatrix} E_n + m_0 \\ p_n \end{pmatrix}, \quad (5.78a)$$

$$v_n = \frac{1}{\sqrt{2E_n(E_n+m_0)}} \begin{pmatrix} p_n \\ E_n + m_0 \end{pmatrix}, \quad (5.78b)$$

where p_n and E_n are given as

$$p_n = \frac{2\pi n}{L}, \quad E_n = \sqrt{p_n^2 + m_0^2}. \quad (5.79)$$

For the γ matrices, we employ the Dirac representation which is convenient for the massive fermion case

$$\gamma_0 = \begin{pmatrix} 1 & 0 \\ 0 & -1 \end{pmatrix}, \quad \gamma_1 = \begin{pmatrix} 0 & 1 \\ -1 & 0 \end{pmatrix}, \quad \gamma_5 \equiv \gamma_0\gamma_1 = \begin{pmatrix} 0 & 1 \\ 1 & 0 \end{pmatrix}. \tag{5.80}$$

5.4.4. Quantized Hamiltonian of QED$_2$

In this case, the quantized Hamiltonian of QED$_2$ can be written as

$$H = \sum_n E_n \left(a_n^\dagger a_n + \mathrm{b}_n^\dagger \mathrm{b}_n\right) + \frac{g^2 L}{8\pi^2} \sum_p \frac{1}{p^2} \tilde{j}_0(-p)\tilde{j}_0(p), \tag{5.81}$$

where $\tilde{j}_0(p)$ is the momentum representation of the current density $j_0(x)$ and is written as

$$\tilde{j}_0(p) = \sum_q \Big[a_{q-p}^\dagger a_q F(q-p,q) + a_{-p-q}^\dagger \mathrm{b}_q^\dagger G(-q-p,q)$$
$$+\mathrm{b}_{p-q} a_q G(p-q,q) - \mathrm{b}_q^\dagger \mathrm{b}_{p+q} F(p+q,q)\Big]. \tag{5.82}$$

Here, $F(p,q)$ and $G(p,q)$ are defined as

$$F(p,q) = u_p^\dagger u_q, \quad G(p,q) = u_p^\dagger v_q$$

and they are written explicitly as

$$\begin{aligned} F(p,q) &= \frac{1}{\sqrt{4E_pE_q(E_p+m_0)(E_q+m_0)}}\left[(E_p+m_0)(E_q+m_0)+\frac{4\pi^2 pq}{L^2}\right], \\ G(p,q) &= \frac{1}{\sqrt{4E_pE_q(E_p+m_0)(E_q+m_0)}}\left[\frac{2\pi q}{L}(E_p+m_0)+\frac{2\pi p}{L}(E_q+m_0)\right]. \end{aligned} \tag{5.83}$$

Interaction Hamiltonian

The interaction part of the Hamiltonian in eq.(5.81) as denoted by H_I can be explicitly written as

$$H_I \equiv \frac{g^2 L}{8\pi^2} \sum_p \frac{1}{p^2} \tilde{j}_0(-p)\tilde{j}_0(p) = H_{MR}^{(1)} + H_{MR}^{(2)} + H_C + H_A + H_R + H_{non}, \tag{5.84}$$

where each term is defined below with some physical meaning.

Mass Renormalization Terms

$$H_{MR}^{(1)} = \frac{g^2 L}{8\pi^2} \sum_{p,q} \frac{1}{p^2} \{a_q^\dagger a_q + \mathrm{b}_q^\dagger \mathrm{b}_q\} F(q,q-p)F(q-p,q), \tag{5.85a}$$

$$H_{MR}^{(2)} = -\frac{g^2 L}{8\pi^2} \sum_{p,q} \frac{1}{p^2} \{a_q^\dagger a_q + \mathrm{b}_q^\dagger \mathrm{b}_q\} G(q,-p-q)G(-q-p,q). \tag{5.85b}$$

Coulomb Interaction Terms

$$H_C = \frac{g^2L}{8\pi^2} \sum_{p,q,q'} \frac{1}{p^2} \{a^\dagger_{q'+p} b^\dagger_q a_{q'} b_{q+p} + h.c.\} F(q'+p,q') F(q+p,q), \tag{5.85c}$$

$$H_A = -\frac{g^2L}{8\pi^2} \sum_{p,q,q'} \frac{1}{p^2} \{a^\dagger_{p-q'} b^\dagger_{q'} a_q b_{p-q} + h.c.\} G(p-q',q') G(p-q,q)\}, \tag{5.85d}$$

$$H_R = -\frac{g^2L}{8\pi^2} \sum_{p,q,q'} \frac{1}{p^2} \left\{ a^\dagger_{q'+p} a^\dagger_{q-p} a_{q'} a_q + b^\dagger_{q'} b^\dagger_q b_{q'-p} b_{q+p} \right\} F(q-p,q) F(q'+p,q'). \tag{5.85e}$$

These Hamiltonians conserve the fermion anti-fermion number.

Fermion Anti-fermion Number Non-conserving Terms

On the other hand, the Hamiltonian H_{non} does not conserve fermion anti-fermion number and can be written as

$$H_{non} = H^{(2)} + H^{(22)} + H^{(4)}, \tag{5.86}$$

where

$$H^{(2)} = -\frac{g^2L}{4\pi^2} \sum_{p,q} \frac{1}{p^2} \{a_q b_{-q} + h.c.\} F(q+p,q) G(-q,q+p), \tag{5.87a}$$

$$H^{(22)} = -\frac{g^2L}{4\pi^2} \sum_{p,q,q'} \frac{1}{p^2} \left\{ a^\dagger_{q'+p} a^\dagger_{-p-q} b^\dagger_q a_{q'} F(q'+p,q') G(-q-p,q) \right.$$

$$\left. + a^\dagger_{-p-q} b^\dagger_{q'} b^\dagger_q b_{-p+q'} F(q'-p,q') G(-q-p,q) + h.c. \right\}, \tag{5.87b}$$

$$H^{(4)} = -\frac{g^2L}{8\pi^2} \sum_{p,q,q'} \frac{1}{p^2} \left\{ a^\dagger_{p-q'} a^\dagger_{-p-q} b^\dagger_{q'} b^\dagger_q G(-q'+p,q') G(-q-p,q) + h.c \right\}. \tag{5.87c}$$

This is all that is needed to evaluate QED_2.

5.4.5. Boson Fock States

Since the Hamiltonian is now an operator, one should prepare some state onto which it can operate. The difficulty of solving QED_2 is that one cannot find the Fock states which are the eigenstate of the QED_2 Hamiltonian. This is because the QED_2 Hamiltonian contains the fermion and anti-fermion number non-conserving interactions and therefore it is practically impossible to find the eigenstate of the Hamiltonian. In general, the Fock state of bosons $|\mathcal{B}\rangle$ should be expanded in terms of the fermion $|q\rangle$ and anti-fermion $|\bar{q}\rangle$ pairs as

$$|\mathcal{B}\rangle = c_1|q\bar{q}\rangle + c_2|q^2\bar{q}^2\rangle + \cdots + c_n|q^n\bar{q}^n\rangle. \tag{5.88}$$

The coefficients c_n can be determined by asking that the Fock state eq.(5.88) must be the eigenstate of the Hamiltonian within the truncated space.

Fermion-Antifermion Fock States

Here, one can prepare a simplest Fock space which is spanned by the fermion and anti-fermion Fock state. However, the reliability of the calculated spectrum of bosons is not justified unless the fermion mass is very large.

5.4.6. Boson Wave Function

In this case, the boson state $|\Psi\rangle$ can be written as

$$|\Psi\rangle = \sum_n f_n a_n^\dagger \mathrm{b}_{-n}^\dagger |0\rangle \tag{5.89}$$

which is the momentum zero state for the boson system. Here, $|0\rangle$ denotes the perturbative (trivial) vacuum state and satisfies the following conditions

$$a_n|0\rangle = 0, \quad \mathrm{b}_n|0\rangle = 0.$$

f_n corresponds to the wave function in momentum space and should be determined by solving the Schrödinger-like equations as obtained below.

5.4.7. Boson Mass

The energy of the bosonic state $\mathcal{M}$ can be evaluated by

$$\mathcal{M} = \langle\Psi|H|\Psi\rangle. \tag{5.90}$$

It should be noted that the boson energy $\mathcal{M}$ is measured at the system where the total momentum of the boson is zero. This is the same as the vacuum state which is always in the momentum zero state. Therefore, the $\mathcal{M}$ corresponds to the boson mass itself in this calculation. The boson mass $\mathcal{M}$ can be easily calculated as

$$\mathcal{M} = \sum_{n,m} f_n f_m \left\{ 2\delta_{n,m}\sqrt{m_0^2 + \left(\frac{2\pi n}{L}\right)^2} + \frac{g^2 L}{4\pi^2}\left[\delta_{n,m}\sum_k \left(\frac{1}{(n-k)^2}|F(n,k)|^2\right)\right.\right.$$
$$-\delta_{n,m}\sum_k \left(\frac{1}{(n-k)^2}|G(n,-k)|^2\right) - \frac{1}{(n-m)^2}F(n,m)F(-n,-m)$$
$$\left.\left. + \lim_{K_0\to 0}\frac{1}{K_0^2}G(K_0-n,n)G(m,K_0-m)\right]\right\}, \tag{5.91}$$

where f_n satisfies

$$\sum |f_n|^2 = 1.$$

Boson Mass in Thermodynamic Limit

In the thermodynamic limit ($L \longrightarrow \infty$), the boson mass $\mathcal{M}$ becomes

$$\mathcal{M} = \int 2\left(E_p - \frac{g^2}{4\pi E_p}\right)|f(p)|^2 dp + \frac{g^2 m_0^2}{8\pi}\int \frac{f(p)f(q)}{E_p^2 E_q^2}dp\,dq$$

$$+\frac{g^2}{4\pi}\int \frac{(f(p)-f(q))^2}{(p-q)^2}\left(\frac{E_pE_q+pq+m_0^2}{2E_pE_q}\right)dp\,dq, \tag{5.92}$$

where $f(p)$ corresponds to f_n, and $E_p = \sqrt{m_0^2+p^2}$.

Boson Mass in Massless Fermion Limit

The energy eigenvalues can be obtained by diagonalizing the above equation in the numerical evaluation, and one can easily find the boson spectrum. However, one sees from the numerical calculations that the boson mass becomes negative when the fermion mass m_0 becomes very small. In fact, the behavior of the small fermion mass region can be found analytically by approximating the wave function in the following way

$$f(p) = \begin{pmatrix} f_0 & \text{for } \ |p| < p_0 \\ 0 & \text{for } \ |p| > p_0 \end{pmatrix}, \tag{5.93}$$

where f_0 is a constant. In this case, one can write the energy of eq.(5.92) in the limit of $m_0 \longrightarrow 0$ as

$$\mathcal{M} \simeq f_0^2\frac{g}{\sqrt{\pi}}\left[\ln\left(\frac{m_0}{g}\right)+2\left(\frac{p_0}{g}\right)^2-\ln\left(\frac{2p_0}{g}\right)+\frac{\pi^2}{2}\right]. \tag{5.94}$$

This obviously becomes negative for very small m_0 because of $\ln(m_0)$ term, that is

$$\ln\left(\frac{m_0}{g}\right) \longrightarrow -\infty \ \text{when } m_0 \longrightarrow 0.$$

Since this is a crudest variational wave function, the realistic eigenvalue must be even lower than that, and therefore eq.(5.92) gives unphysical boson mass for small fermion mass regions.

Improper Vacuum

What is the basic problem of this infrared instability? This turns out to be due to the improper vacuum state (trivial). Clearly, the vacuum structure is most important in the field theory calculations since the boson spectrum is constructed right on this vacuum state. To avoid this infra-red instability, one should improve the vacuum structure. A simple way of improving the vacuum state in QED_2 is to employ the Bogoliubov transformation which gives a special way of summing up the $|q^n\bar{q}^n\rangle$ states in a non-perturbative manner. However, it is not the eigenstate of the Hamiltonian and therefore it is not an exact method but is based on the variational principle which is constructed on the Bogoliubov vacuum. Nevertheless, it can give a good description of the boson spectrum in QED_2.

5.5. Bogoliubov Transformation in QED_2

In this section, we present an interesting technique which is often employed in solid state physics. The method we discuss here is called Bogoliubov transformation method which is equivalent to BCS theory in the treatment of the superconductor theory. Here, we apply the Bogoliubov transformation method to the description of the boson spectrum in QED_2 with massive fermions. The Bogoliubov transformation method is essentially based on the variational principle, and the variables in this case are the Bogoliubov angles θ_n.

5.5.1. Bogoliubov Transformation

In order to treat the interaction that violates the fermion anti-fermion number conservation, one can employ the Bogoliubov transformation to define a new vacuum which is a unitary transformation from one vacuum to the other. The Bogoliubov vacuum state is determined such that the new vacuum energy is minimized with respect to the Bogoliubov angles. Once the Bogoliubov angles are determined, then one obtains a new vacuum state, and thus one can construct the boson state by operating the creation operators onto the vacuum state. With this Fock states, one should diagonalize the Hamiltonian without making any further approximations such as the mean field approximation. In this respect, the Bogoliubov vacuum itself is an approximate vacuum, but otherwise one should solve the dynamics as accurately as possible.

New Fermion Operators

Now, one defines new fermion operators by Bogoliubov transformation.

$$c_n = e^{-\mathcal{A}} a_n e^{\mathcal{A}} = \cos\theta_n a_n - \sin\theta_n \mathrm{b}_{-n}^{\dagger}, \tag{5.95a}$$

$$d_n = e^{-\mathcal{A}} \mathrm{b}_n e^{\mathcal{A}} = \cos\theta_n \mathrm{b}_n - \sin\theta_n a_{-n}^{\dagger}, \tag{5.95b}$$

where the generator of the Bogoliubov transformation is given by

$$\mathcal{A} = \sum_n \theta_n (\mathrm{b}_{-n} a_n - a_n^{\dagger} \mathrm{b}_{-n}^{\dagger}). \tag{5.96}$$

θ_n denotes the Bogoliubov angle which can be taken to be real here and satisfies the following condition,

$$\theta_n = -\theta_{-n}. \tag{5.97}$$

The inverse transformation can also be written as

$$a_n = \cos\theta_n c_n + \sin\theta_n d_{-n}^{\dagger}, \tag{5.98a}$$

$$\mathrm{b}_n = \cos\theta_n d_n + \sin\theta_n c_{-n}^{\dagger}. \tag{5.98b}$$

Bogoliubov Vacuum

In this case, one can obtain the new vacuum state $|\Omega\rangle_B$ by

$$|\Omega\rangle_B = e^{-\mathcal{A}}|0\rangle. \tag{5.99}$$

Clearly the new vacuum state $|\Omega\rangle_B$ satisfies the following conditions

$$c_n|\Omega\rangle_B = 0, \tag{5.100a}$$

$$d_n|\Omega\rangle_B = 0 \tag{5.100b}$$

since, for example,

$$c_n|\Omega\rangle_B = e^{-\mathcal{A}} a_n e^{\mathcal{A}} e^{-\mathcal{A}}|0\rangle = e^{-\mathcal{A}} a_n|0\rangle = 0.$$

Now, the Hamiltonian should be written in terms of the new operators c_n , d_n. In order to do so, one first rewrites the current $\tilde{j}_0(p)$

$$\tilde{j}_0(p) = \sum_q \Big[c^\dagger_{q-p}c_q\tilde{F}(q-p,q) + c^\dagger_{-p-q}d^\dagger_q\tilde{G}(-p-q,q)$$
$$+d_{p-q}c_q\tilde{G}(p-q,q) - d^\dagger_q d_{p+q}\tilde{F}(q+p,q)\Big], \tag{5.101}$$

where $\tilde{F}$ and $\tilde{G}$ can be written as

$$\tilde{F}(p,q) = \cos(\theta_p-\theta_q)F(p,q) + \sin(\theta_p-\theta_q)G(p,-q), \tag{5.102a}$$

$$\tilde{G}(p,q) = -\sin(\theta_p+\theta_q)F(p,-q) + \cos(\theta_p+\theta_q)G(p,q). \tag{5.102b}$$

Bogoliubov Transformed Interaction Hamiltonian

Therefore, the interaction Hamiltonians which conserve the fermion anti-fermion number become

$$H^{(1)}_{MR} = \frac{g^2L}{8\pi^2}\sum_{p,q}\frac{1}{p^2}\{c^\dagger_q c_q + d^\dagger_q d_q\}\tilde{F}(q,q-p)\tilde{F}(q-p,q), \tag{5.103}$$

$$H^{(2)}_{MR} = -\frac{g^2L}{8\pi^2}\sum_{p,q}\frac{1}{p^2}\{c^\dagger_q c_q + d^\dagger_q d_q\}\tilde{G}(q,-p-q)\tilde{G}(-q-p,q), \tag{5.104}$$

$$H_C = \frac{g^2L}{8\pi^2}\sum_{p,q,q'}\frac{1}{p^2}\{c^\dagger_{q'+p}d^\dagger_q c_{q'}d_{q+p} + h.c.\}\tilde{F}(q'+p,q')\tilde{F}(q+p,q), \tag{5.105}$$

$$H_A = -\frac{g^2L}{8\pi^2}\sum_{p,q,q'}\frac{1}{p^2}\{c^\dagger_{p-q'}d^\dagger_{q'}c_q d_{p-q} + h.c.\}\tilde{G}(p-q',q')\tilde{G}(p-q,q), \tag{5.106}$$

$$H_R = -\frac{g^2L}{8\pi^2}\sum_{p,q,q'}\frac{1}{p^2}\Big\{c^\dagger_{q'+p}c^\dagger_{q-p}c_{q'}c_q + d^\dagger_{q'}d^\dagger_q d_{q'-p}d_{q+p}\Big\}\tilde{F}(q-p,q)\tilde{F}(q'+p,q'). \tag{5.107}$$

Further, the $(cd + d^\dagger c^\dagger)$ terms which arise not only from the interaction term but also from the kinetic energy terms become

$$\tilde{H}^{(2)} = \sum_q \left[-E_q \sin 2\theta_q + \sum_p \left\{ A(p,q)\sin 2(\theta_p + \theta_q) + B(p,q)\cos 2(\theta_p + \theta_q) \right\} \right] \times (c_q d_{-q} + d^\dagger_{-q} c^\dagger_q), \tag{5.108}$$

where

$$A(p,q) = -\frac{g^2 L}{8\pi^2} \frac{1}{(p+q)^2} \left(|F(p,-q)|^2 - |G(p,q)|^2 \right), \tag{5.109a}$$

$$B(p,q) = \frac{g^2 L}{8\pi^2} \frac{2}{(p+q)^2} F(p,-q) G(p,q). \tag{5.109b}$$

Determination of Bogoliubov Angle

The condition that the terms proportional to $(cd + d^\dagger c^\dagger)$ should vanish can determine the Bogoliubov angle θ_q . This is equivalent to the condition that the vacuum energy after the Bogoliubov transformation must be minimized, and this means that the Bogoliubov method is based on the variational principle. In this case, the condition for the Bogoliubov angle θ_q is given as

$$\tan 2\theta_q = \frac{\tilde{A}_1(q)}{E_q - \tilde{A}_2(q)}, \tag{5.110}$$

where

$$\tilde{A}_1(q) = \sum_p \left[A(p,q)\sin 2\theta_p + B(p,q)\cos 2\theta_p \right], \tag{5.111a}$$

$$\tilde{A}_2(q) = \sum_p \left[A(p,q)\cos 2\theta_p - B(p,q)\sin 2\theta_p \right]. \tag{5.111b}$$

Eq.(5.110) can be solved numerically by the iteration procedure since $\tilde{A}_1(q)$ and $\tilde{A}_2(q)$ contain θ_p. After a few hundreds of iterations, one can determine the values of the Bogoliubov angles θ_p, and thus $\tilde{F}(p,q)$ and $\tilde{G}(p,q)$ can be calculated. In this way, one can diagonalize the new Hamiltonian and obtain the boson spectrum.

5.5.2. Boson Mass in Bogoliubov Vacuum

The spectrum of bosons in QED_2 can be calculated just in the same way as in the trivial vacuum case. Now, the boson state $|\Psi\rangle$ can be written as

$$|\Psi\rangle = \sum_n f_n c^\dagger_n d^\dagger_{-n} |\Omega\rangle_B. \tag{5.112}$$

The energy of the bosonic state $\mathcal{M}$ can be evaluated to be

$$\mathcal{M} = \langle \Psi | H | \Psi \rangle. \tag{5.113}$$

The diagonalization procedure is just the same as the trivial vacuum case, and one can obtain the spectrum of boson mass as the function of the fermion mass m_0.

It is surprising and interesting to note here that the boson mass calculated by the Bogoliubov method can reproduce the right Schwinger boson mass [106]

$$\mathcal{M} = \frac{g}{\sqrt{\pi}} \tag{5.114}$$

at the massless fermion limit of $m_0 = 0$.

Lowest Boson Mass

In the finite but small fermion mass m_0 region, the lowest boson mass can be expressed as [10]

$$\mathcal{M} \simeq \frac{g}{\sqrt{\pi}} + e^{\gamma} m_0, \tag{5.115}$$

where γ denotes Euler's constant, $\gamma = 0.577216$, and the numerical calculation is just consistent with the result of eq.(5.115) [106]. As the mass increases, the number of the bosonic bound states increases. When the mass is much larger than the Schwinger boson $\frac{g}{\sqrt{\pi}}$, then the system becomes just the same as the nonrelativistic quantum mechanics. The particle and antiparticle are interacting with each other through the linear rising potential. In this case, the vacuum state is just like the perturbative vacuum, and there is no fermion condensate.

In this respect, an interesting region must be the one in which the fermion mass is much smaller than the Schwinger boson $\frac{g}{\sqrt{\pi}}$, but it is still finite. If the fermion mass is finite, then there is no chiral symmetry in the Lagrangian density. This means that there cannot be any symmetry breaking phenomena in the massive fermion QED_2. However, if one evaluates the chiral condensates, then one finds a finite condensate value for the massive fermion case in the Bogoliubov transformation method.

5.5.3. Chiral Condensate

The chiral condensate value for the massive fermion QED_2 can be calculated by the Bogoliubov transformation formalism, and indeed one obtains the absolute value of the fermion condensate as the function of the fermion mass [75, 106]

$$|\langle\Omega|\frac{1}{L}\int \bar{\psi}\psi dx|\Omega\rangle| \simeq \frac{g}{\sqrt{\pi}}\frac{e^{\gamma}}{2\pi} - 0.5m_0 \quad \text{for } m_0 \ll \frac{g}{\sqrt{\pi}}. \tag{5.116}$$

This result shows that the chiral condensate is a continuous function of the fermion mass m_0 in a small fermion mass region. This strongly suggests that the chiral condensate value is not the consequence of the symmetry breaking, but it only shows how many of the virtual particle pairs can be found in the vacuum. Or in other words, the chiral condensate is related to the change of the momentum distributions of the negative energy particles in comparison with the symmetric distributions of the free particles in the vacuum state.

This chiral condensate values are obtained by the Bogoliubov transformation method. Even though it can reproduce the right Schwinger boson at the massless fermion limit, the Bogoliubov vacuum is still not exact. In other words, the Bogoliubov vacuum is not the eigenstate of the Hamiltonian. Nevertheless, the finite value of the chiral condensate in the Bogoliubov vacuum is not inconsistent with the prediction of the condensate value in Section 4.3. In the case of the spontaneous symmetry breaking physics, the axial vector current is always conserved, and therefore the chiral charge Q_5 is a conserved quantity and has the same eigenstate as the Hamiltonian

$$\hat{Q}_5|\Omega\rangle = q_5|\Omega\rangle \quad \text{for spontaneous symmetry breaking.}$$

On the other hand, the axial vector current is not conserved in the Schwinger model due to the chiral anomaly, and therefore $\hat{Q}_5$ does not have the same eigenstate as the Hamiltonian

$$\hat{Q}_5|\Omega\rangle \neq q_5|\Omega\rangle \quad \text{for Schwinger model.}$$

Thus, one cannot make use of the identity equation of eq.(4.17). This indicates that the chiral condensate value in the Schwinger model does not have to vanish and indeed it is found to be finite.

5.6. QED$_2$ in Light Cone

It is well known that the light cone calculation reproduces an exact result of the boson mass in the Schwinger model. However, it is also confirmed that the vacuum of QED$_2$ should possess a finite condensate value even in the light cone vacuum. In this respect, it is interesting to know any reason why the light cone calculation with the trivial vacuum gives a correct Schwinger boson even though it predicts no fermion condensate.

5.6.1. Light Cone Quantization

In this section, we briefly describe the light cone quantization which is first introduced by Dirac [27]. In particular, we present the evaluation of the boson mass in QED$_2$ in the light cone calculations [18, 29, 74].

Light Cone Coordinate

In the light cone in two dimensions, one introduces new variables by

$$x^+ = t + x, \quad x^- = t - x \tag{5.117}$$

and one identifies x^+ as a light cone time and x^- as a light cone coordinate.

Light Cone Hamiltonian

Accordingly, the light cone momentum P^+ and energy P^- can be written by the energy momentum tensor in the light cone coordinate as

$$P^+ = \int dx^- \psi_+^\dagger(-i\partial_-)\psi_+, \tag{5.118a}$$

$$P^- = m_0^2 \int dx^- \psi_+^\dagger \frac{1}{(-i\partial_-)} \psi_+ + 2g^2 \int dx^- \psi_+^\dagger \psi_+ \frac{1}{(-i\partial_-)^2} \psi_+^\dagger \psi_+, \tag{5.118b}$$

where the differential operator $i\partial_-$ is kept in the denominator since it should become convenient for the momentum representation. The invariant mass squared is therefore written as

$$\mathcal{M}^2 = P^+P^-. \tag{5.119}$$

Now, one defines K and H by

$$P^+ = \frac{2\pi}{L}K, \tag{5.120a}$$

$$P^- = \frac{L}{2\pi}H, \tag{5.120b}$$

where L is the size of the box in the light cone coordinate. The K is called the harmonic resolution and the H the light cone Hamiltonian.

Now, one can quantize the field ψ_+ and express the Hamiltonian by creation and annihilation operators. In this case, H is written as

$$H = \sum_{n=1}^{\infty} \frac{m_0^2}{n}\left(a_n^\dagger a_n + b_n^\dagger b_n\right) + \frac{g^2}{\pi}\sum_{n=1}^{\infty} c_n^\dagger c_n, \tag{5.121}$$

where c_n is defined as

$$c_n = \frac{1}{\sqrt{n}}\left[\sum_{m=0}^{n-1} a_m^\dagger b_{n-m}^\dagger + \sum_{m=0}^{\infty} a_{n+m}^\dagger a_m - \sum_{m=1}^{\infty} b_{n+m}^\dagger b_m\right]. \tag{5.122}$$

These $c_n^\dagger$ and c_n represent the creation and annihilation operators of bosons in massless Schwinger model and obey the usual boson commutation relations if they operate on any physical states $|\mathrm{phys}\rangle$

$$[c_n, c_{n'}^\dagger]|\mathrm{phys}\rangle = \delta_{\mathrm{n,n'}}|\mathrm{phys}\rangle. \tag{5.123}$$

All the other commutators vanish. Here, it should be noted that, in light cone quantization, all the momenta are nonnegative. This is very important from the computational point of view.

Fock Space and Wave Function

In order to obtain the eigenvalues of the Hamiltonian H, one prepares Fock spaces. If one takes the $q\bar{q}$ subspace which is the simplest Fock space for the charge zero sector, then one can write the $q\bar{q}$ state $|\Phi_K\rangle$ of the Fock space which has a total momentum K,

$$|\Phi_K\rangle = \sum_{n=\frac{1}{2}}^{K} f_n a_n^\dagger b_{K-n}^\dagger |0\rangle, \tag{5.124}$$

where f_n is a wave function in momentum space. $|0\rangle$ denotes a perturbative vacuum state. The Hamiltonian H is evaluated with $|\Phi_K\rangle$ by introducing the momentum fraction x as

$$n = xK$$

and one obtains the following equation

$$\mathcal{M}^2 = \int_0^1 \frac{m_0^2}{x(1-x)} |f(x)|^2 \, dx$$

$$+\frac{g^2}{\pi}\left[\left(\int_0^1 f(x)\,dx\right)^2 + \frac{1}{2}\int_0^1\int_0^1 \frac{(f(x)-f(y))^2}{(x-y)^2}\,dx\,dy\right], \qquad (5.125)$$

where f_n is now written as $f(x)$. The wave function $f(x)$ satisfies the normalization condition

$$\int_0^1 |f(x)|^2 \, dx = 1. \qquad (5.126)$$

From eq.(5.125), one clearly sees that one can find an exact solution when the fermion is massless

$$f(x) = 1 \text{ with } m_0 = 0$$

which can indeed satisfy eq.(5.125). In this case, the boson mass becomes

$$\mathcal{M} = \frac{g}{\sqrt{\pi}}$$

which is just the Schwinger boson [96]. On the other hand, the fermion condensate value of QED_2 vanishes in the light cone calculation unless one modifies the light cone vacuum state.

Chapter 6

Quantum Chromodynamics

Physics of the strong interactions is described by quantum chromodynamics (QCD), and this is by now well established. Many experimental observations support that the number of the color must be three, and interactions between quarks should be mediated by gluons which are gauge bosons with colors. In addition to colors, quarks have six flavors of up, down, strangeness, charm, bottom and top.

However, it is extremely difficult to solve QCD in a non-perturbative fashion and obtain any reasonable spectrum of hadrons from QCD since quantum field theory has infinite degrees of freedom. At the present stage, one should make some kind of approximations in order to obtain physical observables. The perturbative treatment is most popular. However, there is a serious problem in the unperturbed QCD Hamiltonian since there are no free quark and gluon states in physical space, and indeed the unperturbed Fock space is gauge dependent. In addition, we present an inherent problem connected to the gauge non-invariance of the unperturbed and interaction Lagrangian densities due to the non-conservation of the quark color current. Therefore there is a basic difficulty of carrying out the perturbative expansion unless one can overcome the difficulty of the gauge dependences of both the unperturbed and the interaction Lagrangian densities in some way or the other.

Therefore, in the first part of this chapter, we discuss the basic field theory aspects of the non-abelian gauge theory. In particular, we clarify what are the physical observables in QCD since some of known quantities are not gauge invariant and thus they cannot be observed. Secondly, we discuss the non-perturbative aspects of QCD and, here, we treat the bound state problem in QCD in two dimensions in terms of the total Hamiltonian which is gauge invariant. Two dimensional field theory is, of course, not realistic, but one can learn some important physics of the non-perturbative nature. In two dimensions, one can calculate observables non-perturbatively to a sufficiently reliable degree. In particular, we show the calculation of the boson mass in QCD_2 based on the Bogoliubov transformation method.

6.1. Properties of QCD with $SU(N_c)$ Colors

In this section, we explain some properties of QCD since the basic part of discussions concerning the properties of the Lagrangian density is common to the two and four dimensional

QCD. The essential difference between the two dimensional QCD and four dimensional QCD is the absence of the transverse vector fields in two dimensional QCD.

6.1.1. Lagrangian Density of QCD

The Lagrangian density of QCD with $SU(N_c)$ color is described as

$$\mathcal{L} = \bar{\psi}(i\gamma^\mu\partial_\mu - g\gamma^\mu A_\mu - m_0)\psi - \frac{1}{2}\mathrm{Tr}\{G_{\mu\nu}G^{\mu\nu}\}, \tag{6.1}$$

where $G_{\mu\nu}$ is written as

$$G_{\mu\nu} = \partial_\mu A_\nu - \partial_\nu A_\mu + ig[A_\mu, A_\nu], \tag{6.2}$$

$$A_\mu = A_\mu^a T^a \equiv \sum_{a=1}^{N_c^2-1} A_\mu^a T^a, \tag{6.3}$$

where T^a denotes the generator of $SU(N_c)$ group and satisfies the following commutation relations

$$[T^a, T^b] = iC^{abc}T^c, \tag{6.4}$$

where C^{abc} denotes the structure constant of the group generators. For $SU(2)$ case, the structure constant C^{abc} becomes just the anti-symmetric symbol ε_{abc}. In eq.(6.1), Tr { } means the trace of the group generators of $SU(N_c)$, and the generators T^a are normalized according to

$$\mathrm{Tr}\{T^aT^b\} = \frac{1}{2}\delta^{ab}. \tag{6.5}$$

Therefore, the last term of eq.(6.1) can be rewritten as

$$\frac{1}{2}\mathrm{Tr}\{G_{\mu\nu}G^{\mu\nu}\} = \frac{1}{4}G_{\mu\nu}^a G^{a,\mu\nu},$$

where $G_{\mu\nu}^a$ is described as

$$G_{\mu\nu}^a = \partial_\mu A_\nu^a - \partial_\nu A_\mu^a - gC^{abc}A_\mu^b A_\nu^c. \tag{6.6}$$

m_0 denotes the fermion mass, and at the massless limit, the Lagrangian density has a chiral symmetry.

6.1.2. Infinitesimal Local Gauge Transformation

QCD Lagrangian density is invariant under the following infinitesimal local gauge transformation

$$\psi' = (1 - ig\chi)\psi = (1 - igT^a\chi^a)\psi, \quad \text{with } \chi = T^a\chi^a, \tag{6.7}$$

$$A'_\mu = A_\mu + ig[A_\mu, \chi] + \partial_\mu\chi \quad \text{or} \tag{6.8a}$$

$$A'^a_\mu = A_\mu^a - gC^{abc}A_\mu^b\chi^c + \partial_\mu\chi^a, \tag{6.8b}$$

where χ is infinitesimally small. By defining the covariant derivative D_μ by

$$D_\mu = \partial_\mu + igT^a A_\mu^a$$

one can see

$$D'_\mu \psi' = \left[\partial_\mu + igT^a\left(A_\mu^a - gC^{abc}A_\mu^b \chi^c + \partial_\mu \chi^a\right)\right](1 - igT^a\chi^a)\psi$$

$$= (1 - igT^a\chi^a)D_\mu \psi. \tag{6.9}$$

Therefore, one can prove that

$$\bar{\psi}' i\gamma^\mu D'_\mu \psi' = \bar{\psi} i\gamma^\mu D_\mu \psi, \tag{6.10}$$

$$G'_{\mu\nu} = (1 - igT^a\chi^a)G_{\mu\nu}(1 + igT^a\chi^a). \tag{6.11}$$

and one obtains

$$\mathrm{Tr}\{G'_{\mu\nu}G'^{\mu\nu}\} = \mathrm{Tr}\left\{(1 - igT^a\chi^a)G_{\mu\nu}G^{\mu\nu}(1 + igT^a\chi^a)\right\} = \mathrm{Tr}\{G_{\mu\nu}G^{\mu\nu}\}. \tag{6.12}$$

ThereforeCone sees that the Lagrangian density of eq.(6.1) is invariant under the infinitesimal local gauge transformation.

6.1.3. Local Gauge Invariance

Now, the local gauge transformation with finite χ is defined as

$$A'_\mu = U(\chi)A_\mu U^\dagger(\chi) - \frac{i}{g}U(\chi)\partial_\mu U^\dagger(\chi), \tag{6.13}$$

$$\psi' = U(\chi)\psi, \tag{6.14}$$

where $U(\chi)$ is described in terms of χ as

$$U(\chi) = e^{-ig\chi}. \tag{6.15}$$

Here, one can easily prove the following equations

$$\bar{\psi}' i\gamma^\mu D'_\mu \psi' = \bar{\psi} i\gamma^\mu D_\mu \psi, \tag{6.16}$$

$$G'_{\mu\nu} = U(\chi)G_{\mu\nu}U(\chi)^{-1} \tag{6.17}$$

and by making use of the following identity

$$\sum_{a=1}^{n_a} G'^a_{\mu\nu}G'^{a,\mu\nu} = 2\mathrm{Tr}\{G'_{\mu\nu}G'^{\mu\nu}\} = 2\mathrm{Tr}\{G_{\mu\nu}G^{\mu\nu}\} = \sum_{a=1}^{n_a} G^a_{\mu\nu}G^{a,\mu\nu} \tag{6.18}$$

the gauge invariance of the Lagrangian density is easily seen.

6.1.4. Noether Current in QCD

The QCD Lagrangian density is invariant under the following infinitesimal global gauge transformation

$$\psi' = (1 - igT^a\theta^a)\psi, \tag{6.19a}$$

$$A'^a_\nu = A^a_\nu - gC^{abc}A^b_\nu\theta^c, \tag{6.19b}$$

where θ^a is an infinitesimally small constant. In this case, one finds

$$\delta\mathcal{L} = \mathcal{L}(\psi', \partial_\mu\psi', A'^a_\nu, \partial_\mu A'^a_\nu) - \mathcal{L}(\psi, \partial_\mu\psi, A^a_\nu, \partial_\mu A^a_\nu) = 0. \tag{6.20}$$

By making use of the equations of motion, one obtains

$$\delta\mathcal{L} = \Big[- ig(i\partial_\mu\bar{\psi}\gamma^\mu T^a\psi + i\bar{\psi}\gamma^\mu T^a\partial_\mu\psi)$$

$$-g(\partial_\mu G^{\mu\nu,c}C^{bca}A^b_\nu + G^{\mu\nu,c}C^{bca}\partial_\mu A^b_\nu)\Big]\theta^a = 0. \tag{6.21}$$

Therefore, one easily finds that

$$\partial_\mu\left(\bar{\psi}\gamma^\mu T^a\psi + C^{abc}G^{\mu\nu,b}A^c_\nu\right) = 0. \tag{6.22}$$

This means that the Noether current

$$I^{\mu,a} \equiv j^{\mu,a} + C^{abc}G^{\mu\nu,b}A^c_\nu \tag{6.23}$$

is indeed conserved. That is,

$$\partial_\mu I^{\mu,a} = 0, \tag{6.24}$$

where the quark color current j^a_μ is defined as

$$j^a_\mu = \bar{\psi}\gamma_\mu T^a\psi.$$

Thus, the quark color current alone cannot be conserved, and therefore there is no conservation of the quark color charge. This is consistent with the fact that the color current of quarks is not a gauge invariant quantity.

6.1.5. Conserved Charge of Color Octet State

From eqs.(6.22) and (6.23), one sees that the color octet vector current of one quark and one gluon state $I^{\mu,a}$ is conserved. Since $\partial_\mu I^{\mu,a}$ is a gauge invariant quantity, one can integrate it over all space

$$\int \partial_\mu I^{\mu,a}\,d^3r = \frac{d}{dt}\int I^a_0\,d^3r + \int \boldsymbol{\nabla}\cdot\boldsymbol{I}^a\,d^3r = \frac{dQ^a_I}{dt} + \int \boldsymbol{I}^a\cdot d\boldsymbol{S} = \frac{dQ^a_I}{dt} = 0,$$

where the color charge Q^a_I is defined as

$$Q^a_I = \int I^{0,a}\,d^3r. \tag{6.25}$$

Therefore, the color charge Q^a_I is indeed a conserved quantity, and there may be some chance that the color charge Q^a_I becomes a physical observable.

6.1.6. Gauge Non-invariance of Interaction Lagrangian

The interaction Lagrangian density of QCD that involves quark currents is written as

$$\mathcal{L}_I = -g j^a_\mu A^{\mu,a}. \tag{6.26}$$

Now, the interaction Lagrangian density $\mathcal{L}_I$ is not gauge invariant, and therefore if one wishes to make any perturbation calculations involving the quark color currents, then one should check it in advance whether one can make the gauge invariant quark-quark interactions.

The interaction Lagrangian density is transformed into a new shape under the infinitesimal local gauge transformation

$$\mathcal{L}_I = -g j^a_\mu (A^{\mu,a} + \partial^\mu \chi^a), \tag{6.27}$$

where the second term is a gauge dependent term. In the same way as QED case, one can rewrite the second term by making use of the conserved current as

$$-g j^a_\mu \partial^\mu \chi^a = -g\partial^\mu (j^a_\mu \chi^a) + gC^{abc}\chi^a \partial^\mu G^b_{\mu\nu} A^{\nu,c}. \tag{6.28}$$

The first term is a total derivative and thus does not contribute. However, there is no way to erase the second term.

Therefore, one sees that one cannot make any simple-minded perturbative calculations of quark-quark interactions in QCD, contrary to the QED case where the electron-electron interaction is well defined and calculated. This means that there is a difficulty of defining any potential between quarks, and this is of course consistent with the picture that the color charge of quarks are gauge dependent and is not a conserved quantity.

6.1.7. Equations of Motion

The Lagrangian equations of motion now become

$$(i\gamma^\mu \partial_\mu - g\gamma^\mu A_\mu - m_0)\psi = 0, \tag{6.29}$$

$$\partial_\mu G^{\mu\nu,a} = gJ^{\nu,a} = g\left(j^{\nu,a} + C^{abc} G^{\nu\rho,b} A^c_\rho\right). \tag{6.30}$$

One can see that the equation of motion for the gauge fields has gauge field source terms in addition to the quark color current. Even though the equation of motion looks similar to that of QED, physics of QCD must be very different from the QED case. Now, one can introduce the color electric field $\boldsymbol{E}^a$ and the color magnetic field $\boldsymbol{B}^a$ by

$$\boldsymbol{E}^a = -\dot{\boldsymbol{A}}^a - \boldsymbol{\nabla} A^a_0 - gC^{abc}\boldsymbol{A}^b A^c_0, \tag{6.31a}$$

$$\boldsymbol{B}^a = \boldsymbol{\nabla} \times \boldsymbol{A}^a + \frac{1}{2} gC^{abc}\boldsymbol{A}^b \times \boldsymbol{A}^c. \tag{6.31b}$$

It should be noted that the fields $\boldsymbol{E}^a$ and $\boldsymbol{B}^a$ themselves are not gauge invariant, contrary to the QED case.

Now, eq.(6.30) can be rewritten in terms of $\boldsymbol{E}^a$ and $\boldsymbol{B}^a$ as

$$\nabla\cdot\boldsymbol{E}^a = gj_a^0 - gC^{abc}\boldsymbol{A}^b\cdot\boldsymbol{E}^c, \tag{6.32a}$$

$$\nabla\times\boldsymbol{B}^a - \frac{\partial\boldsymbol{E}^a}{\partial t} = g\boldsymbol{j}^a - gC^{abc}\left(A_0^b\boldsymbol{E}^c + \boldsymbol{A}^b\times\boldsymbol{B}^c\right). \tag{6.32b}$$

From eq.(6.30), one sees that the current $I^{\nu,a}$ is a conserved quantity, $\partial_\mu I^{\mu,a} = 0$. In order to solve the dynamics of QCD, it should be inevitable to take into account the conservation of this current $I^{\nu,a}$. A question is of course in which way one should consider this effect of the current conservation in QCD dynamics, and this is still an open question which cannot be answered in this textbook.

6.1.8. Hamiltonian Density of QCD

Now, one can construct the Hamiltonian density of QCD just in the same way as the QED case. The Hamiltonian density $\mathcal{H}$ can be defined by the energy momentum tensor $\mathcal{T}^{\mu\nu}$ as

$$\begin{aligned}\mathcal{H} \equiv \mathcal{T}^{00} &= \sum_i\left(\frac{\partial\mathcal{L}}{\partial\dot{\psi}_i^a}\dot{\psi}_i^a + \frac{\partial\mathcal{L}}{\partial\dot{\psi}_i^{a\dagger}}\dot{\psi}_i^{a\dagger}\right) + \sum_k\left(\frac{\partial\mathcal{L}}{\partial\dot{A}_k^a}\dot{A}_k^a\right) - \mathcal{L}\\ &= \sum_i\left(\Pi_{\psi_i^a}\dot{\psi}_i + \Pi_{\psi_i^{a\dagger}}\dot{\psi}_i^{a\dagger}\right) + \sum_k \Pi_{A_k^a}\dot{A}_k^a - \mathcal{L},\end{aligned} \tag{6.33}$$

where the conjugate fields $\Pi_{\psi_i^a}$, $\Pi_{\psi_i^{a\dagger}}$ and $\Pi_{A_k^a}$ are defined as

$$\Pi_{\psi_i^a} = \frac{\partial\mathcal{L}}{\partial\dot{\psi}_i^a}, \quad \Pi_{\psi_i^{a\dagger}} = \frac{\partial\mathcal{L}}{\partial\dot{\psi}_i^{a\dagger}}, \quad \Pi_{A_k^a} = \frac{\partial\mathcal{L}}{\partial\dot{A}_k^a}. \tag{6.34}$$

Now, the conjugate vector field $\Pi_{A_\mu^a}$ can be evaluated to be

$$\Pi_{A_0^a}^a = 0, \quad \Pi_{A_k^a}^a = \frac{\partial\mathcal{L}}{\partial\dot{A}_k^a} = \dot{A}_k^a + \frac{\partial A_0^a}{\partial x_k} + gC^{abc}A_k^bA_0^c = -E_k^a.$$

Note that there is no corresponding conjugate field for A_0^a, and there is no kinetic energy term present for A_0^a. In this way, the Hamiltonian density of QCD is now written as

$$\mathcal{H} = \bar{\psi}[-i\boldsymbol{\gamma}\cdot\nabla + m_0]\psi + gj_0^aA_0^a - g\boldsymbol{j}^a\cdot\boldsymbol{A}^a - \boldsymbol{E}^a\cdot\dot{\boldsymbol{A}}^a - \frac{1}{2}[\boldsymbol{E}^a\cdot\boldsymbol{E}^a - \boldsymbol{B}^a\cdot\boldsymbol{B}^a]. \tag{6.35a}$$

By employing the equation of motion [eq.(6.30)], one obtains

$$\mathcal{H} = \bar{\psi}[-i\boldsymbol{\gamma}\cdot\nabla + m_0]\psi - g\boldsymbol{j}^a\cdot\boldsymbol{A}^a + \frac{1}{2}[\boldsymbol{E}^a\cdot\boldsymbol{E}^a + \boldsymbol{B}^a\cdot\boldsymbol{B}^a] \tag{6.35b}$$

which is, of course, gauge invariant.

6.1.9. Hamiltonian of QCD

The Hamiltonian can be obtained by integrating the Hamiltonian density over all space

$$H = \int d^3r \left\{ \bar{\psi}[-i\boldsymbol{\gamma}\cdot\boldsymbol{\nabla} + m_0]\psi - g\boldsymbol{j}^a\cdot\boldsymbol{A}^a + \frac{1}{2}[\boldsymbol{E}^a\cdot\boldsymbol{E}^a + \boldsymbol{B}^a\cdot\boldsymbol{B}^a] \right\}. \tag{6.36}$$

In order to calculate the spectrum emerged from this Hamiltonian, one should quantize the fields A_μ^a by making use of the equations of motion

$$(i\gamma^\mu\partial_\mu - g\gamma^\mu A_\mu - m_0)\psi = 0,$$

$$\partial_\mu G^{\mu\nu,a} = g\left(j^{\nu,a} + C^{abc}G^{\nu\rho,b}A_\rho^c\right).$$

After obtaining the quantized Hamiltonian, one should prepare Fock spaces and then evaluate the Hamiltonian to obtain the mass of hadrons, assuming finite quark masses. This is a difficult task, but it should be done in future.

6.2. Hamiltonian of QCD in Two Dimensions

When one wishes to solve QCD in four dimensions, one should prepare Fock states. Suppose one calculates the spectrum of bosons from QCD. Then, the Fock state of boson $|\mathcal{B}\rangle$ may be composed of

$$|\mathcal{B}\rangle = a_0|q\bar{q}\rangle + b_1|q\bar{q}g\rangle + b_2|q\bar{q}g^2\rangle + \cdots + c_1|q^2\bar{q}^2\rangle + c_2|q^2\bar{q}^2g\rangle + \cdots, \tag{6.37}$$

where q and g denote the quark and gluon states, respectively. Since the QCD Hamiltonian contains the fermion and anti-fermion number non-conserving interactions, it is practically impossible to find any eigenstate of the Hamiltonian, contrary to the Thirring model which does not contain the fermion and anti-fermion number non-conserving interactions.

Therefore, one should find some approximate scheme of solving QCD and obtaining the spectrum of bosons in QCD. The simplest way of solving QCD must be to truncate the Fock space by hand. For example, if one considers only the first term in eq.(6.37), then there is some chance to find the approximate energy eigenvalue of the QCD Hamiltonian. However, there is no reason to believe that gluons do not play any important role in bosons in QCD.

As the first step, however, one should consider the case without gluons and this is possible if one treats the two dimensional QCD which does not have any transverse gluon degree of freedoms. In this section, therefore, we discuss the boson spectrum in two dimensional QCD which has no transverse gluons. This means that, in QCD_2, one can describe the Hamiltonian in terms of the fermion fields only, and indeed the Hamiltonian of QCD_2 becomes four fermion interactions.

6.2.1. Gauge Fixing

In order to construct the Hamiltonian, one can make use of the equation of motions in QCD_2. In this case, there is a gauge freedom, and therefore one should fix the gauge to solve the equation of motions. Here, the axial gauge fixing

$$A_1^a = 0$$

is chosen for simplicity, and in this gauge, the equation of motion for A_0^a becomes

$$\partial_1^2 A_0^a = -g j_0^a \tag{6.38}$$

because of the following relations due to the two dimensionality of the fields

$$G^{0\rho,b} A_\rho^c = G^{01,b} A_1^c = 0.$$

Since there is no transverse gluon part in QCD$_2$, there is no chromomagnetic field ($\boldsymbol{B}^a = 0$). The chromo-electric field E_1^a is simply written as

$$E_1^a = -\partial_1 A_0^a$$

and therefore, one can write the Hamiltonian of QCD$_2$ [eq.(6.36)] as

$$H = \int dx \left[-i\bar{\psi}\gamma^1 \partial_1 \psi + m_0 \bar{\psi}\psi + \frac{1}{2} g j_0^a A_0^a \right], \tag{6.39a}$$

where the following equation is employed

$$\frac{1}{2} g \int (E_1^a)^2 \, dx = \frac{1}{2} g \int (\partial_1 A_0^a)^2 \, dx = -\frac{1}{2} g \int (\partial_1^2 A_0^a) A_0^a \, dx = \frac{1}{2} g \int j_0^a A_0^a \, dx.$$

Also, eq.(6.38) can be solved for A_0^a, and one obtains

$$A_0^a(x) = -\frac{1}{2} g \int |x - x'| j_0^a(x') \, dx'.$$

Therefore, one can write the Hamiltonian in terms of fermion fields only

$$H = \int dx \left[-i\bar{\psi}\gamma^1 \partial_1 \psi + m_0 \bar{\psi}\psi \right] - \frac{g^2}{4} \int dx \, dx' \, j_0^a(x) |x - x'| j_0^a(x') \tag{6.39b}$$

which is just similar to the QED Hamiltonian of eq.(5.75).

6.2.2. Quantization of Fields

The classical field Hamiltonian eq.(6.39) is described only by the fermion fields without involving the vector field A_k^a. Therefore, in this sense, it is a closed system. However, the classical field Hamiltonian itself cannot give us much information on the dynamics of the system.

In quantum field theory, one should quantize fields, and together with the quantized fields, one obtains the quantized Hamiltonian. The quantization of the fermion field $\psi(x)$ can be done by expanding it in terms of creation and annihilation operators in two dimensional field theory

$$\psi(x) = \frac{1}{\sqrt{L}} \sum_{n,\alpha} \begin{pmatrix} a_{n,\alpha} \\ b_{n,\alpha} \end{pmatrix} e^{ip_n x}, \quad \text{with } p_n = \frac{2\pi n}{L}, \tag{6.40}$$

where $a_{n,\alpha}$ and $b_{n,\alpha}$ denote the annihilation operators of the fermions in the chiral representation. They satisfy the following anti-commutation relations

$$\{a_{n,\alpha}^\dagger, a_{m,\beta}\} = \{b_{n,\alpha}^\dagger, b_{m,\beta}\} = \delta_{n,m}\delta_{\alpha,\beta}. \tag{6.41}$$

All other anti-commutation relations vanish. The color index of $SU(N_c)$ has been, until now, described by roman letters $a,\ b$, but hereafter, α and β are employed and they run as $\alpha = 1, \dots, N_c$.

In this case, the current density $j_0^\alpha(x)$ is related to the momentum representation current density $\tilde{j}_{0,n}^\alpha$ as

$$j_0^\alpha(x) = \frac{1}{L}\sum_n \tilde{j}_{0,n}^\alpha e^{ip_n x}. \tag{6.42}$$

Here, $\tilde{j}_{0,n}^\alpha$ can be written in terms of creation and annihilation operators of $a_{m,\alpha}$ and $b_{m,\alpha}$ as

$$\tilde{j}_{0,n}^\alpha = \tilde{j}_{a,n,\alpha\beta} + \tilde{j}_{b,n,\alpha\beta},$$

where $\tilde{j}_{a,n,\alpha\beta}$ and $\tilde{j}_{b,n,\alpha\beta}$ are expressed as

$$\tilde{j}_{a,n,\alpha\beta} = \sum_m a_{m,\alpha}^\dagger a_{m+n,\beta}, \tag{6.43a}$$

$$\tilde{j}_{b,n,\alpha\beta} = \sum_m b_{m,\alpha}^\dagger b_{m+n,\beta}. \tag{6.43b}$$

6.2.3. Quantized Hamiltonian of QCD$_2$ with $SU(N_c)$

Therefore, one can express the Hamiltonian of QCD$_2$ with $SU(N_c)$ colors in terms of creation and annihilation operators [76]

$$\begin{aligned}\hat{H} = &\sum_{n,\alpha} p_n \left(a_{n,\alpha}^\dagger a_{n,\alpha} - b_{n,\alpha}^\dagger b_{n,\alpha}\right) + m_0 \sum_{n,\alpha}\left(a_{n,\alpha}^\dagger b_{n,\alpha} + b_{n,\alpha}^\dagger a_{n,\alpha}\right)\\ &-\frac{g^2}{4N_cL}\sum_{n\neq 0,\alpha,\beta}\frac{1}{p_n^2}\left(\tilde{j}_{a,n,\alpha\alpha}+\tilde{j}_{b,n,\alpha\alpha}\right)\left(\tilde{j}_{a,-n,\beta\beta}+\tilde{j}_{b,-n,\beta\beta}\right)\\ &+\frac{g^2}{4L}\sum_{n\neq 0,\alpha,\beta}\frac{1}{p_n^2}\left(\tilde{j}_{a,n,\alpha\beta}+\tilde{j}_{b,n,\alpha\beta}\right)\left(\tilde{j}_{a,-n,\beta\alpha}+\tilde{j}_{b,-n,\beta\alpha}\right).\end{aligned} \tag{6.44}$$

Now, the Hamiltonian is an operator, and therefore one should prepare the Fock states like those given in eq.(6.37). However, as one sees in the QED$_2$ case, the Fock states which are built on the perturbative (trivial) vacuum cannot describe the boson spectrum. In particular, the boson spectrum with the trivial vacuum state becomes unphysical when the fermion mass is sufficiently small. Therefore, one should employ more realistic vacuum states than the trivial one. Here, the Bogoliubov vacuum state can be employed since it is simple and tractable.

6.2.4. Bogoliubov Transformed Hamiltonian

New fermion operators are defined by making the Bogoliubov transformation [39, 43]

$$c_{n,\alpha} = e^{\mathcal{A}} a_{n,\alpha} e^{-\mathcal{A}} = \cos\theta_{n,\alpha} a_{n,\alpha} - \sin\theta_{n,\alpha} b_{n,\alpha}, \tag{6.45a}$$

$$d_{-n,\alpha}^\dagger = e^{\mathcal{A}} b_{n,\alpha} e^{-\mathcal{A}} = \sin\theta_{n,\alpha} a_{n,\alpha} + \cos\theta_{n,\alpha} b_{n,\alpha}, \tag{6.45b}$$

where $\theta_{n,\alpha}$ denotes the Bogoliubov angle. The generator of the Bogoliubov transformation is given by

$$\mathcal{A} = \sum_{n,\alpha} \theta_{n,\alpha} \left(a^{\dagger}_{n,\alpha} b_{n,\alpha} - b^{\dagger}_{n,\alpha} a_{n,\alpha} \right) \tag{6.46}$$

which transforms the free vacuum state $|0\rangle$ into the Bogoliubov vacuum $|\Omega\rangle_B$

$$|\Omega\rangle_B = e^{\mathcal{A}} |0\rangle. \tag{6.47}$$

The vacuum state $|0\rangle$ is filled with massess fermions with the negative energy states. In this case, the Hamiltonian of QCD$_2$ can be rewritten as

$$H = \sum_{n,\alpha} \mathcal{E}_{n,\alpha} (c^{\dagger}_{n,\alpha} c_{n,\alpha} + d^{\dagger}_{-n,\alpha} d_{-n,\alpha}) + H', \tag{6.48}$$

where

$$\mathcal{E}^2_{n,\alpha} = \left\{ p_n + \frac{g^2}{4N_c L} \sum_{m,\beta} \frac{(N_c \cos 2\theta_{m,\beta} - \cos 2\theta_{m,\alpha})}{(p_m - p_n)^2} \right\}^2 + \left\{ m_0 + \frac{g^2}{4N_c L} \sum_{m,\beta} \frac{(N_c \sin 2\theta_{m,\beta} - \sin 2\theta_{m,\alpha})}{(p_m - p_n)^2} \right\}^2 \tag{6.49}$$

with

$$p_n = \frac{2\pi}{L} n.$$

H' denotes the interaction Hamiltonian in terms of the new operators but is quite complicated, and therefore it is given at the end of this chapter.

6.2.5. Determination of Bogoliubov Angle

The conditions that the vacuum energy is minimized give the constraint equations which can determine the Bogoliubov angles

$$\tan 2\theta_{n,\alpha} = \frac{m_0 + \frac{g^2}{4N_c L} \sum_{m,\beta} \frac{(N_c \sin 2\theta_{m,\beta} - \sin 2\theta_{m,\alpha})}{(p_m - p_n)^2}}{p_n + \frac{g^2}{4N_c L} \sum_{m,\beta} \frac{(N_c \cos 2\theta_{m,\beta} - \cos 2\theta_{m,\alpha})}{(p_m - p_n)^2}}. \tag{6.50}$$

This is coupled equations for $\theta_{n,\alpha}$ and can be solved numerically. It should be noted that eq.(6.50) can be also obtained by requiring that the coefficient of the term

$$(c^{\dagger}_{n,\alpha} d^{\dagger}_{-n,\alpha} + d_{-n,\alpha} c_{n,\alpha})$$

should vanish. In this procedure, any mean field type of approximations should not be made for the interaction terms. In this respect, the treatment itself is exact even though the Bogoliubov method is an approximate scheme.

6.2.6. Fermion Condensate

The fermion condensate value C_{N_c} is written as

$$C_{N_c} = \langle\Omega|\frac{1}{L}\int dx\bar{\psi}\psi|\Omega\rangle_B = \frac{1}{L}\sum_{n,\alpha}\sin 2\theta_{n,\alpha}. \tag{6.51}$$

The values of $\theta_{n,\alpha}$ can be obtained by numerically solving eq.(6.50).

6.2.7. Boson Mass

Now, one can calculate the boson mass for the $SU(N_c)$ color. First, one defines the wave function for the color singlet boson as

$$|\Psi_K\rangle = \frac{1}{\sqrt{N_c}}\sum_{n,\alpha} f_n c^\dagger_{n,\alpha} d^\dagger_{K-n,\alpha}|\Omega\rangle_B, \tag{6.52}$$

where n runs from $-n_{max}$ to n_{max}, and n_{max} is taken to be around 1000. In this case, the boson mass can be described as

$$\begin{aligned}
\mathcal{M} &= \langle\Psi_K|H|\Psi_K\rangle = \frac{1}{N_c}\sum_{n,\alpha}(E_{n,\alpha}+E_{n-K,\alpha})\,|f_n|^2 \\
&+\frac{g^2}{2N_c^2 L}\sum_{l,m,\alpha}\frac{f_l f_m}{(p_l-p_m)^2}\cos(\theta_{l,\alpha}-\theta_{m,\alpha})\cos(\theta_{l-K,\alpha}-\theta_{m-K,\alpha}) \\
&-\frac{g^2}{2N_c L}\sum_{l,m,\alpha,\beta}\frac{f_l f_m}{(p_l-p_m)^2}\cos(\theta_{l,\alpha}-\theta_{m,\beta})\cos(\theta_{l-K,\alpha}-\theta_{m-K,\beta}) \\
&+\frac{g^2}{2N_c^2 L}\sum_{l,m,\alpha,\beta}\frac{f_l f_m}{K^2}\sin(\theta_{l-K,\alpha}-\theta_{l,\alpha})\sin(\theta_{m,\beta}-\theta_{m-K,\beta}) \\
&-\frac{g^2}{2N_c L}\sum_{l,m,\alpha}\frac{f_l f_m}{K^2}\sin(\theta_{l-K,\alpha}-\theta_{l,\alpha})\sin(\theta_{m,\alpha}-\theta_{m-K,\alpha}),
\end{aligned} \tag{6.53}$$

where $K \to 0$ should be taken. Here, it should be noted that the last two terms should be carefully estimated since the apparent divergence at $K = 0$ is well defined and finite. Since the Bogoliubov angles are determined by solving eq.(6.50), the Hamiltonian of eq.(6.53) can be diagonalized numerically, and one can obtain the energy eigenvalues of the Hamiltonian. From the diagonalization procedure, one can determine the wave function f_n at the same time as the energy eigenvalues.

Conditions for Numerical Calculations

For the maximum momentum in eq.(6.53), one can take

$$p_{max} \simeq 15\frac{g}{\sqrt{\pi}}$$

which should be sufficiently large compared to a typical boson mass scale of $\frac{g}{\sqrt{\pi}}$. The number of the states n_{max} in eq.(6.53) is varied from 500 to 1500, and one can confirm that the calculated energy eigenvalues are well converged already when the number of the states is around 1000. Since the box length L is given as

$$L = \frac{2\pi n_{max}}{p_{max}}$$

the finite size effect is well controlled. The calculated results show that the boson masses as well as the condensate values do not depend on the value of p_{max} once it is larger than $10\frac{g}{\sqrt{\pi}}$.

6.2.8. Condensate and Boson Mass in $SU(N_c)$

The calculations of the condensate and the boson mass in eq.(6.53) can be carried out numerically for arbitrary numbers of N_c value [39]. The calculated condensate and boson mass can be expressed by analytical formula for the large N_c values of $SU(N_c)$ where numerical calculations are carried out up to $N_c = 50$.

Massless Fermions

First, one can calculate the condensate and boson mass for the massless fermion case. The calculated condensate values for the large N_c values of $SU(N_c)$ agree with the prediction of the $1/N_c$ expansion [19, 20, 115] as the function of the N_c

$$C_{N_c} = -\frac{N_c}{\sqrt{12}}\sqrt{\frac{N_c g^2}{2\pi}}. \tag{6.54}$$

Further, the calculated boson masses can be described by the following phenomenological formula for the large N_c values [39, 43]

$$\mathcal{M}_{N_c} = \frac{2}{3}\sqrt{\frac{N_c g^2}{3\pi}} \tag{6.55}$$

which is a fit to the calculated result of the boson mass with massless fermions. Indeed, the calculated boson masses for N_c larger than $N_c = 10$ perfectly agree with the predicted value of eq.(6.55). This suggests that that the boson mass of $SU(N_c)$ QCD_2 may well have some analytical solution, though it is not known yet how to derive it analytically.

From the calculations, it is found that the second excited state energy for $SU(N_c)$ color is higher than the twice of the boson mass at the massless fermion. Therefore, there is only one bound state in QCD_2 with the $SU(N_c)$ since there is no bound state below the two boson continuum states. In order to confirm that the continuum states start right above the two boson state, one has to carry out the calculation with four fermion Fock space. However, the four fermion Fock space calculation is very complicated, even though it should be done in future.

Masssive Fermions

For the finite fermion mass, one can carry out the numerical calculations of the boson mass just in the same way as the massless case. From the calculated result of the boson mass, one sees that the boson mass can be well expressed in the following phenomenological formula

$$\mathcal{M}_{N_c} \approx \left(\frac{2}{3}\sqrt{\frac{2}{3}} + \frac{10}{3}\frac{m_0}{\sqrt{\frac{N_c g^2}{\pi}}} \right) \sqrt{\frac{N_c g^2}{2\pi}} \tag{6.56}$$

for the finite but small fermion mass m_0 regions. This phenomenological expression is obtained by fitting the calculated results as the function of m_0, and this must be valid for small m_0 regions. At least, if the fermion mass m_0 is much smaller than the Schwinger boson mass

$$m_0 \ll \frac{g}{\sqrt{\pi}}$$

then the expression can reproduce well the calculated results.

6.3. 't Hooft Model

Here, we discuss the boson mass of QCD_2 with $SU(N_c)$ color in the large N_c limit. The large N_c expansion is beyond the perturbation theory and gives a bound state mass of boson.

6.3.1. $1/N_c$ Expansion

't Hooft presented an interesting method in which some of the leading Feynman diagrams of perturbative calculations can be summed up to all orders in the $1/N_c$ expansion [70], [71]. From this sum of the non-perturbative contributions, one can obtain coupled equations which can determine the boson mass.

In the evaluation of summing up the Feynman diagrams, there is one important assumption which may well be justified. That is, the gluon propagator is replaced by the propagator of the $q-\bar{q}$ states which should have N_c^2 degrees of freedom. On the other hand, the gluon propagator has the degrees of freedom of N_c^2-1. However, it is obvious that, in the large N_c limit, there is no difference between them.

In this case, one can prove that those planar diagrams which can be written on the same plane with no topological numbers can be summed up non-perturbatively. Here, there is one important constraint, that is, the quark propagators should not appear in the intermediate states since they are smaller than the gluon propagators by $1/N_c$.

In this way, one can build up coupled equations for the boson states in the self-consistent way. In principle, the 't Hooft equations must be exact up to the order of $1/N_c$. Therefore, it is expected that the right boson mass can be obtained from the equations at the order of $1/N_c$. In fact, 't Hooft calculated the boson mass in the large N_c limit, and showed that the boson mass vanishes at the massless fermion limit

$$\mathcal{M}_{(N_c=\infty)} = 0 \text{ at } m_0 = 0.$$

Even though this result is inconsistent with the theorem that there should not exist any massless boson in two dimensions, the argument against 't Hooft result is ended without understanding the problem in depth.

On the other hand, the calculations of the boson mass with the $SU(N_c)$ colors in terms of Bogoliubov transformation method show that the boson mass can be well described by

$$\mathcal{M}_{N_c} = \frac{2}{3}\sqrt{\frac{N_c g^2}{3\pi}}$$

as the function of N_c for the large values of N_c. In the 't Hooft model, the boson mass should be proportional to $\sqrt{\frac{N_c g^2}{2\pi}}$, and therefore, the above expression of the boson mass is indeed consistent with the 't Hooft evaluation as far as the expansion parameter is concerned. Therefore, the boson mass calculation by the planar diagram evaluations of 't Hooft must be reasonable. In this respect, the boson mass prediction of 't Hooft should be reexamined from the point of view of the light cone procedure [86]. The 't Hooft equations in the light cone have lost one important information which is related to the θ_p variables [9]. The equations without the θ_p variables should correspond to the trivial vacuum, and therefore, if one can recover the θ_p variable constraints in the 't Hooft equations in the light cone, then one may obtain the right boson mass from the 't Hooft model.

6.3.2. Examination of 't Hooft Model

In the $1/N_c$ expansion procedure, 't Hooft assumed one important constraint, that is, $g^2 N_c$ must be fixed. This constraint itself has no problem since this is just the way in which the $1/N_c$ expansion is well defined.

Mass Dimension

However, one should be careful for this constraint since g has a mass dimension in two dimensional QCD. Suppose one writes

$$g^2 N_c = C_0. \tag{6.57}$$

In this case, C_0 should be a constant which has a mass square dimension. Where can one find the constant with dimensions in QCD_2 with massless fermions? This is rather a difficult question, and it has never been seriously discussed.

Here, one may postulate that the dimension of this constant C_0 should be given by the boson mass $\mathcal{M}$ as

$$C_0 = a_0 \mathcal{M}^2, \tag{6.58}$$

where a_0 is a dimensionless constant. This is somewhat puzzling since the boson mass can be given only after the dynamics of QCD_2 is solved. Nevertheless, this should be the only candidate if the $1/N_c$ expansion in 't Hooft model has any physical meaning. This indicates that the boson mass must be finite, and if one finds a massless boson only, then the calculation must have some intrinsic problems. This is also a good reason why there must be a boson with the finite mass even in the large N_c limit.

6.4. Spontaneous Symmetry Breaking in QCD_2

The Lagrangian density of QCD_2 has a chiral symmetry when the fermion mass m_0 is set to zero. In this case, there should be no condensate for the vacuum state if the symmetry is preserved in the vacuum state. However, as we saw above, the physical vacuum state in QCD_2 has a finite condensate value, and thus the chiral symmetry is broken. In contrast to the Schwinger model, there is no anomaly in QCD_2 since the field strength in QCD_2 has a color while the axial vector current is a color singlet object. Therefore, the axial vector current is conserved without anomaly, and thus, this symmetry breaking is *spontaneous*. However, there appears no massless boson. Even though no appearance of the Goldstone boson is very reasonable in two dimension, this means that the Goldstone theorem does not hold for the fermion field theory model of QCD_2.

Further, it seems that the chiral anomaly does not play an important role in the symmetry breaking business though it has been believed that the Schwinger model breaks the chiral symmetry due to the anomaly. However, the massless limit in QED_2 is not singular. The condensate value and the boson mass are smooth as the function of the fermion mass m_0. This means that the vacuum structure is smoothly connected from the massive case to the massless one.

This is just in contrast to the Thirring model where the massless limit is a singular point. The structure of the vacuum is completely different from the massive case to the massless one in the Thirring model. Further, the boson mass in the Thirring model is not smooth function of the fermion mass m_0. For the massive Thirring model, there is one boson state, and the boson mass is proportional to the fermion mass m_0. On the other hand, for the massless Thirring model, excitation energies are proportional to the cut-off Λ by which all of the physical observables are measured.

On the other hand, QED_2 and QCD_2 are very different from the Thirring model in that the coupling constant of the models have the mass scale dimensions, and all of the physical quantities are described by the coupling constant g even at the massless limit. The super-renormalizability for QED_2 and QCD_2 must be quite important in this respect, while the Thirring model has no dimensional quantity, and this makes the vacuum structure very complicated when the fermion mass is zero.

Table 6.1 summarizes the physical quantities of the chiral symmetry breaking for QED_2, QCD_2 and Thirring models which are evaluated by employing the Bogoliubov transformation technique even though it is not exact. All the condensates and the masses are measured in units of $\frac{g}{\sqrt{\pi}}$ for QED_2 and QCD_2. The Λ and g_0 in the Thirring model denote the cut-off parameter and the coupling constant, respectively. Also, the value of $\alpha(g_0)$ can be obtained by solving the equation for bosons in the Thirring model.

For QED_2, there is an anomaly, and therefore, the axial vector current is not conserved while, for QCD_2 and the Thirring model, the chiral current is conserved. From Table 6.1, one sees that the symmetry breaking mechanism is just the same for QED_2 and QCD_2. However, the Thirring model has a singularity at the massless fermion limit, and this gives rise to somewhat different behaviors from the gauge theory. In Chapter 7, we will discuss the vacuum structure and some physical properties of the Thirring model in terms of the Bethe ansatz solution which is exact. The exact solutions of the Thirring model show that there is no condensate and no boson but a finite gap in the excitation spectrum.

Table 6.1

	Condensate		Boson Mass	
	$m_0 = 0$	$m_0 \neq 0$	$m_0 = 0$	$m_0 \neq 0$
QED$_2$	-0.283	$-0.283 + O(m_0)$	1	$1 + O(m_0)$
QCD$_2$	$-\frac{N_c}{\sqrt{12}}\sqrt{\frac{N_c}{2}}$	$-\frac{N_c}{\sqrt{12}}\sqrt{\frac{N_c}{2}} + O(m_0)$	$\frac{2}{3}\sqrt{\frac{N_c}{3}}$	$\left(\frac{2}{3}\sqrt{\frac{2}{3}} + \frac{10}{3}\frac{m_0}{\sqrt{N_c}}\right)\sqrt{\frac{N_c}{2}}$
Thirring	$\frac{\Lambda}{g_0 \sinh\left(\frac{\pi}{g_0}\right)}$	0	$\frac{\alpha(g_0)\Lambda}{\sinh\left(\frac{\pi}{g_0}\right)}$	$\alpha(g_0) m_0$

6.5. Explicit Expression of H'

$$H' = H_C + H_A + H_R + H^{(4)} + H^{(22)}, \tag{6.59}$$

$$\begin{aligned} H_C = \frac{g^2}{4L} \sum_{n,m,l,\alpha,\beta} \frac{1}{p_n^2} \Bigg[& \frac{1}{N_c} \cos(\theta_{m,\alpha} - \theta_{m+n,\alpha}) \cos(\theta_{l,\beta} - \theta_{l-n,\beta}) \\ & \times (c^\dagger_{m,\alpha} c_{m+n,\alpha} d^\dagger_{-l+n,\beta} d_{-l,\beta} + c^\dagger_{l,\beta} c_{l-n,\beta} d^\dagger_{-m-n,\alpha} d_{-m,\alpha}) \\ & - \cos(\theta_{m,\alpha} - \theta_{m+n,\beta}) \cos(\theta_{l,\beta} - \theta_{l-n,\alpha}) \\ & \times (c^\dagger_{m,\alpha} c_{m+n,\beta} d^\dagger_{-l+n,\alpha} d_{-l,\beta} + c^\dagger_{l,\beta} c_{l-n,\alpha} d^\dagger_{-m-n,\beta} d_{-m,\alpha}) \Bigg], \end{aligned} \tag{6.60a}$$

$$\begin{aligned} H_A = \frac{g^2}{4L} \sum_{n,m,l,\alpha,\beta} \frac{1}{p_n^2} \Bigg[& \frac{1}{N_c} \sin(\theta_{m+n,\alpha} - \theta_{m,\alpha}) \sin(\theta_{l-n,\beta} - \theta_{l,\beta}) \\ & \times (c^\dagger_{m,\alpha} c_{l-n,\beta} d^\dagger_{-m-n,\alpha} d_{-l,\beta} + c^\dagger_{l,\beta} c_{m+n,\alpha} d^\dagger_{-l+n,\beta} d_{-m,\alpha}) \\ & - \sin(\theta_{m+n,\beta} - \theta_{m,\alpha}) \sin(\theta_{l-n,\alpha} - \theta_{l,\beta}) \\ & \times (c^\dagger_{m,\alpha} c_{l-n,\alpha} d^\dagger_{-m-n,\beta} d_{-l,\beta} + c^\dagger_{l,\beta} c_{m+n,\beta} d^\dagger_{-l+n,\alpha} d_{-m,\alpha}) \Bigg], \end{aligned} \tag{6.60b}$$

$$\begin{aligned} H_R = \frac{g^2}{4L} \sum_{n,m,l,\alpha,\beta} \frac{1}{p_n^2} \Bigg[& \frac{1}{N_c} \cos(\theta_{m,\alpha} - \theta_{m+n,\alpha}) \cos(\theta_{l,\beta} - \theta_{l-n,\beta}) \\ & \times (c^\dagger_{m,\alpha} c^\dagger_{l,\beta} c_{m+n,\alpha} c_{l-n,\beta} + d^\dagger_{-m-n,\alpha} d^\dagger_{-l+n,\beta} d_{-m,\alpha} d_{-l,\beta}) \\ & - \cos(\theta_{m,\alpha} - \theta_{m+n,\beta}) \cos(\theta_{l,\beta} - \theta_{l-n,\alpha}) \\ & \times (c^\dagger_{m,\alpha} c^\dagger_{l,\beta} c_{m+n,\beta} c_{l-n,\alpha} + d^\dagger_{-m-n,\beta} d^\dagger_{-l+n,\alpha} d_{-m,\alpha} d_{-l,\beta}) \Bigg], \end{aligned} \tag{6.60c}$$

$$H^{(4)} = \frac{g^2}{4L} \sum_{n,m,l,\alpha,\beta} \frac{1}{p_n^2} \left[\frac{1}{N_c} \sin(\theta_{m+n,\alpha} - \theta_{m,\alpha}) \sin(\theta_{l-n,\beta} - \theta_{l,\beta}) \right.$$
$$\times (c^\dagger_{m,\alpha} c^\dagger_{l,\beta} d^\dagger_{-m-n,\alpha} d^\dagger_{-l+n,\beta} + c_{m+n,\alpha} c_{l-n,\beta} d_{-m,\alpha} d_{-l,\beta})$$
$$- \sin(\theta_{m+n,\beta} - \theta_{m,\alpha}) \sin(\theta_{l-n,\alpha} - \theta_{l,\beta})$$
$$\left. (c^\dagger_{m,\alpha} c^\dagger_{l,\beta} d^\dagger_{-m-n,\beta} d^\dagger_{-l+n,\alpha} + c_{m+n,\beta} c_{l-n,\alpha} d_{-m,\alpha} d_{-l,\beta}) \right], \quad (6.60d)$$

$$H^{(22)} = \frac{g^2}{4L} \sum_{n,m,l,\alpha,\beta} \frac{1}{p_n^2} \left[\frac{1}{N_c} \left\{ \cos(\theta_{m,\alpha} - \theta_{m+n,\alpha}) \sin(\theta_{l-n,\beta} - \theta_{l,\beta}) \right. \right.$$
$$\times (c^\dagger_{m,\alpha} c^\dagger_{l,\beta} c_{m+n,\alpha} d^\dagger_{-l+n,\beta} - c^\dagger_{l,\beta} d^\dagger_{-m-n,\alpha} d^\dagger_{-l+n,\beta} d_{-m,\alpha}$$
$$- c^\dagger_{m,\alpha} c_{m+n,\alpha} c_{l-n,\beta} d_{-l,\beta} + c_{l-n,\beta} d^\dagger_{-m-n,\alpha} d_{-m,\alpha} d_{-l,\beta})$$
$$+ \sin(\theta_{m+n,\alpha} - \theta_{m,\alpha}) \cos(\theta_{l,\beta} - \theta_{l-n,\beta})$$
$$\times (-c^\dagger_{m,\alpha} c^\dagger_{l,\beta} c_{l-n,\beta} d^\dagger_{-m-n,\alpha} + c^\dagger_{m,\alpha} d^\dagger_{-m-n,\alpha} d^\dagger_{-l+n,\beta} d_{-l,\beta}$$
$$\left. + c^\dagger_{l,\beta} c_{m+n,\alpha} c_{l-n,\beta} d_{-m,\alpha} - c_{m+n,\alpha} d^\dagger_{-l+n,\beta} d_{-m,\alpha} d_{-l,\beta}) \right\}$$
$$- \left\{ \cos(\theta_{m,\alpha} - \theta_{m+n,\beta}) \sin(\theta_{l-n,\alpha} - \theta_{l,\beta}) \right.$$
$$\times (c^\dagger_{m,\alpha} c^\dagger_{l,\beta} c_{m+n,\beta} d^\dagger_{-l+n,\alpha} - c^\dagger_{l,\beta} d^\dagger_{-m-n,\beta} d^\dagger_{-l+n,\alpha} d_{-m,\alpha}$$
$$- c^\dagger_{m,\alpha} c_{m+n,\beta} c_{l-n,\alpha} d_{-l,\beta} + c_{l-n,\alpha} d^\dagger_{-m-n,\beta} d_{-m,\alpha} d_{-l,\beta})$$
$$+ \sin(\theta_{m+n,\beta} - \theta_{m,\alpha}) \cos(\theta_{l,\beta} - \theta_{l-n,\alpha})$$
$$\times (-c^\dagger_{m,\alpha} c^\dagger_{l,\beta} c_{l-n,\alpha} d^\dagger_{-m-n,\beta} + c^\dagger_{m,\alpha} d^\dagger_{-m-n,\beta} d^\dagger_{-l+n,\alpha} d_{-l,\beta}$$
$$\left. \left. + c^\dagger_{l,\beta} c_{m+n,\beta} c_{l-n,\alpha} d_{-m,\alpha} - c_{m+n,\beta} d^\dagger_{-l+n,\alpha} d_{-m,\alpha} d_{-l,\beta}) \right\} \right]. \quad (6.60e)$$

Chapter 7

Thirring Model

Since quantum field theory has infinite degrees of freedoms, it is extremely difficult to solve it exactly. Therefore, it is important to have an exactly solvable field theory model. The Thirring model is exactly solvable, and presents a nontrivial field theory model. The solution is given in terms of the Bethe ansatz technique and one can learn a lot about the essence of the vacuum structure in the nontrivial field theory model which can otherwise be impossible to understand.

In this chapter, we first explain the Bethe ansatz technique which enables to solve the massive Thirring model in an exact fashion. In this method, one postulates the shape of the exact eigenstates of the Hamiltonian and then finds conditions and confirms that the states which are prepared are indeed the eigenstates of the Hamiltonian. By imposing the periodic boundary conditions on the wave functions, one obtains physical observables from the Bethe ansatz equations.

Then, we present the Bethe ansatz solution of the massless Thirring model and see how the spectrum emerges from the Bethe ansatz equations. In addition, we learn the most important physics of the spontaneous symmetry breaking since the vacuum of the massless Thirring model breaks the chiral symmetry spontaneously. From the exact solution of the vacuum state, we can understand why there should not exist any Goldstone boson.

Further, we discuss the correspondence between the massive Thirring model and the sine-Gordon field theory since we can also learn a lot from the fermion and boson models which give the same structure of an arbitrary order of the correlation functions.

Finally, we treat the Bogoliubov transformation method in the Thirring model. Even though it is an approximate scheme, it should be important to understand in which way the approximation enters into the method. By comparing the result of the Bogoliubov method with the exact solution of the Thirring model, one can understand from where the approximation comes in and to which level the approximation is reliable.

7.1. Bethe Ansatz Method for Massive Thirring Model

In this section, we describe the Bethe ansatz technique for the massive Thirring model [13, 105]. First, we start from the quantized Hamiltonian of the massive Thirring model in the chiral representation

$$\hat{H} = \int dx \left[-i \left(\psi_a^\dagger \frac{\partial}{\partial x} \psi_a - \psi_b^\dagger \frac{\partial}{\partial x} \psi_b \right) + m_0 (\psi_a^\dagger \psi_b + \psi_b^\dagger \psi_a) + 2g \psi_a^\dagger \psi_a \psi_b^\dagger \psi_b \right], \quad (7.1)$$

where $\psi_a(x)$ and $\psi_b(x)$ denote the fermion operators in coordinate space and can be written in terms of creation and annihilation operators a_k and b_k in momentum space

$$\begin{pmatrix} \psi_a(x) \\ \psi_b(x) \end{pmatrix} = \frac{1}{\sqrt{L}} \sum_k \begin{pmatrix} a_k \\ b_k \end{pmatrix} e^{ikx}. \quad (7.2)$$

The creation and annihilation operators should satisfy the following anti-commutation relations

$$\{a_k, a_{k'}^\dagger\} = \{b_k, b_{k'}^\dagger\} = \delta_{k,k'}, \quad \{a_k, a_{k'}\} = \{b_k, b_{k'}\} = \{a_k, b_{k'}\} = 0.$$

Bogoliubov Transformation

Now, new operators A_k and B_k are introduced by

$$A_k = \cos\theta_k \, a_k + \sin\theta_k \, b_k, \quad (7.3a)$$

$$B_k = -\sin\theta_k \, a_k + \cos\theta_k \, b_k \quad (7.3b)$$

and this is just the same as the Bogoliubov transformation from (a_k, b_k) to (A_k, B_k). It is easy to check that the same anti-commutation relations should hold for A_k, B_k

$$\{A_k, A_{k'}^\dagger\} = \{B_k, B_{k'}^\dagger\} = \delta_{k,k'}, \quad \{A_k, A_{k'}\} = \{B_k, B_{k'}\} = \{A_k, B_{k'}\} = 0.$$

Here, the Bogoliubov transformation corresponds to the unitary transformation from the massless basis to the massive one, and the free Hamiltonian can be diagonalized by the massive fermion basis as will be seen below.

7.1.1. Free Fermion System

Before going to the Thirring model case, we first treat the free massive fermion system. In this case, the Hamiltonian $\hat{H}_0$ of massive free fermions can be written as

$$\hat{H}_0 = \int dx \left[-i \left(\psi_a^\dagger \frac{\partial}{\partial x} \psi_a - \psi_b^\dagger \frac{\partial}{\partial x} \psi_b \right) + m_0 (\psi_a^\dagger \psi_b + \psi_b^\dagger \psi_a) \right]$$

$$= \sum_k \left[k \left(a_k^\dagger a_k - b_k^\dagger b_k \right) + m_0 \left(a_k^\dagger b_k + b_k^\dagger a_k \right) \right]. \quad (7.4)$$

This Hamiltonian can be diagonalized in terms of eqs.(7.3) as

$$\hat{H}_0 = \sum_k E_k (A_k^\dagger A_k - B_k^\dagger B_k), \ \text{with} \ E_k = \sqrt{m_0^2 + k^2}, \quad (7.5)$$

where θ_k is chosen to be

$$\tan 2\theta_k = \frac{m_0}{k}. \quad (7.6)$$

For example, the two particle eigenstate $|k_1,k_2\rangle_0$ for free fermion states with the momenta k_1 and k_2 is given as

$$|k_1,k_2\rangle_0 = A_{k_1}^\dagger A_{k_2}^\dagger |0\rangle, \tag{7.7}$$

where the vacuum state $|0\rangle$ is defined as

$$a_k|0\rangle = 0, \quad b_k|0\rangle = 0. \tag{7.8}$$

It should be noted that the new fermion operators A_k and B_k also satisfy

$$A_k|0\rangle = 0, \quad B_k|0\rangle = 0$$

because of eqs.(7.3). The Bogoliubov transformation [eqs.(7.3)] of a_k and b_k to A_k and B_k corresponds to the transformation from the massless basis to the massive basis. The generator of the Bogoliubov transformation is given as

$$\mathcal{A} = \sum_k \theta_k (a_k^\dagger b_{-k} - b_{-k}^\dagger a_k)$$

and the new vacuum $|\Omega\rangle$

$$|\Omega\rangle = e^{-\mathcal{A}}|0\rangle = |0\rangle$$

is just the same as the original vacuum $|0\rangle$.

7.1.2. Bethe Ansatz State in Two Particle System

Here, one should find a state which is the eigenstate of the Hamiltonian H of the massive Thirring model. In order to diagonalize the Hamiltonian of eq.(7.1), one first considers two particle system since it is straightforward to extend the result to N-particle system.

Fock State

In the Bethe ansatz method, one postulates that the two particle states (positive energy state) with their momenta k_1 and k_2 can be written as

$$|k_1,k_2\rangle = \int dx_1\, dx_2 e^{ik_1x_1+ik_2x_2}\left[1+i\lambda\varepsilon(x_1-x_2)\right]\psi_A^\dagger(x_1,k_1)\psi_A^\dagger(x_2,k_2)|0\rangle, \tag{7.9}$$

where $\psi_A(x,k)$ is defined as

$$\psi_A(x,k) = \cos\theta_k\ \psi_a(x) + \sin\theta_k\ \psi_b(x) \tag{7.10}$$

which should correspond to A_k in eq.(7.3a) in coordinate space representation. $\varepsilon(x)$ denotes a step function as defined

$$\varepsilon(x) = \begin{pmatrix} -1 & \text{for } x<0 \\ 1 & \text{for } x>0 \end{pmatrix}. \tag{7.11}$$

λ is a parameter which should be determined such that the two particle state $|k_1,k_2\rangle$ can be the eigenstate of the Hamiltonian. Here, the basic point is that the derivative of the step function produces a delta function

$$\frac{d\varepsilon(x)}{dx} = 2\delta(x) \tag{7.12}$$

which comes from the kinetic energy term, and this term should be cancelled by the interaction term that also contains the $\delta(x)$ function type term due to the four fermion interaction. It should be noted that, when $\lambda = 0$, then the state $|k_1, k_2\rangle$ of eq.(7.9) is just reduced to

$$|k_1, k_2\rangle = A^\dagger_{k_1} A^\dagger_{k_2} |0\rangle = |k_1, k_2\rangle_0 \ \ (\lambda = 0) \tag{7.13}$$

which is the eigenstate of the free Hamiltonian $\hat{H}_0$.

Eigenvalue Equation of Hamiltonian

Now, the Hamiltonian $\hat{H}$ can be operated on $|k_1, k_2\rangle$ and one finds

$$\hat{H}|k_1, k_2\rangle = (E_{k_1} + E_{k_2})|k_1, k_2\rangle \tag{7.14}$$

if the following condition is satisfied

$$\int dx e^{i(k_1+k_2)x} \left[4\lambda \sin(\theta_{k_1} + \theta_{k_2}) - 2g \sin(\theta_{k_1} - \theta_{k_2})\right] \psi^\dagger_a(x) \psi^\dagger_b(x)|0\rangle = 0. \tag{7.15}$$

Therefore, one obtains the following condition for λ

$$\lambda = \frac{1}{2} g \frac{\sin(\theta_{k_1} - \theta_{k_2})}{\sin(\theta_{k_1} + \theta_{k_2})}. \tag{7.16}$$

One can easily extend the result of the two particle system to N particle system.

7.1.3. Bethe Ansatz State in N Particle System

The eigenstate of the N particle system can be written as

$$|k_1, \ldots, k_N\rangle = \int dx_1 \cdots dx_N \Psi(x_1, \ldots, x_N) \prod_{i=1}^{N} \psi^\dagger_A(x_i, k_i)|0\rangle, \tag{7.17}$$

where the Bethe ansatz wave function $\Psi(x_1, \ldots, x_N)$ for N particles is assumed to have the following shape

$$\Psi(x_1, \ldots, x_N) = \exp\left(i m_0 \sum_{i=1}^{N} x_i \sinh\beta_i \right) \prod_{1 \le i < j \le N} \left[1 + i\lambda\varepsilon(x_i - x_j)\right]. \tag{7.18}$$

Rapidity β_i

In eq.(7.18), rapidity β_i is introduced, for convenience, which is related to the momentum k_i and the energy E_i of i-th particle as

$$k_i = m_0 \sinh\beta_i, \tag{7.19a}$$

$$E_i = m_0 \cosh\beta_i. \tag{7.19b}$$

In the same way as the two particle system, one requires that the state $|k_1, \ldots, k_N\rangle$ should be the eigenstate of the Hamiltonian $\hat{H}$, and this determines the λ which can be written as

$$\lambda = -\frac{1}{2} g \tanh\frac{1}{2}(\beta_i - \beta_j). \tag{7.20}$$

Phase Shift Function

By defining the phase shift function $\phi(\beta_i, \beta_j)$ as

$$\exp\left[i\phi(\beta_i, \beta_j)\right] \equiv \frac{1+i\lambda}{1-i\lambda} \tag{7.21}$$

and by making use of the following identity

$$e^{i\phi} = \frac{1+i\tan\frac{\phi}{2}}{1-i\tan\frac{\phi}{2}} \tag{7.22}$$

one can write the phase shift function $\phi(\beta_i, \beta_j)$ explicitly as

$$\phi(\beta_i, \beta_j) = -2\tan^{-1}\left[\frac{1}{2}g\tanh\frac{1}{2}(\beta_i - \beta_j)\right]. \tag{7.23}$$

In this case, denoting the state $|k_1, \ldots, k_N\rangle$ by $|\beta_1, \ldots, \beta_N\rangle$, one obtains the eigenvalue equation for the Hamiltonian

$$\hat{H}|\beta_1, \ldots, \beta_N\rangle = \left(\sum_{i=1}^{N} m_0 \cosh\beta_i\right)|\beta_1, \ldots, \beta_N\rangle. \tag{7.24}$$

Periodic Boundary Conditions

The Bethe ansatz wave functions satisfy the eigenvalue equation [eq.(7.24)]. However, they still do not have proper boundary conditions. The best and simplest way to define field theoretical models is to put the theory into a box of length L and impose periodic boundary conditions (PBC) on the states. Therefore, it is demanded that $\Psi(x_1, .., x_N)$ should be periodic in each argument x_i. This gives the boundary condition

$$\Psi(x_i = 0) = \Psi(x_i = L) \tag{7.25}$$

which leads to the following PBC equations,

$$\exp\left(im_0 L \sinh\beta_i\right) = \exp\left(-i\sum_{j=1}^{N}\phi(\beta_i, \beta_j)\right) \quad (i = 1, .., N). \tag{7.26}$$

Taking the logarithm of eq.(7.26), one obtains

$$m_0 L \sinh\beta_i = 2\pi n_i - \sum_{j=1}^{N}\phi(\beta_i, \beta_j) \quad (i = 1, \ldots, N), \tag{7.27}$$

where n_i runs as

$$n_i = 0, \pm 1, \pm 2, \ldots.$$

This is the basic Bethe ansatz equations which can determine all the physical quantities in the Thirring model.

7.2. Bethe Ansatz Method for Field Theory

Now, we discuss the field theoretical treatment of the Bethe ansatz solutions in the massive Thirring model. The difficulty of quantum field theory is mostly concerned with the vacuum state which is composed of infinite many negative energy particles. Here, we consider the N-particle system in the vacuum state and later we let the value of N as large as required.

7.2.1. Vacuum State of Massive Thirring Model

Now, one can solve the PBC equations of eqs.(7.27) and construct the vacuum state. The box length L and particle number N are the parameters one should first choose. Also, one knows the maximum momentum or cut-off in this field theory, that is , Λ which is defined as

$$\Lambda = \frac{2\pi}{L} N_0, \quad N_0 = \frac{1}{2}(N-1). \tag{7.28}$$

First, one should make a vacuum which is filled with negative energy particles whose rapidity is written as

$$\beta_i = i\pi - \alpha_i.$$

This is clear since the energy E_i is given as

$$E_i = m_0 \cosh\beta_i = m_0 \cosh(i\pi - \alpha_i) = -m_0 \cosh\alpha_i$$

which shows the negative energy of the particle state. In this case, one can write the PBC equations for the vacuum as

$$\sinh\alpha_i = \frac{2\pi n_i}{L_0} - \frac{2}{L_0}\sum_{j\neq i} \tan^{-1}\left[\frac{1}{2} g \tanh\frac{1}{2}(\alpha_i - \alpha_j)\right] \quad (i = 1, \ldots, N), \tag{7.29}$$

where L_0 is defined as

$$L_0 = m_0 L$$

which is made dimensionless. By fixing the values of L_0 and N, one can solve eq.(7.29) and determine the values of the rapidities β_i. Therefore, the vacuum energy E_v can be written as

$$E_v = -\sum_{i=-N_0}^{N_0} m_0 \cosh\alpha_i. \tag{7.30}$$

The vacuum energy is determined without any ambiguity. The physical vacuum can be obtained by making $N_0 \to \infty$, which means that the value of N_0 should be made sufficiently large such that any physical observables should become stable.

7.2.2. Excited States

The lowest excited state must be made by one particle-one hole (1p-1h) state. That is, one takes out one negative energy particle (n_{i_0}-th particle) and puts it into a positive energy state. In this case, the PBC equations become

$i \neq i_0$

$$\sinh\alpha_i = -\frac{2\pi n_i}{L_0} - \frac{2}{L_0}\tan^{-1}\left[\frac{1}{2}g\coth\frac{1}{2}(\alpha_i+\beta_{i_0})\right]$$
$$-\frac{2}{L_0}\sum_{j\neq i,i_0}\tan^{-1}\left[\frac{1}{2}g\tanh\frac{1}{2}(\alpha_i-\alpha_j)\right]; \qquad (7.31a)$$

$i = i_0$

$$\sinh\beta_{i_0} = -\frac{2\pi n_{i_0}}{L_0} + \frac{2}{L_0}\sum_{j\neq i_0}\tan^{-1}\left[\frac{1}{2}g\coth\frac{1}{2}(\beta_{i_0}+\alpha_j)\right]. \qquad (7.31b)$$

These PBC equations determine the energy of the one particle-one hole states which is denoted by $E^{i_0}_{1p1h}$,

$$E^{i_0}_{1p1h} = m_0\cosh\beta_{i_0} - \sum_{\substack{i=-N_0\\ i\neq i_0}}^{N_0} m_0\cosh\alpha_i. \qquad (7.32)$$

It is important to notice that the momentum allowed for the positive energy state must be determined by the PBC equations. Also, the momenta occupied by the negative energy particles are different from the vacuum case even though the rapidity difference in the negative energy particles is indeed very small.

7.2.3. Lowest Excited State (Boson)

The lowest configuration one can consider is the case in which one takes out $n_i = 0$ particle and puts it into the positive energy state. This must be the first excited state since it has a discrete symmetry of $\alpha_i = \alpha_{-i}$. Indeed, by numerical calculations, one finds that there is one isolated state between the vacuum state and other excited states. This corresponds to a boson in the massive Thirring model.

7.2.4. Higher Excited States

One can calculate the two particle-two hole (2p-2h) states by putting two negative energy particles into positive energy states. This calculation is complicated and is not shown here. According to the numerical calculations, one sees that there is no bosonic state in the spectrum of the massive Thirring model [55]. In terms of analytical evaluations, one can prove that the interactions between bosons are always repulsive, and therefore there is no bound state between bosons.

7.2.5. Continuum States

Next, the following configurations are considered in which one takes out particles with their momentum

$$k_i = \frac{2\pi n_i}{L_0}, \quad \text{with } n_i = \pm 1, \pm 2, \ldots.$$

and puts them into the positive energy states. They become continuum states and therefore, the lowest energy state of the one particle-one hole continuum state should correspond to the twice of the physical fermion mass, that is, $2m$. From this point, one can measure the mass of the boson and indeed the first excited state with the discrete symmetry of $\alpha_i = \alpha_{-i}$ becomes a boson. Numerically, it is confirmed that there is only one boson state from the Bethe ansatz solutions. This should be related to the fact that there is only one state which has the discrete symmetry of $\alpha_i = \alpha_{-i}$ in the one particle-one hole states, and all other states do not have this symmetry.

It is interesting to notice that the state with some discrete symmetry should become lower than the other states without this symmetry. On the other hand, the spontaneous symmetry breaking of the vacuum state is just opposite to this discrete symmetry preservation. In the spontaneous symmetry breaking, the vacuum state which breaks the continuous symmetry (chiral symmetry) becomes lower than the state which preserves the chiral symmetry. As we discussed in Chapter 4, the reason why the vacuum in quantum field theory prefers the symmetry broken state is simply because there is a negative sign in front of the total energy of the particles.

On the other hand, the state with the discrete symmetry of $\alpha_i = \alpha_{-i}$ is found in the positive energy state and therefore it is natural that the symmetry preserving state becomes the lowest state in this system.

7.3. Bethe Ansatz Method for Massless Thirring Model

Now, we discuss the massless Thirring model whose Lagrangian density is given as

$$\mathcal{L} = i\bar{\psi}\gamma_\mu\partial^\mu\psi - \frac{1}{2}gj^\mu j_\mu \tag{7.33}$$

and its Hamiltonian can be written in the chiral representation as

$$\hat{H} = \int dx \left\{ -i\left(\psi_a^\dagger \frac{\partial}{\partial x}\psi_a - \psi_b^\dagger \frac{\partial}{\partial x}\psi_b\right) + 2g\psi_a^\dagger\psi_b^\dagger\psi_b\psi_a \right\}. \tag{7.34}$$

In the same way as the massive Thirring model, the Hamiltonian can be diagonalized by the Bethe ansatz wave function for N particles [5, 40, 52, 95]. In this case, the condition for λ is written as

$$\lambda = -\frac{g}{2}S_{ij}, \tag{7.35}$$

where S_{ij} is written as

$$S_{ij} = \frac{k_iE_j - k_jE_i}{k_ik_j - E_iE_j - \varepsilon^2} \tag{7.36}$$

which is obtained from eq.(7.16). Here, ε denotes a infra-red regulator which should be infinitesimally small. Concerning the infra-red regulator ε, it is important to note that the physical observables like momentum k_i and energy E_i should not depend on the infra-red regulator ε. If any of physical observables depended on the regulator ε, then it would mean that the system is not solved properly. The solutions presented below do not depend on the regulator ε at all.

Hamiltonian and PBC Equations

In this case, the eigenvalue equation becomes

$$\hat{H}|k_1,\ldots,k_N\rangle = \sum_{i=1}^{N} E_i|k_1,\ldots,k_N\rangle. \tag{7.37}$$

From the periodic boundary conditions (PBC), one obtains the following PBC equations,

$$k_i = \frac{2\pi n_i}{L} + \frac{2}{L}\sum_{j\neq i}^{N} \tan^{-1}\left(\frac{g}{2}S_{ij}\right), \tag{7.38}$$

where n_i's runs as

$$n_i = 0, \pm 1, \pm 2, \ldots, \pm N_0 \text{ with } N_0 = \frac{1}{2}(N-1).$$

7.3.1. Vacuum State of Massless Thirring Model

First, one should construct a vacuum and write the PBC equations for the vacuum which is filled with N negative energy particles

$$k_i = \frac{2\pi n_i}{L} - \frac{2}{L}\sum_{\substack{i\neq j \\ k_i\neq k_j}}^{N} \tan^{-1}\left(\frac{g}{2}\frac{k_i|k_j| - k_j|k_i|}{k_ik_j - |k_i||k_j| - \varepsilon^2}\right). \tag{7.39}$$

Now, one should fix the maximum momentum of the negative energy particles which is denoted by the cut off momentum Λ. In this case, the box length L is expressed as

$$L = \frac{2\pi N_0}{\Lambda}. \tag{7.40}$$

If one solves eq.(7.39), then one can determine all the values of k_i in the vacuum state, and the vacuum energy E_v can be written as

$$E_v = -\sum_{i=1}^{N} |k_i|. \tag{7.41}$$

It should be noted that physical observables are obtained by taking the thermodynamic limit where one makes $L \to \infty$ and $N \to \infty$, keeping Λ finite. If there is other scale like the mass, then one should take the Λ which is sufficiently larger than the other scaleful parameters. However, there is no other scale in the massless Thirring model and therefore all the physical observables are measured by the Λ.

7.3.2. Symmetric Vacuum State

The solution of eq.(7.39) is easily found as [5, 95]

$$k_1 = 0 \text{ for } n_1 = 0, \tag{7.42a}$$

$$k_i = \frac{2\pi n_i}{L} + \frac{2N_0}{L}\tan^{-1}\left(\frac{g}{2}\right) \quad \text{for } n_i = 1, 2, \ldots, N_0, \tag{7.42b}$$

$$k_i = \frac{2\pi n_i}{L} - \frac{2N_0}{L}\tan^{-1}\left(\frac{g}{2}\right) \quad \text{for } n_i = -1, -2, \ldots, -N_0. \tag{7.42c}$$

This gives a symmetric vacuum state, and the vacuum energy E_v^{sym} can be written as

$$E_v^{\text{sym}} = -\Lambda\left\{N_0 + 1 + \frac{2N_0}{\pi}\tan^{-1}\left(\frac{g}{2}\right)\right\}. \tag{7.43}$$

7.3.3. True Vacuum (Symmetry Broken) State

Eq.(7.39) has completely different solutions from the above symmetric solutions, and one finds that the vacuum energy of the new solution is lower than the symmetric vacuum energy.

Analytical Solutions of True Vacuum

The new solutions can be analytically written

$$k_1 = \frac{2N_0}{L}\tan^{-1}\left(\frac{g}{2}\right) \quad \text{for } n_1 = 0, \tag{7.44a}$$

$$k_i = \frac{2\pi n_i}{L} + \frac{2N_0}{L}\tan^{-1}\left(\frac{g}{2}\right) \quad \text{for } n_i = 1, 2, \ldots, N_0, \tag{7.44b}$$

$$k_i = \frac{2\pi n_i}{L} - \frac{2(N_0+1)}{L}\tan^{-1}\left(\frac{g}{2}\right) \quad \text{for } n_i = -1, -2, \ldots, -N_0. \tag{7.44c}$$

The new vacuum has no $k_i = 0$ solution, and breaks the left-right symmetry. Instead, all of the momenta of the negative energy particles become non-zero. The energy E_v^{true} of the true vacuum state can be written as

$$E_v^{\text{true}} = -\Lambda\left\{N_0 + 1 + \frac{2(N_0+1)}{\pi}\tan^{-1}\left(\frac{g}{2}\right)\right\}. \tag{7.45}$$

One can easily see that the energy E_v^{true} is lower than the energy E_v^{sym}, and therefore, this is the true vacuum state.

Effective Fermion Mass

From the distributions of the negative energy particles, one sees that this solution breaks the chiral symmetry. This situation can be easily seen from the analytical solutions since the absolute value of the momentum of the negative energy particles is larger than $\frac{\Lambda}{\pi}\tan^{-1}\left(\frac{g}{2}\right)$. Therefore, one can define the effective fermion mass M_N by

$$M_N = \frac{\Lambda}{\pi}\tan^{-1}\left(\frac{g}{2}\right). \tag{7.46}$$

This suggests that the real vacuum state of the Thirring model has somewhat a similarity with a massive fermion vacuum state.

Chiral Charge

Further, from the solutions of the symmetry broken vacuum (true vacuum), one sees that the chiral charge q_5 has

$$q_5 = \pm 1.$$

This means that the vacuum of the Thirring model has two-fold degeneracy.

7.3.4. $1p-1h$ State

Next, one particle-one hole $(1p-1h)$ states can be evaluated. There, one takes out one negative energy particle (i_0-th particle) and puts it into a positive energy state. In this case, the PBC equations become

$$k_i = \frac{2\pi n_i}{L} - \frac{2}{L}\tan^{-1}\left(\frac{g}{2}\frac{k_i|k_{i_0}|+k_{i_0}|k_i|}{k_ik_{i_0}+|k_i||k_{i_0}|+\varepsilon^2}\right)$$

$$-\frac{2}{L}\sum_{\substack{j\neq i,i_0\\ k_j\neq k_i,k_{i_0}}}^{N}\tan^{-1}\left(\frac{g}{2}\frac{k_i|k_j|-k_j|k_i|}{k_ik_j-|k_i||k_j|-\varepsilon^2}\right) \quad \text{for } i\neq i_0, \tag{7.47a}$$

$$k_{i_0} = \frac{2\pi n_{i_0}}{L} - \frac{2}{L}\sum_{\substack{j\neq i_0\\ k_j\neq -k_{i_0}}}^{N}\tan^{-1}\left(\frac{g}{2}\frac{k_{i_0}|k_j|+k_j|k_{i_0}|}{k_{i_0}k_j+|k_{i_0}||k_j|+\varepsilon^2}\right) \quad \text{for } i=i_0. \tag{7.47b}$$

Therefore, the energy of the one particle-one hole states $E^{1p1h}_{(i_0)}$ is given as

$$E^{1p1h}_{(i_0)} = |k_{i_0}| - \sum_{\substack{i=1\\ i\neq i_0}}^{N}|k_i|. \tag{7.48}$$

It turns out that the solutions of eqs.(7.47) can be found at the specific value of n_{i_0} and then from this n_{i_0} value on, one finds the continuum spectrum of the $1p-1h$ states.

Analytical Solutions of $1p-1h$ State

Here, the analytical solutions of eqs.(7.47) are shown for the lowest $1p-1h$ state

$$k_{i_0} = \frac{2\pi n_{i_0}}{L} - \frac{2N_0}{L}\tan^{-1}\left(\frac{g}{2}\right) \quad \text{for } n_{i_0}, \tag{7.49a}$$

$$k_i = \frac{2\pi n_i}{L} + \frac{2(N_0+1)}{L}\tan^{-1}\left(\frac{g}{2}\right) \quad \text{for } n_i = 0,1,2,\ldots,N_0, \tag{7.49b}$$

$$k_i = \frac{2\pi n_i}{L} - \frac{2N_0}{L}\tan^{-1}\left(\frac{g}{2}\right) \quad \text{for } n_i = -1,-2,\ldots,-N_0, \tag{7.49c}$$

where n_{i_0} is given by the following equation

$$n_{i_0} = \left[\frac{N_0}{\pi}\tan^{-1}\left(\frac{g}{2}\right)\right], \tag{7.50}$$

where $\lceil X \rceil$ denotes the smallest integer value which is larger than X. In this case, one can express the lowest $1p-1h$ state energy analytically

$$E_0^{1p-1h} = -\Lambda\left\{(N_0+1) - \frac{2n_{i_0}}{N_0} + \frac{2(N_0+1)}{\pi}\tan^{-1}\left(\frac{g}{\pi}\right)\right\}. \tag{7.51}$$

Therefore, the lowest excitation energy ΔE_0^{1p-1h} with respect to the true vacuum state becomes

$$\Delta E_0^{1p-1h} \equiv E_0^{1p-1h} - E_v^{\text{true}} = \frac{2\Lambda}{N_0} n_{i_0}. \tag{7.52}$$

If one takes the thermodynamic limit, that is, $N \to \infty$ and $L \to \infty$, then eq.(7.52) can be reduced to

$$\Delta E_0^{1p-1h} = \frac{2\Lambda}{\pi}\tan^{-1}\left(\frac{g}{2}\right) = 2M_N \tag{7.53}$$

which is just the twice of the effective fermion mass.

7.3.5. Momentum Distribution of Negative Energy States

Fig. 7.1 shows the momentum distribution of symmetric and true (symmetry broken) vacuum states. Note that the true vacuum state is degenerate due to the $k \leftrightarrow -k$ symmetry, which is always the case with Bethe ansatz solutions. There is no zero mode ($k = 0$) in the true vacuum while there exists the zero mode in the symmetric vacuum. The energy of the true vacuum state $\mathcal{E}_v^{true}$ can be written as

$$\mathcal{E}_v^{true} = -\left[N_0 + 1 + \frac{2(N_0+1)}{\pi}\tan^{-1}\left(\frac{g}{2}\right)\right]. \tag{7.54}$$

From the figure, one can see that the true vacuum state breaks the chiral symmetry or left-right symmetry of the momentum distribution.
Fig. 7.2 shows the dispersion relation of the symmetry broken vacuum predicted by the Bethe ansatz, and it can be fit by the following function

$$E_k = -\sqrt{k^2 + \xi^2}, \tag{7.55}$$

where ξ^2 is a constant. Therefore, the Bogoliubov transformation method may present a good approximate scheme for the analysis of the Thirring model. In particular, the vacuum property may be described well by the Bogoliubov transformation method. However, the Bogoliubov transformation tends to overestimate the attraction between fermions and anti-fermions since it predicts a bosonic bound state while the exact solution shows no bosonic bound state.

7.4. Bosonization of Thirring Model

Here, we briefly review the bosonization procedure in the massless Thirring models and show that the massless Thirring model cannot be bosonized properly due to the lack of the zero mode of the boson field.

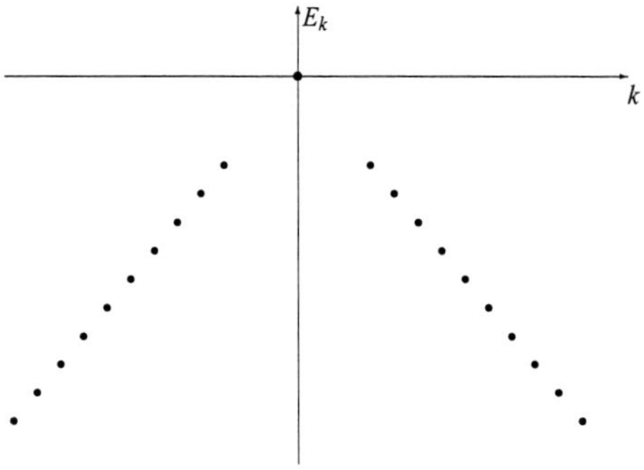

(a)

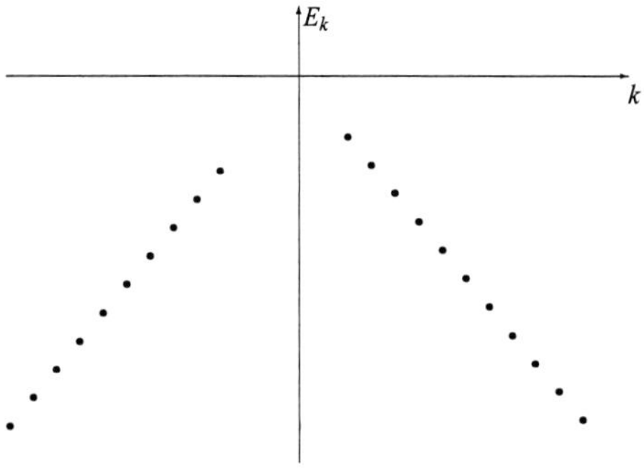

(b)

Figure 7.1. The momentum distributions of the symmetry preserved vacuum (a) and the true vacuum (b) are shown.

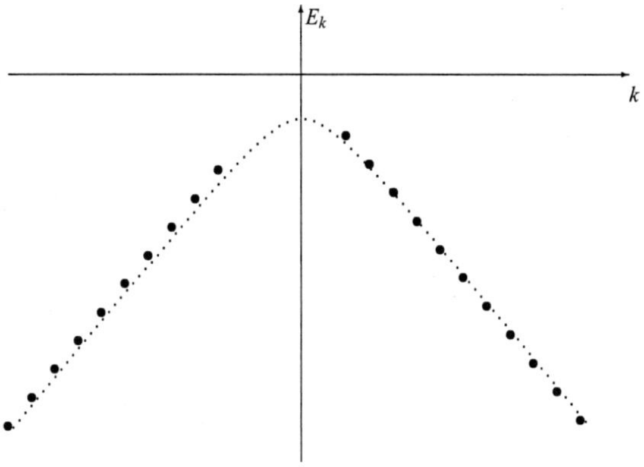

Figure 7.2. The bullets (•) show the Bethe ansatz solutions for the true vacuum of the Thirring model. The dot line shows the dispersion relation ($E_k = -\sqrt{k^2 + \xi^2}$, where ξ^2 is a constant.) which is based on the Bogoliubov transformation calculation.

7.4.1. Massless Thirring Model

The Hamiltonian of the massless Thirring model can be written in the chiral representation in terms of the creation and annihilation operators

$$\hat{H} = \sum_p \left(\frac{2\pi}{L}p\right)\left(a_p^\dagger a_p - b_p^\dagger b_p\right) + \frac{g}{2L}\sum_p \Big(\tilde{j}_0(p)\tilde{j}_0(-p) - \tilde{j}_1(p)\tilde{j}_1(-p)\Big), \tag{7.56}$$

where $\tilde{j}_0(p)$ and $\tilde{j}_1(p)$ are given in eqs.(5.56) and (5.57) in Chapter 5

$$\tilde{j}_0(p) = \sum_k \left[a_{k+p}^\dagger a_k + b_{k+p}^\dagger b_k\right],$$

$$\tilde{j}_1(p) = \sum_k \left[a_{k+p}^\dagger a_k - b_{k+p}^\dagger b_k\right].$$

Kinetic Energies of Fermion and Boson

In the same way as the Schwinger model bosonization, the kinetic energy part of the fermion Hamiltonian is written in terms of the kinetic energy of the boson Hamiltonian in eq.(5.73) in Chapter 5

$$\sum_p \left(\frac{2\pi p}{L}\right)\left(a_p^\dagger a_p - b_p^\dagger b_p\right) = \frac{1}{2}\sum_{p\neq 0}\left\{\Pi^\dagger(p)\Pi(p) + \left(\frac{2\pi p}{L}\right)^2 \Phi^\dagger(p)\Phi(p)\right\}.$$

Boson Fields and Currents

In addition, the momentum representation of the currents $\tilde{j}_0(p)$ and $\tilde{j}_1(p)$ are described in terms of the boson fields $\Phi(p)$ and $\Pi(p)$ in eqs.(5.61) in Chapter 5

$$\tilde{j}_0(p) = ip\sqrt{\frac{4\pi}{L}}\Phi(p) \quad \text{for } p \neq 0,$$

$$\tilde{j}_1(p) = \sqrt{\frac{L}{\pi}}\Pi(p) \quad \text{for } p \neq 0.$$

It should be important to note that the zero mode $p = 0$ is not included for the boson fields.

Hamiltonian in Terms of Boson Fields

In this way, one can rewrite the Thirring model Hamiltonian as

$$\hat{H} = \frac{1}{2}\sum_{p\neq 0}\left\{\left(1-\frac{g}{\pi}\right)\Pi^\dagger(p)\Pi(p) + \left(1+\frac{g}{\pi}\right)p^2\Phi^\dagger(p)\Phi(p)\right\} + \hat{H}(0), \tag{7.57}$$

where $\hat{H}(0)$ is the zero mode part and is given as

$$\hat{H}(0) = K_{vac} + \frac{g}{2L}\left(\tilde{j}_0(0)\tilde{j}_0(0) - \tilde{j}_1(0)\tilde{j}_1(0)\right). \tag{7.58}$$

Here, K_{vac} denotes the vacuum energy of the fermion kinetic energy part, and it is related to the gauge field A_1 in the Schwinger model. In the massless Thirring model, there is no corresponding field that can describe the K_{vac} and indeed, in the massless Thirring model, one finds

$$K_{vac} = 0.$$

It has been long believed that the massless Thirring model can be bosonized and its Hamiltonian is given by eq.(7.57) without the $H(0)$ term. However, as one sees, the zero mode ($p = 0$ part) is missing in the bosonized Hamiltonian. In fact, there is a serious problem in the definition of the boson field $\Phi(0)$ and $\Pi(0)$ at the zero momentum $p = 0$. In the Schwinger model, one finds the $\Phi(0)$ due to the anomaly equation. However, the Thirring model has no anomaly, and therefore the $\Phi(0)$ identically vanishes. That is,

$$\Phi(0) = 0.$$

There is no way to find the corresponding zero mode of the boson field in the massless Thirring model since the axial vector current is always conserved. Therefore, the Hamiltonian of the massless Thirring model [eq.(7.56)] does not correspond to a massless boson. It is interesting to notice that the problem is closely related to the zero mode which exhibits the infra-red property of the Hamiltonian. This is just consistent with the non-existence of the massless boson due to the infra-red singularity of the propagator in two dimensions. Further, as discussed in the previous section, the Bethe ansatz solutions confirm the finite gap of the massless Thirring spectrum, and this rules out a possibility of any excuse of the massless boson in the massless Thirring model.

7.4.2. Massive Thirring Model

On the other hand, it is well known that the massive Thirring model is equivalent to the sine-Gordon field theory. The proof of the equivalence is based on the observation that the arbitrary number of the correlation functions between the two models agree with each other if some constants and the fields of the two models are properly identified between them. This bosonization technique is explained in the next section.

For the bosonization in terms of the fermion currents, one can see how one finds the zero mode of the boson field. This is connected to the axial vector current conservation which is in fact violated by the mass term in the massive Thirring model

$$\partial_\mu j_5^\mu = 2im\bar{\psi}\gamma_5\psi. \tag{7.59a}$$

It should be noted that j_0^5 is equal to j_1 in two dimensions. Therefore, one can always define the $\dot{Q}_5$ by

$$\dot{Q}_5 = 2im \int \bar{\psi}\gamma_5\psi\, dx \tag{7.59b}$$

and one obtains the field $\Phi(0)$ of the boson

$$\Phi(0) = \frac{2im\pi}{g\sqrt{L}} \int \bar{\psi}\gamma_5\psi\, dx. \tag{7.60}$$

7.4.3. Physics of Zero Mode

What is the physics behind the Hamiltonian without the zero mode? Here, we discuss the effect of the zero mode and the eigenvalues of the Hamiltonian in a simplified way. The Hamiltonian [eq.(7.57)] can be rewritten as

$$\hat{H} = \hat{H}_B - \frac{1}{2}\left(1 - \frac{g}{\pi}\right)\Pi^\dagger(0)\Pi(0), \tag{7.61}$$

where the $\Pi(0)$ field is introduced by hand, and the existence of the $\Pi(0)$ and $\Phi(0)$ fields is assumed. Here, $\hat{H}_B$ denotes the free boson Hamiltonian and is written as

$$\hat{H}_B = \frac{1}{2}\sum_p \left\{\left(1 - \frac{g}{\pi}\right)\Pi^\dagger(p)\Pi(p) + \left(1 + \frac{g}{\pi}\right)p^2\Phi^\dagger(p)\Phi(p)\right\}. \tag{7.62}$$

Now, one can introduce the following eigenstates for $\hat{H}_B$ and $\Pi^\dagger(0)\Pi(0)$ by

$$\hat{H}_B|p\rangle = E_p|p\rangle, \tag{7.63a}$$

$$\Pi^\dagger(0)\Pi(0)|\Lambda_0\rangle = \Lambda_0|\Lambda_0\rangle, \tag{7.63b}$$

where

$$E_p = \frac{2\pi}{L}p \text{ with } p = 0, 1, 2, \ldots$$

and Λ_0 is related to the box length L by

$$\Lambda_0 = \frac{c_0}{L}$$

with c_0 constant. Eq.(7.63a) is just the normal eigenvalue equation for the massless boson and its spectrum. On the other hand, eq.(7.63b) is somewhat artificial since the state $|\Lambda_0\rangle$ is introduced by hand. The zero mode state of the Hamiltonian H_B should couple with the state $|\Lambda_0\rangle$, and therefore new states can be made by the superposition of the two states

$$|v\rangle = c_1|\Lambda_0\rangle + c_2|0\rangle, \tag{7.64}$$

where c_1 and c_2 are constants. Further, the overlapping integral between the $|0\rangle$ and the $|\Lambda_0\rangle$ states is assumed to be small and is given by ε

$$\langle 0|\Lambda_0\rangle = \varepsilon. \tag{7.65}$$

In this case, the energy eigenvalues $\langle v|\hat{H}|v\rangle$ of eq.(7.61) become at the order of $O(\varepsilon)$

$$E_{\Lambda_0} = \langle\Lambda_0|\hat{H}_B|\Lambda_0\rangle - \frac{1}{2}\left(1 - \frac{g}{\pi}\right)\Lambda_0, \tag{7.66a}$$

$$E_0 = -\frac{1}{2}\left(1 - \frac{g}{\pi}\right)\langle 0|\Pi^\dagger(0)\Pi(0)|0\rangle. \tag{7.66b}$$

Since the magnitude of the $\langle\Lambda_0|\hat{H}_B|\Lambda_0\rangle$ and $\langle 0|\Pi^\dagger(0)\Pi(0)|0\rangle$ should be appreciably smaller than the Λ_0

$$\langle\Lambda_0|\hat{H}_B|\Lambda_0\rangle \ll \Lambda_0, \tag{7.67a}$$

$$\langle 0|\Pi^\dagger(0)\Pi(0)|0\rangle \ll \Lambda_0 \tag{7.67b}$$

the spectrum of the Hamiltonian eq.(7.61) should have a finite gap, and the continuum states of the massless excitations start right above the gap. This is just the same as the spectrum obtained from the Bethe ansatz solutions discussed in the previous section.

7.5. Massive Thirring vs Sine-Gordon Models

In the previous section, we saw that the massive Thirring model can be bosonized. Indeed, there is a nice way of bosonizing the massive Thirring model which is presented by Coleman [23], and therefore we explain briefly the procedure of bosonizing the fermion field theory model in two dimensions and show that the massive Thirring model is equivalent to the sine-Gordon field theory model.

The proof of the equivalence between two field theory models is based on the observation that any orders of the correlation functions have the same structure between the two models once the correspondence relations of the coupling constant and the mass parameter between the two models are properly chosen.

7.5.1. Sine-Gordon Field Theory Model

The sine-Gordon field theory model is a boson field theory model in two dimensions. The Lagrangian density of the sine-Gordon model is written as

$$\mathcal{L} = \frac{1}{2}\partial_\mu\phi\partial^\mu\phi + \frac{m^2}{\beta^2}\cos\beta\phi, \tag{7.68}$$

where m and β denotes the mass scale parameter and the coupling constant, respectively. The field equation of the sine-Gordon model becomes

$$\left(\frac{\partial^2}{\partial t^2} - \frac{\partial^2}{\partial x^2}\right)\phi + \frac{m^2}{\beta}\sin\beta\phi = 0. \tag{7.69}$$

By rescaling ϕ by

$$\beta\phi \to \phi$$

one obtains the equation for the sine-Gordon model

$$\left(\frac{\partial^2}{\partial t^2} - \frac{\partial^2}{\partial x^2}\right)\phi + m^2\sin\phi = 0. \tag{7.70}$$

Soliton Solution

This has a soliton solution as a classical field theory and is given as

$$\phi(t,x) = 4\tan^{-1}\left[\exp\left(\frac{m(x-vt)}{\sqrt{1-v^2}}\right)\right], \tag{7.71}$$

where v is a constant.

This sine-Gordon field is a real field, and therefore there is no finite current density. However, this field has a soliton solution which behaves like a particle, and therefore, it may be that the zero current density is somehow understandable since the field is confined in the limited area of space and time.

7.5.2. Correlation Functions

In order to see the equivalence between the sine-Gordon field theory and the massive Thirring model, we should evaluate the n-point functions in both of the models.

Sine-Gordon Model

In the case of the sine-Gordon field theory, one defines

$$A_\pm = e^{\pm i\beta\phi}. \tag{7.72}$$

Now, one can calculate the expectation value of $\prod_{i=1}^{n} [A_+(x_i)A_-(y_i)]$ with the vacuum state, and one obtains

$$\langle 0|T\left\{\prod_{i=1}^{n}\left[A_+(x_i)A_-(y_i)\right]\right\}|0\rangle = \frac{\prod_{i>j}\left[(x_i-x_j)^2(y_i-y_j)^2c^2m^4\right]^{\frac{\beta^2}{4\pi}}}{\prod_{i,j}\left[(x_i-y_j)^2cm^2\right]^{\frac{\beta^2}{4\pi}}}, \tag{7.73}$$

where c and m denote a numerical constant and a mass parameter, respectively.

Massive Thirring Model

For the massive Thirring model, one defines $\sigma_\pm$ in terms of the fermion operators by

$$\sigma_\pm = \frac{1}{2}\bar{\psi}(1\pm\gamma_5)\psi. \tag{7.74}$$

In this case, the expectation value of $\prod_{i=1}^{n} [\sigma_+(x_i)\sigma_-(y_i)]$ with the vacuum state can be evaluated by making use of Klaiber's formula [81] as

$$\langle 0|T\left\{\prod_{i=1}^{n}\left[\sigma_+(x_i)\sigma_-(y_i)\right]\right\}|0\rangle = \left(\frac{1}{2}\right)^{2n}\frac{\prod_{i>j}\left[(x_i-x_j)^2(y_i-y_j)^2M^4\right]^{\frac{1}{1+\frac{g}{\pi}}}}{\prod_{i,j}\left[(x_i-y_j)^2M^2\right]^{\frac{1}{1+\frac{g}{\pi}}}}, \tag{7.75}$$

where g and M denote the coupling constant and a mass parameter in the Thirring model, respectively.

Equivalence

By comparing eq.(7.73) with eq.(7.75), one notices that the two perturbation theories are identical if one makes the following identifications

$$\sigma_\pm = \frac{1}{2}A_\pm, \quad M^2 = cm^2, \quad \frac{1}{1+\frac{g}{\pi}} = \frac{\beta^2}{4\pi}. \tag{7.76}$$

This implies that the two field theory models of sine-Gordon and massive Thirring (charge zero sector) must be equivalent to each other as long as the perturbation expansion converges. Since the Lagrangian density of the massive Thirring model is given as

$$\mathcal{L} = i\bar{\psi}\gamma_\mu\partial^\mu\psi - m\bar{\psi}\psi - \frac{1}{2}gj^\mu j_\mu \tag{7.77}$$

one sees the correspondence between the sine-Gordon field theory and the massive Thirring model in the following way.

7.5.3. Correspondence

Firstly, the kinetic energy term of the sine-Gordon model should correspond to the kinetic energy part plus the interaction term in the massive Thirring model,

$$\left\{\frac{1}{2}\partial_\mu\phi\partial^\mu\phi\right\} \Longrightarrow \left\{i\bar{\psi}\gamma_\mu\partial^\mu\psi - \frac{1}{2}gj^\mu j_\mu\right\}. \tag{7.78}$$

Further, the interaction term of the sine-Gordon model should correspond to the mass term in the massive Thirring model

$$\left\{\frac{m_0^2}{\beta^2}\cos\beta\phi\right\} \Longrightarrow \{-m\bar{\psi}\psi\}. \tag{7.79}$$

However, it is important to note that the correspondence is proved for the expectation values of the above terms, and the correspondence between these terms as operators is not necessarily shown here.

This indicates that the equivalence between the Thirring and the sine-Gordon field theory models holds true in the sense that any physical observables of the charge zero sector must be the same between the two models. However, their equivalence may not be justified in the sense of operator equations. This is in contrast to the Schwinger model which can be properly written in terms of the massive boson field, and this should be connected to the fact that the Schwinger model has no free fermion state due to the confinement force. In other words, the Schwinger model has only bosons as physical particles, and therefore it is natural that it becomes equivalent to the massive boson field theory.

7.6. Bogoliubov Method for Thirring Model

As we saw in the Section 7.1, the massless Thirring model is solved exactly in terms of the Bethe ansatz technique. The vacuum of the Thirring model has a chiral symmetry broken state, and there is no bosonic excitation in the spectrum. However, there is a finite gap, and there is no massless state in the Thirring model. This is consistent with the observation that there should be no massless boson in two dimensions since a massless boson in two dimension has an infra-red singularity and thus cannot exist as a physical state. In this section, we present the Bogoliubov transformation method [68]. The Bogoliubov method is often employed in many fields of research. In particular, it was applied to the understanding of the superconductor phenomena.

7.6.1. Massless Thirring Model

First, we treat the massless Thirring model in the Bogoliubov transformation method.

Hamiltonian in Chiral Representation

The Hamiltonian of the massless Thirring model can be written in the chiral representation

$$\hat{H} = \sum_n \left[p_n(a_n^\dagger a_n - b_n^\dagger b_n) + \frac{2g}{L}\left(\sum_l a_l^\dagger a_{l+n}\right)\left(\sum_m b_m^\dagger b_{m+n}\right)\right], \tag{7.80}$$

where

$$p_n = \frac{2\pi}{L} n.$$

7.6.2. Bogoliubov Transformation

Now, new fermion operators by the Bogoliubov transformation are defined

$$c_n = e^{\mathcal{A}} a_n e^{-\mathcal{A}} = \cos\left(\frac{\theta_n}{2} - \frac{\pi}{4}\right) a_n - \sin\left(\frac{\theta_n}{2} - \frac{\pi}{4}\right) b_n, \tag{7.81a}$$

$$d_{-n}^{\dagger} = e^{\mathcal{A}} b_n e^{-\mathcal{A}} = \cos\left(\frac{\theta_n}{2} - \frac{\pi}{4}\right) b_n + \sin\left(\frac{\theta_n}{2} - \frac{\pi}{4}\right) a_n, \tag{7.81b}$$

where the generator of the Bogoliubov transformation is given by

$$\mathcal{A} = \sum_n \left(\frac{\theta_n}{2} - \frac{\pi}{4}\right) (a_n^{\dagger} b_n - b_n^{\dagger} a_n). \tag{7.82}$$

θ_n denotes the Bogoliubov angle which can be determined by the condition that the vacuum energy is minimized. In this case, the new vacuum state is obtained as

$$|\Omega\rangle_B = e^{-\mathcal{A}} |0\rangle, \tag{7.83}$$

where $|0\rangle$ denotes the free fermion vacuum which is filled with the negative energy particles.

7.6.3. Bogoliubov Transformed Hamiltonian

Now, one can obtain the new Hamiltonian under the Bogoliubov transformation,

$$\hat{H} = \sum_n \left[\left\{ p_n \sin\theta_n + \frac{g}{L}\mathcal{B} \cos\theta_n \right\} (c_n^{\dagger} c_n + d_{-n}^{\dagger} d_{-n}) \right]$$
$$+ \sum_n \left\{ -p_n \cos\theta_n + \frac{g}{L}\mathcal{B} \sin\theta_n \right\} (c_n^{\dagger} d_{-n}^{\dagger} + d_{-n} c_n) + \hat{H}', \tag{7.84}$$

where $\hat{H}'$ denotes the interaction terms of the Bogoliubov transformed state.

Definition of Effective Mass

Here, $\mathcal{B}$ is defined as

$$\mathcal{B} = \sum_m \cos\theta_m \tag{7.85}$$

and the term $\frac{g}{L}\mathcal{B}$ should correspond to an effective mass M which is thus defined as

$$M = \frac{g}{L}\mathcal{B}. \tag{7.86}$$

Determination of Bogoliubov Angles

The Bogoliubov angle θ_n can be determined by imposing the condition that the vacuum energy must be minimized. Equivalently, one sees that the term proportional to $(c_n^\dagger d_{-n}^\dagger + d_{-n}c_n)$ in eq.(7.84) should vanish, and therefore, one obtains

$$\tan\theta_n = \frac{p_n}{M}. \tag{7.87}$$

Determination of Effective Mass

In this case, one can express the self-consistency condition for M

$$M = \frac{g}{L}\mathcal{B} = \frac{g}{L}\sum_m \cos\theta_m = \frac{g}{2\pi}\int dp\,\cos\theta_p = \frac{g}{\pi}\int_0^\Lambda \frac{M}{\sqrt{M^2+p^2}}dp$$

which leads to the following constraint

$$M = \frac{g}{\pi}M\ln\left(\frac{\Lambda}{M} + \sqrt{1+\left(\frac{\Lambda}{M}\right)^2}\right), \tag{7.88}$$

where Λ denotes the cut-off momentum. Since the massless Thirring model has no scale, one should measure all the physical observables in terms of Λ. Therefore, one can express the effective mass M in terms of Λ,

$$M = \frac{\Lambda}{\sinh(\frac{\pi}{g})}. \tag{7.89}$$

Vacuum Energy

Further, the vacuum energy E_{vac} as measured from the trivial vacuum is given

$$E_{vac} = -\frac{L}{2\pi}\frac{\Lambda^2}{\sinh(\frac{\pi}{g})}e^{-\frac{\pi}{g}}. \tag{7.90}$$

From this value of the vacuum energy, one sees that the new vacuum energy is indeed lower than the trivial one. Therefore, the chiral symmetry is spontaneously broken in the new vacuum state since the fermion becomes massive.

7.6.4. Eigenvalue Equation for Boson

Now, one can carry out the calculations of the spectrum of the bosons in the Fock space expansion. The boson state $|B\rangle$ can be expressed as

$$|B\rangle = \sum_n f_n c_n^\dagger d_{-n}^\dagger |\Omega\rangle_B, \tag{7.91}$$

where f_n is a wave function in momentum space, and $|\Omega\rangle_B$ denotes the Bogoliubov vacuum state. The energy eigenvalue equation of the Hamiltonian for the large L limit can be written as

$$\mathcal{M}\, f(p) = 2E_p f(p) - \frac{g}{2\pi}\int dq\, f(q)\left(1+\frac{M^2}{E_pE_q}+\frac{pq}{E_pE_q}\right), \tag{7.92}$$

where $\mathcal{M}$ denotes the boson mass. E_p is given as

$$E_p = \sqrt{M^2+p^2}\,.$$

7.6.5. Solution of Separable Interactions

Eq.(7.92) can be solved exactly since the interaction is a separable type. This is a well known method in the scattering theory if the interaction kernel has the separable type

$$\sum_{i=1}^{N}\int dq\, f(q)G_i(q)H_i(p),$$

where $G_i(p)$ and $H_i(p)$ are arbitrary functions. Suppose one wishes to solve the following equation for $f(p)$

$$C(p)f(p) = \sum_{i=1}^{N}\int dq\, f(q)G_i(q)H_i(p), \tag{7.93}$$

where $C(p)$ denotes a known function. Then, one can solve the above equation in an exact fashion. The explicit way of solving eq.(7.93) is just simple. First, one defines

$$A_i = \int dq\, f(q)G_i(q)$$

which is a constant. In this case, one can easily solve eq.(7.93) for $f(p)$

$$f(p) = \frac{1}{C(p)}\sum_{i=1}^{N} A_iH_i(p).$$

Then, one can put this $f(p)$ back to A_i and obtain

$$A_i = \sum_{i,j=1}^{N} V_{ij}A_j,$$

where V_{ij} is defined as

$$V_{ij} = \int dq\, \frac{1}{C(q)}H_j(q)G_i(q).$$

Since V_{ij} is just a constant, this is a matrix equation for A_i and can be solved exactly. The energy eigenvalues can be obtained from the determinant

$$\det(\delta_{ij}-V_{ij}) = 0.$$

7.6.6. Boson Spectrum

Eq.(7.92) can be solved in the same way as the above separable type interactions. Now, one defines A and B by

$$A = \int_{-\Lambda}^{\Lambda} dp\, f(p), \tag{7.94a}$$

$$B = \int_{-\Lambda}^{\Lambda} dp \frac{f(p)}{E_p}. \tag{7.94b}$$

Using A and B, one can solve eq.(7.92) for $f(p)$ and obtain

$$f(p) = \frac{g}{2\pi(2E_p - \mathcal{M})}\left(A + \frac{m^2}{E_p}B\right). \tag{7.95}$$

Putting this $f(p)$ back into eqs.(7.94), one obtains the matrix equations

$$A = \frac{g}{2\pi}\int_0^{\Lambda} \frac{2dp}{2E_p - \mathcal{M}}\left(A + \frac{m^2}{E_p}B\right), \tag{7.96a}$$

$$B = \frac{g}{2\pi}\int_0^{\Lambda} \frac{2dp}{(2E_p - \mathcal{M})E_p}\left(A + \frac{m^2}{E_p}B\right). \tag{7.96b}$$

This is a matrix equation for A and B, and therefore by requiring that the determinant of this matrix should be zero, one can obtain the boson spectrum as the function of the coupling constant g/π. From this calculation, one sees that there is always one massive boson, and there is no massless boson in this spectrum even though the boson mass for the very small coupling constant g is exponentially small.

7.6.7. Axial Vector Current Conservation

Now, we discuss the axial vector current conservation. In order to examine it, we evaluate the following equation,

$$i\dot{Q}_5 = [\hat{H}, Q_5], \tag{7.97}$$

where Q_5 is defined as

$$Q_5(t) = \int j_0^5(x,t)\,dx = \int \bar{\psi}\gamma_0\gamma_5\psi\, dx. \tag{7.98}$$

Clearly, one can show that the right hand side of eq.(7.97) vanishes for the massless Thirring Hamiltonian. However, since the Bogoliubov transformation is unitary, $[\hat{H}, Q_5]$ is invariant and thus it remains to be zero. Therefore, the axial vector current is still conserved after the Bogoliubov transformation.

7.6.8. Fermion Condensate

Here, one can also calculate the fermion condensate C for this vacuum state and obtain

$$|C| = \langle\Omega|\frac{1}{L}\int\bar{\psi}\psi\, dx|\Omega\rangle_B = \frac{M}{g} \tag{7.99}$$

which is a finite value. Therefore, the massless Thirring model has a chiral symmetry broken vacuum, but, the axial vector current is conserved. Therefore, this situation contradicts the Goldstone theorem since there appears no massless boson in this symmetry breaking phenomena.

7.6.9. Massive Thirring Model

The massive Thirring model can be calculated just in the same ways as the massless case. The Hamiltonian of the massive Thirring model can be written as

$$\hat{H} = \sum_n \left[p_n \left(a_n^\dagger a_n - b_n^\dagger b_n\right) + m_0 \left(a_n^\dagger b_n + b_n^\dagger a_n\right) + \frac{2g}{L}\tilde{j}_{a,p_n}\tilde{j}_{b,-p_n}\right]. \tag{7.100}$$

Now, one can carry out the Bogoliubov transformation just in the same way as the massless case, and obtain

$$\hat{H} = \sum_n \left[\left\{ p_n \sin\theta_n + \left(m_0 + \frac{g}{L}\mathcal{B}\right)\cos\theta_n \right\} \left(c_n^\dagger c_n + d_{-n}^\dagger d_{-n}\right)\right]$$
$$+ \sum_n \left\{ -p_n \cos\theta_n + \left(m_0 + \frac{g}{L}\mathcal{B}\right)\sin\theta_n \right\} \left(c_n^\dagger d_{-n}^\dagger + d_{-n} c_n\right) + \hat{H}'. \tag{7.101}$$

Renormalized Fermion Mass

Therefore, one can define the effective mass M by

$$M = m_0 + \frac{g}{L}\mathcal{B}. \tag{7.102}$$

which corresponds to the renormalization of the fermion mass m_0. In this case, all of the procedures of carrying out the boson mass are just the same as the massless Thirring case, except that the effective fermion mass M is not determined by the cut-off momentum Λ any more. Therefore, one has to introduce the cut-off momentum Λ independently from the effective fermion mass M.

Boson Spectrum in Massive Thirring Model

The calculated boson mass $\mathcal{M}$ is always massive, and there is no massless boson. At the strong coupling limit, the boson mass $\mathcal{M}$ goes to zero. It should be also noted that there is only one boson in the massive Thirring model [38, 54]. The semi-classical treatment of the massive Thirring model predicted many bosonic excitations [24], but the spectrum is obtained with the approximation that considers only the second order fluctuations in the path integral in the semiclassical method. However, apart from the higher excitations, the boson mass predicted by the semi-classical approximation agrees rather well with that calculated by the Bogoliubov transformation.

In addition, the boson spectrum in the massive Thirring model is obtained by the Bethe ansatz technique [44, 55]. The numerical results show that there is only one boson in the massive Thirring model since the interaction between two bosons are always repulsive. Also, the boson mass calculated by the Bethe ansatz method is consistent with the boson mass calculated by the Bogoliubov transformation method as the function of the coupling constant g.

7.6.10. NJL Model

We have not discussed the spectrum of the Nambu-Jona-Lasinio (NJL) model which is a four dimensional current current interaction model. The Lagrangian density is given in eq.(1.34) in Chapter 1

$$\mathcal{L} = i\bar{\psi}\gamma_\mu\partial^\mu\psi - m\bar{\psi}\psi + \frac{1}{2}G\left[(\bar{\psi}\psi)^2 + (\bar{\psi}i\gamma_5\psi)^2\right].$$

In this textbook, we do not treat the NJL model mainly because it is not a renormalizable field theory model in the perturbation theory.

However, it should be noted that the NJL model has been playing some important role for the symmetry breaking physics even though there is a wrong belief that the model possesses a massless boson.

Concerning the bound state problem of bosons, there is a similarity between the Thirring model and the NJL model with massless fermions. Both of the models predict a massive boson if one employs the Bogoliubov transformation method [69]. However, there is no boson in the Thirring model if it is solved exactly since the Bogoliubov vacuum overestimates the attraction between fermion and anti-fermion. In this respect, there should be no boson in the massless NJL model either since the interaction in four dimensions must be much weaker than in the two dimensions. In other words, it is generally true that bound states in four dimensions requires much stronger interactions than those in two dimensions if the shape of the interactions is the same.

Chapter 8

Lattice Field Theory

In field theory, space and time are ranged from $-\infty$ to ∞. Since it is normally difficult to solve field theory in this real space and time, one puts the theory into a box with its length of L. In addition, the space and time may be discretized into a lattice. This lattice field theory becomes fashionable mainly because it can be solved by computer simulations. Unfortunately, however, it is still difficult to judge how much one can trust the calculated results of the lattice simulations if the number of the lattice sites is less than 100. Normally, one has to carry out calculations with more than 1000 of lattice sites if one wishes to obtain any reasonable results for physical observables which crucially depend on the finite size effects.

In this chapter, we discuss two aspects of the lattice field theory. The first one is connected to the specialty of the two dimensional field theory, and one sees that the Heisenberg spin chain of XYZ in one space dimension is shown to be equivalent to the massive Thirring model in the continuum limit.

The second one is the lattice gauge theory and, in particular, we treat the gauge fields on a lattice, which is developed by Wilson [112]. However, there are some unsolved problems related to his action in evaluating the Wilson loop integral even though the lattice formulation itself is interesting.

Before going to the discussion of concrete examples of lattice field theory models, we should make some general comments on the connection and difference between the continuum field theory and the lattice field theory.

8.1. General Remark on Discretization of Space

When one wishes to discretize space in one dimension which now ranges from $-\frac{L}{2}$ to $\frac{L}{2}$, then one usually divides the L into N sites. Therefore, one sees that the lattice constant a must be given as

$$a = \frac{L}{N}.$$

This is just a normal way of making the discretization of L in one space dimension.

8.1.1. Equal Spacing

Now, a question may arise as to why one can make an equal division of space. In other words, the discretization of equal spacing is simply one of many ways of dividing space. For example, if the interaction fields are concentrated on some area of space, then the discretization of equal spacing should be far from the best way. It should be better and efficient to solve the field theory model if one divides a limited region of space with a small lattice constant where most of the interactions may take place and the rest of area with a large lattice constant.

Discretization of Space

It is by now clear that the discretization of space has infinitely many ways and there is no reason to believe that the equal spacing with the lattice constant a is the only way to discretize space. Therefore, we may write the coordinate x_n in the discretized space as

$$x_n = F_n(x),$$

where the mapping function F_n should have an arbitrary functional shape. In the equal spacing case, it becomes

$$x_n = -\frac{L}{2} + an, \quad n = 1, 2, \ldots, N.$$

This is, of course, the simplest case of the discretization and seems to be most natural. However, the discretization of space itself is not a natural thing at all, and therefore one should judge the discretization method in terms of the efficiency in solving the field theory model.

8.1.2. Continuum Limit

If one works in the lattice field theory, then one has to make the continuum limit, that is,

$$a \to 0$$

at the end of the calculation since one should recover the real space. In the process of making the continuum limit, one may find a different continuum field theory model, depending on the way of keeping relevant terms in the continuum limit. However, this is connected to the approximation one has made, and it is not due to the complexity of the lattice field theory model.

As one sees above, the continuum field theory may lead to many different discretized field theory models, depending on the way of dividing space, namely, depending on the functional dependence of $F_n(x)$ itself. In this respect, it is, of course, clear that the continuum field theory has much more information than the lattice version.

Relevant Scale

The continuum limit in physics means that the lattice constant a should be much smaller than any of relevant scales in the field theory model one considers. However, what should be the guiding principle when there is no relevant scale in the model? This is indeed the case in the massless Thirring model or QED and QCD with massless fermions in four dimensions, and all of physical observables must be measured in units of the lattice constant a. In this case, one should take the site number N and the box length L as large as required

$$N \to \infty, \ L \to \infty.$$

However, one keeps the lattice constant a finite

$$a = \frac{L}{N} \to \text{finite}.$$

The important point is that any physical observables should not depend on the values of N and L. Instead, physical quantities should be described in terms of the lattice constant a or the cutoff momentum Λ. Normally, they are related to each other by

$$\Lambda \simeq \frac{\pi}{a}.$$

For example, the spectrum of the system should be described in terms of the cutoff momentum Λ as

$$E_n = c_n \Lambda.$$

In this case, physically meaningful quantities should be the ratio to the lowest energy state E_0. In the case of the massless Thirring model, there is a gap between the vacuum state E_0 and the first excited state E_1 and then the continuum states start. In this respect, physically the existence of the gap itself is most important and interesting, but the absolute magnitude of the gap is not a meaningful quantity. This becomes important when one introduces a new scale like a fermion mass m, and in this case, the spectrum is described in terms of the mass m and therefore, the magnitude of the gap itself can be a physical observable.

8.2. Bethe Ansatz Method in Heisenberg Model

The Hamiltonian of the Heisenberg XXZ model with spin $\frac{1}{2}$ is constructed by the spin spin interactions between neighboring sites. The spins are defined on the lattice, and the lattice constant a in this model is always finite since it corresponds to the distance between two lattice sites in solid state physics. This model is most frequently studied among all models in spin systems and is well understood by various methods. In particular, it is solved by the Bethe ansatz technique [11]. Here, we briefly describe the Heisenberg model and the Bethe ansatz solutions. The model is described by the following Hamiltonian

$$H = J \sum_{i=1}^{N} \left(S_i^x S_{i+1}^x + S_i^y S_{i+1}^y + \Delta S_i^z S_{i+1}^z \right), \tag{8.1}$$

where J denotes the coupling constant, and Δ is a anisotropy parameter. Here, one can set $J = 1$ without loss of generality.

This Hamiltonian can be diagonalized by the Bethe ansatz method in which one assumes the shape of the eigenstate in advance and finds conditions so that the eigenstates can indeed satisfy the following eigenvalue equation of the Hamiltonian

$$H\Psi^{(m)} = E_m \Psi^{(m)}, \tag{8.2}$$

where the Bethe ansatz wave function $\Psi^{(m)}$ is assumed to have m down spin states while all the rest should have up spin states.

8.2.1. Exchange Operator $P_{i,j}$

Here, the exchange operator $P_{i,j}$ can be introduced which is defined as

$$P_{i,j}\phi(i,j) = \phi(j,i).$$

Denoting the symmetric and anti-symmetric states by

$$\text{symmetric} : \ \phi_s(i,j), \quad \text{anti}-\text{symmetric} : \ \phi_a(i,j)$$

one finds

$$P_{i,j}\phi_s(i,j) = \phi_s(i,j), \quad P_{i,j}\phi_a(i,j) = -\phi_a(i,j).$$

Therefore, $\phi_s(i,j)$ and $\phi_a(i,j)$ are the eigenfunction of the exchange operator $P_{i,j}$. On the other hand, the scalar product of the spin 1/2 operator $\boldsymbol{S}_i \cdot \boldsymbol{S}_j$ has the following properties

$$\left(2\boldsymbol{S}_i \cdot \boldsymbol{S}_j + \frac{1}{2}\right)\phi_s(i,j) = \phi_s(i,j), \quad \left(2\boldsymbol{S}_i \cdot \boldsymbol{S}_j + \frac{1}{2}\right)\phi_a(i,j) = -\phi_a(i,j).$$

Therefore, the scalar product of the spin operator $\boldsymbol{S}_i \cdot \boldsymbol{S}_j$ can be written in terms of the exchange operator $P_{i,j}$

$$\boldsymbol{S}_i \cdot \boldsymbol{S}_j = \frac{1}{2}P_{i,j} - \frac{1}{4} \tag{8.3}$$

and thus the Hamiltonian eq.(8.1) can be rewritten as

$$H = \sum_{i=1}^{N}\left(\frac{1}{2}P_{i,j} + (\Delta - 1)S_i^z S_{i+1}^z\right) - \frac{1}{4}' N. \tag{8.1$'$}$$

8.2.2. Heisenberg XXZ for One Magnon State

First, we consider the simplest case in which there is only one spin down state at the z_i site and all the rest should have spin up states. This is called one magnon state ($m = 1$), and one can write the Bethe ansatz wave function by the superposition of the down spin wave function $\phi(z_i)$ as

$$\Psi^{(1)} = \sum_{i=1}^{N} A_i \phi(z_i)\Phi_{N-1}(\text{up}), \tag{8.4}$$

where the coefficient A_i should be determined such that the state $\Psi^{(1)}$ is the eigenfunction of the Hamiltonian of eq.(8.1$'$).

Periodic Boundary Condition

Here, z_i satisfies the following conditions

$$z_1 < z_2 < \cdots < z_N$$

and $\phi(z_i)$ satisfies the periodic boundary condition

$$\phi(z_{N+1}) = \phi(z_1).$$

Eigenvalue Equation

Since one can easily prove the following equations for $m = 1$ down spin states

$$\left(\sum_{j=1}^{N} P_{j,j+1}\right)\phi(z_i) = \phi(z_{i-1}) + \phi(z_{i+1}) + (N-2)\phi(z_i), \tag{8.5a}$$

$$\left(\sum_{j=1}^{N} S_j^z S_{j+1}^z\right)\phi(z_i) = \frac{1}{4}(N-4)\phi(z_i) \tag{8.5b}$$

the eigenvalue equation [eq.(8.2)] becomes

$$\frac{1}{2}\sum_{i=1}^{N}\left[(A_{i+1}+A_{i-1})\phi(z_i) + (N-2)A_i\phi(z_i) + \frac{1}{2}(\Delta-1)(N-4)A_i\phi(z_i)\right]$$

$$= \left(E_1 + \frac{1}{4}N\right)\sum_{i=1}^{N} A_i\phi(z_i). \tag{8.6}$$

Therefore, eq.(8.6) should hold for arbitrary $\phi(z_i)$, and thus one obtains

$$\frac{1}{2}(A_{i+1}+A_{i-1}) + \left\{\frac{1}{4}\Delta(N-4) + \frac{1}{4}N\right\}A_i = \left(E_1 + \frac{1}{4}N\right)A_i. \tag{8.7}$$

Solutions

Eq.(8.7) can be satisfied when A_i is assumed to be of the following shape

$$A_i = e^{ikz_i}, \tag{8.8}$$

where k is called *rapidity* and should be determined by the periodic boundary conditions (PBC). In this case, the energy eigenvalue with $m = 1$ becomes

$$E_1 = \cos k + \left(\frac{1}{4}N - 1\right)\Delta \tag{8.9}$$

apart from a constant. Since the PBC for A_i is written as

$$A_i = A_{i+N}$$

one finds an equation which determines the rapidity k

$$e^{ikN} = 1. \tag{8.10a}$$

Therefore, k is written as

$$k = \frac{2\pi}{N}\lambda, \quad \lambda = 1, 2, \ldots, N-1. \tag{8.10b}$$

8.2.3. Heisenberg XXZ for Two Magnon States

Now, we discuss the two down-spin ($m = 2$, two magnon) case, and denote the wave function $\phi(z_{i_1}, z_{i_2})$ for the two down-spin sites where z_{i_1}, z_{i_2} are the coordinates of the down-spin site. All the rest are in the spin up states.

Wave Functions

In this case, the Hamiltonian can be diagonalized by the superposition of the wave functions $\phi(z_{i_1}, z_{i_2})$ as

$$\Psi^{(2)} = \sum_P A_{i_1,i_2}\phi(z_{i_1}, z_{i_2})\Phi_{N-2}(\text{up}), \tag{8.11}$$

where P means all possible permutations of the i_1, i_2. $\Phi_{N-2}(\text{up})$ denotes the wave function of the $N-2$ sites which are in the up spin states.

Solutions

Further, the coefficient A_{i_1,i_2} is assumed to be of the following shape,

$$A_{i_1,i_2} = e^{\frac{i}{2}\varphi} e^{ik_1 z_{i_1} + ik_2 z_{i_2}} + e^{-\frac{i}{2}\varphi} e^{ik_2 z_{i_1} + ik_1 z_{i_2}}, \tag{8.12}$$

where k_1 and k_2 denote the rapidity of the down-spin site. In this case, one can easily confirm that the wave function [eq.(8.11)] is indeed the eigenstate of the Hamiltonian of the Heisenberg model

$$\left[\sum_{i=1}^{N} \left(S_i^x S_{i+1}^x + S_i^y S_{i+1}^y + \Delta S_i^z S_{i+1}^z\right)\right] \Psi^{(2)} = E_2 \Psi^{(2)}. \tag{8.13}$$

Periodic Boundary Condition

Now, one imposes the periodic boundary conditions on $\phi(z_{i_1}, z_{i_2})$, and in this case, A_{i_1,i_2} should satisfy the following condition

$$A_{i_1,i_2} = A_{i_2,i_1+N}. \tag{8.14a}$$

Therefore, one obtains the following equations

$$k_1 N + \varphi = 2\pi\lambda_1, \quad k_2 N - \varphi = 2\pi\lambda_2, \tag{8.14b}$$

where λ_1 and λ_2 are integers running between 0 and $N-1$. Further, the phase shift φ should satisfy the following equation,

$$\cot\frac{\varphi}{2} = \frac{\Delta \sin(\frac{k_1-k_2}{2})}{\cos(\frac{k_1+k_2}{2}) - \Delta\cos(\frac{k_1-k_2}{2})}, \tag{8.15}$$

where φ is taken to be

$$-\pi \le \varphi \le \pi.$$

In this case, the energy eigenvalue of the Hamiltonian can be written as

$$E_2 = \left(\frac{N}{4} - 2\right)\Delta + \cos k_1 + \cos k_2. \tag{8.16}$$

8.2.4. Heisenberg XXZ for m Magnon States

Now, we treat the m down spin case which is called m magnon state. The Hamiltonian can be diagonalized by the superposition of the wave functions $\phi(z_{n_1},\dots,z_{n_m})$ for m down spin case as

$$\Psi^{(m)} = \sum_P A_{n_1,\dots,n_m}\phi(z_{n_1},\dots,z_{n_m})\Phi_{N-m}(\mathrm{up}), \tag{8.17}$$

where P means all possible permutations of the $n_1,\dots,n_m$.

Solutions

Further, the coefficient $A_{n_1,\dots,n_m}$ is assumed to be of the following shape [98],

$$A_{n_1,\dots,n_m} = \sum_{P_\mu}\sum_P \exp\left(i\sum_j^m k_{P_j}n_{\mu_j} + \frac{i}{2}\sum_{j<\ell}\varphi_{P_jP_\ell}\right). \tag{8.18}$$

From the periodic boundary conditions, one obtains the following equations

$$Nk_j = 2\pi\lambda_j + \sum_{\ell=1,\ell\neq j}^m \varphi_{j\ell}, \tag{8.19}$$

where λ_j are integers running between 0 and $N-1$ with the condition

$$\lambda_1 \leq \lambda_2 \leq \cdots \leq \lambda_m.$$

The equation for $\varphi_{j\ell}$ becomes

$$\cot\frac{\varphi_{j\ell}}{2} = \frac{\Delta\sin(\frac{k_j-k_\ell}{2})}{\cos(\frac{k_j+k_\ell}{2}) - \Delta\cos(\frac{k_j-k_\ell}{2})}. \tag{8.20}$$

In this case, one can express the energy eigenvalue E_m as

$$E_m = \left(\frac{N}{4} - m\right)\Delta + \sum_{j=1}^m \cos k_j. \tag{8.21}$$

If one solves the Bethe ansatz equations [(8.19) and (8.20)], then one can determine the energy of the Heisenberg model Hamiltonian.

It should be noted that the solutions of eqs.(8.19) and (8.20) cannot be easily obtained analytically. For the value of m larger than 3, one has to carry out the numerical calculations to obtain the energy eigenvalues.

8.3. Equivalence between Heisenberg XYZ and Massive Thirring Models

The Heisenberg XYZ spin chain in one dimension is known to be equivalent to the massive Thirring model in the continuum limit. Here, we describe the procedure how these two models are related to each other when the lattice spacing is set to zero [88]. The Hamiltonian of the Heisenberg XYZ model is written as

$$H = \sum_{n=1}^{N} \left(J_x S_n^x S_{n+1}^x + J_y S_n^y S_{n+1}^y + J_z S_n^z S_{n+1}^z \right), \tag{8.22}$$

where J_x, J_y, J_z denote the coupling constant. The periodicity for the spin operators is assumed, that is,

$$\boldsymbol{S}_{n+N} = \boldsymbol{S}_n. \tag{8.23}$$

8.3.1. Jordan-Wigner Transformation

The spin operators $\boldsymbol{S}_n$ can be described in terms of the fermion operators ψ_n, $\psi_n^\dagger$. First, one defines

$$S_n^{\pm} = S_n^x \pm i S_n^y.$$

In this case, one can show that $S_n^{\pm}$ and S_n^z can be expressed as

$$S_n^- = \exp\left(-i\pi \sum_{j=1}^{n-1} \psi_j^\dagger \psi_j \right) \psi_n,$$
$$S_n^+ = \psi_n^\dagger \exp\left(i\pi \sum_{j=1}^{n-1} \psi_j^\dagger \psi_j \right), \quad S_n^z = \psi_n^\dagger \psi_n - \frac{1}{2}. \tag{8.24}$$

This is a non-linear transformation which is invented by Jordan and Wigner [78], and it is indeed quite useful. Here, it should be noted that the operators ψ_n, $\psi_n^\dagger$ are defined on the n-th lattice site. Therefore, the index n of the operator ψ_n corresponds to the coordinate x. In this respect, ψ_n is an operator in the coordinate representation. Therefore, they satisfy the following anticommutation relations

$$\{\psi_n^\dagger, \psi_m\} = \delta_{n,m}, \quad \{\psi_n, \psi_m\} = 0, \quad \{\psi_n^\dagger, \psi_m^\dagger\} = 0.$$

Now, one can employ the following equations,

$$S_n^+ S_{n+1}^- = \psi_n^\dagger e^{-in\pi\psi_n^\dagger\psi_n} \psi_{n+1} = \psi_n^\dagger \psi_{n+1}, \quad S_n^+ S_{n+1}^+ = \psi_n^\dagger \psi_{n+1}^\dagger \tag{8.25a}$$

$$S_n^- S_{n+1}^+ = \psi_n^\dagger e^{in\pi\psi_n^\dagger\psi_n} \psi_{n+1} = \psi_n \psi_{n+1}^\dagger, \quad S_n^- S_{n+1}^- = \psi_n \psi_{n+1}, \tag{8.25b}$$

where the relations

$$\psi_n^\dagger \psi_n^\dagger = 0, \quad \psi_n \psi_n = 0$$

should always hold due to the anti-commutation relations. In this case, one can rewrite the Hamiltonian of the Heisenberg XYZ model under the Jordan–Wigner transformation

$$H = \sum_n \left[\frac{v}{2} (\psi_n^\dagger \psi_{n+1} + \psi_{n+1}^\dagger \psi_n) + \frac{J_T}{2} (\psi_n^\dagger \psi_{n+1}^\dagger - \psi_n \psi_{n+1}) + J_z \psi_n^\dagger \psi_n \psi_{n+1}^\dagger \psi_{n+1} \right], \tag{8.26}$$

where only relevant terms are written. Here, v and J_T are defined as

$$v = \frac{1}{2}(J_x + J_y), \quad J_T = \frac{1}{2}(J_x - J_y). \tag{8.27}$$

8.3.2. Continuum Limit

Now, one can prove that the Hamiltonian eq.(8.26) can be reduced to the massive Thirring model Hamiltonian in the continuum limit. In order to see it explicitly, one first makes a transformation

$$\psi_n \to (i)^n \psi_n. \tag{8.28}$$

Next, one defines the following new operators by even and odd numbers of fermion operators at the $2n$ and $2n+1$ sites, separately

$$\chi_n = \psi_{2n}, \quad \xi_n = \psi_{2n+1}. \tag{8.29}$$

In this case, one obtains a new Hamiltonian

$$\begin{aligned} H = & \frac{iv}{2}\sum_n^{N/2}\left[\chi_{2n}^\dagger(\xi_{2n+1}-\xi_{2n-1})+\xi_{2n+1}^\dagger(\chi_{2n+2}-\chi_{2n})\right]. \\ & +\frac{iJ_T}{2}\sum_n^{N/2}\left[(-)^{2n}(\chi_{2n}\xi_{2n+1}-\xi_{2n+1}^\dagger\chi_{2n}^\dagger)+(-)^{2n+1}(\xi_{2n+1}\chi_{2n+2}-\chi_{2n+2}^\dagger\xi_{2n+1}^\dagger)\right] \\ & +J_z\sum_n^{N/2}\left[\chi_{2n}^\dagger\chi_{2n}\xi_{2n+1}^\dagger\xi_{2n+1}+\xi_{2n+1}^\dagger\xi_{2n+1}\chi_{2n+2}^\dagger\chi_{2n+2}\right]. \end{aligned} \tag{8.30}$$

Now, one takes the continuum limit $N \to \infty$ and defines the differential operators

$$\chi_{2n+2}-\chi_{2n} = 2a\sqrt{2a}\frac{\partial\psi_1}{\partial x}, \quad \xi_{2n+2}-\xi_{2n} = 2a\sqrt{2a}\frac{\partial\psi_2}{\partial x} \quad (a \to 0), \tag{8.31a}$$

where a denotes a lattice constant, and $\psi_1(x)$ and $\psi_2(x)$ are introduced as

$$\psi_1(x) = \frac{1}{\sqrt{2a}}\chi_n, \quad \psi_2(x) = \frac{1}{\sqrt{2a}}\xi_n. \tag{8.31b}$$

In addition, x and the summation of n are written in terms of the lattice constant a and the integral

$$x = 2an, \quad \sum_n \longrightarrow \frac{1}{2a}\int dx. \tag{8.31c}$$

In this case, the Hamiltonian becomes

$$\begin{aligned} H = 2a & \left[\frac{v}{2}\int dx\,(i\psi_1^\dagger\partial_x\psi_2 + i\psi_2^\dagger\partial_x\psi_1) + \frac{J_T}{2a}\int dx\,(i\psi_1\psi_2 - i\psi_2^\dagger\psi_1^\dagger)\right. \\ & \left. +2J_z\int dx\psi_1^\dagger\psi_1\psi_2^\dagger\psi_2\right]. \end{aligned} \tag{8.32}$$

This is still not very convenient for comparing it with the fermion field theory model. Therefore, one can make a further transformation

$$\psi_1 = \frac{1}{\sqrt{2}}(\psi_a^\dagger + i\psi_b), \quad \psi_2 = \frac{1}{\sqrt{2}}(i\psi_a - \psi_b^\dagger). \tag{8.33}$$

Apart from some numerical constants, one can write the Hamiltonian of eq.(8.32)

$$H = \int dx \left[-i(\psi_a^\dagger \partial_x \psi_b + \psi_b^\dagger \partial_x \psi_a) + \frac{J_T}{av}(\psi_a^\dagger \psi_a + \psi_b^\dagger \psi_b) + \frac{4J_z}{v} \psi_a^\dagger \psi_a \psi_b^\dagger \psi_b \right]. \quad (8.34)$$

Therefore, by identifying the mass m and the coupling constant g by

$$m = \frac{J_T}{av}, \quad g = \frac{2J_z}{v} \quad (8.35)$$

one can compare eq.(8.34) with the Thirring model Hamiltonian in the Dirac representation which is given as

$$H = \int dx \left[-i(\psi_a^\dagger \partial_x \psi_b + \psi_b^\dagger \partial_x \psi_a) + m(\psi_a^\dagger \psi_a - \psi_b^\dagger \psi_b) + 2g \psi_a^\dagger \psi_a \psi_b^\dagger \psi_b \right].$$

However, it should be noted that the sign in front of $\psi_b^\dagger \psi_b$ in the mass term is opposite to the Thirring model Hamiltonian and it is still not yet clear what it physically means by the sign difference.

8.3.3. Heisenberg XXZ and Massless Thirring Models

The equivalence between the Heisenberg XYZ model and the massive Thirring model is proved at the Hamiltonian level. If one calculates the n-point function between the two models, then one can also prove that the two models have the same structure of the n-point function [88]. This is similar to the case where the sine-Gordon field theory and the massive Thirring model have the same n-point functions between them once one establishes the relations between the coupling constants of the two models and the correspondence between the two mass scales.

Preservation of Symmetry

Now, when the mass term is set to zero in eq.(8.34) which is the massless Thirring model, it has the chiral symmetry as we discussed in chapters 2, 4 and 7. The chiral symmetry corresponds to the change of ψ in the Dirac representation as

$$\psi'_a = \cos\alpha\, \psi_a + i \sin\alpha\, \psi_b, \quad \psi'_b = \cos\alpha\, \psi_b + i \sin\alpha\, \psi_a \quad (8.36)$$

which should keep the Hamiltonian of the massless Thirring model invariant. In the Heisenberg XXZ model, there is no continuous symmetry which should correspond to the chiral symmetry in the massless Thirring model. Therefore, the ground state of the Heisenberg XXZ model cannot keep this symmetry, and the spectrum emerged from the Heisenberg XXZ model has nothing to do with this symmetry. On the other hand, the vacuum state of the massless Thirring model is realized by breaking the chiral symmetry. In this case, the spectrum of the massless Thirring model has a finite gap and then the continuum spectrum starts from the gap. This spectrum is completely different from that of the Heisenberg XXZ model which has a gapless excitation. This should be similar to the spectrum of the massless Thirring model if it was solved with the symmetry preserving vacuum state, which is not a true vacuum state [51].

Non-equivalence

Therefore, the two models (Heisenberg XXZ and massless Thirring models) give the different spectrum and therefore they are not equivalent to each other [52]. This is an interesting observation and is closely related to the fact that the massless limit in the Thirring model is a singular point, and one cannot smoothly connect the massive Thirring model to the massless Thirring model. This singular behavior at the massless limit is, however, simple to understand. The massive Thirring model has a natural scale which is the mass m while the massless Thirring model has no mass scale and therefore all of the physical observables must be measured by the cut-off parameter Λ as we saw in Chapter 7, and therefore the spectrum between the two models cannot be connected to each other.

8.4. Gauge Fields on Lattice

Last twenty years, people have been making every effort to try to solve QCD in terms of the lattice gauge field theory. Space and time are defined on the lattice sites, and gauge fields are defined as link variables. Further, the path integral formulation with imaginary time projection is employed in order that the numerical calculations of the field theory can be carried out. In this section, we briefly review the lattice gauge field theory, and discuss some problems in connection with the Wilson loop integral and Wilson's action.

8.4.1. Discretization of Space

When one wishes to solve quantum field theory, space is, of course, from $-\infty$ to ∞. However, it is practically impossible to solve it in this real space. Therefore, one first puts the theory into a box with its length of L. If one can solve the field theory model in this continuous space, then the next thing one should do is to make the box length L much larger than any scales that appear in the model. This is called *thermodynamic limit*, and in this limit, one can obtain physical quantities which can be compared with experiments.

Now, sometimes people want to solve quantum field theory in the discretized space. In this case, one should divide the box length L by the site number N, and therefore the lattice constant a becomes

$$a = \frac{L}{N}. \tag{8.37}$$

It should be important to note that the lattice constant a here is not a physical variable. This is clear since it depends on as to how one wishes to discretize the space which is in fact related to the site number N.

8.4.2. Wilson's Action

It has been believed that the confining potential between quark and anti-quark is found from the Monte Carlo simulations of the Wilson loop integral with Wilson's action. It seems that the lattice simulations can give a Coulomb type plus a linear rising potential. However, at the present stage, it is extremely difficult to accept this claim since the number of the lattice sites is less than 100. In this case, one knows that the finite size effects are appreciably large, and if at all the finite size effects are insignificant in their calculations, then one should understand why the finite size effects are not very important in QCD.

In this section, we discuss the lattice gauge theory which is developed by Wilson. Since Wilson presented the abelian gauge field theory case, we treat here the lattice QED. First, the path integral function Z_+ with Wilson's action can be written as

$$Z_+ = \prod_m \prod_\mu \int_{-\pi}^{\pi} dB_{m\mu} \exp\left(\frac{1}{2g^2}\sum_{n\mu\nu} e^{if_{n\mu\nu}}\right), \tag{8.38}$$

where the dimensionless quantity $B_{n\mu}$ is related to the discretized vector potential $A_{n\mu}$ as

$$B_{n\mu} = agA_{n\mu}, \tag{8.39}$$

where a is a lattice constant. A dimensionless form of the field strength $f_{n\mu\nu}$ is defined in terms of the discretized field strength $F_{n\mu\nu}$ as

$$f_{n\mu\nu} = a^2 gF_{n\mu\nu} = B_{n\mu} + B_{n+\hat{\mu},\nu} - B_{n+\hat{\nu},\mu} - B_{n\nu}. \tag{8.40}$$

Wilson's Action in Continuum Limit

In eq.(8.38), the gauge field action can be expressed as

$$S_+ = \frac{1}{2g^2}\sum_{n\mu\nu} e^{if_{n\mu\nu}} \tag{8.41}$$

instead of the conventional form of

$$S_0 = -\frac{1}{4}a^4\sum_{n\mu\nu} F_{n\mu\nu}^2. \tag{8.42}$$

For small a, the S_+ can be expanded as

$$S_+ = \frac{1}{2g^2}\sum_{n\mu\nu} e^{if_{n\mu\nu}} \approx \frac{1}{2g^2}\sum_{n\mu\nu}\left(1 + if_{n\mu\nu} - \frac{1}{2}f_{n\mu\nu}^2 \cdots\right). \tag{8.43}$$

The first term is irrelevant and the linear term in $f_{n\mu\nu}$ gives no contribution to the integral. Therefore, Wilson's action S_+ becomes, up to the order of the quadratic term

$$S_+ = -\frac{1}{4}a^4\sum_{n\mu\nu} F_{n\mu\nu}^2 + C, \tag{8.44}$$

which is just the conventional expression of the gauge field action S_0 apart from the irrelevant constant. Therefore, S_+ can be reduced to the physical gauge field action in the continuum limit of $a \to 0$.

Continuum Limit

The evaluation of the path integral formulation is based on the non-perturbative calculation, and therefore there appears no external momentum. In this respect, the continuum limit should be taken just in a simple way, and the coupling constant in this formalism should be taken as a constant which does not depend on the lattice constant a.

8.4.3. Wilson Loop

In order to treat the potential between fermions, Wilson proposed the Wilson loop integral which is defined as

$$W_L = \exp\left(ig\oint A_\mu dx^\mu\right). \tag{8.45}$$

This can be written in the discretized form as

$$W_L = \exp\left(i\sum_P B_{n\mu}\right). \tag{8.46}$$

This integral is important when one considers a scalar current

$$\bar{\psi}(x)\psi(y)$$

in which each $\psi(x)$ is defined in a different space and time point.

Gauge Invariant Scalar Current

In this case, the scalar current $\bar{\psi}(x)\psi(y)$ is not gauge invariant. In order to make it gauge invariant, Schwinger proposed the point splitting regularization as defined

$$\bar{\psi}(x)\psi(y) \longrightarrow \bar{\psi}(x)\exp\left(ig\int_x^y A_\mu dx^\mu\right)\psi(y), \tag{8.47}$$

which is indeed gauge invariant. This is just the same as the gauge invariant transition amplitude and therefore its path integral formula may be related to

$$\langle\bar{\psi}(x')\psi(y')|e^{ig\oint A_\mu dx^\mu}e^{-iHt}|\bar{\psi}(x)\psi(y)\rangle \sim \sum e^{-iE_nt} \sim e^{-iVt}, \tag{8.48}$$

where the first step is obtained only by the analogy with the quantum mechanics case, and the closed loop integral $\oint A_\mu dx^\mu$ as well as the last step in eq.(8.48) are obtained by assuming that the mass of quarks is sufficiently large compared to their motion.

Expectation Value of Wilson Loop

The expectation value of the Wilson loop in terms of the path integral function is written as

$$\langle\ \exp\left(i\sum_P B_{n\mu}\right)\ \rangle_+ \equiv Z_+^{-1}\int[\mathcal{D}B_\mu]W_L e^{iS_+}$$

$$= Z_+^{-1}\prod_m\prod_\mu\int_{-\pi}^{\pi} dB_{m\mu}\exp\left(i\sum_P B_{n\mu} + \frac{1}{2g^2}\sum_{n\mu\nu} e^{if_{n\mu\nu}}\right). \tag{8.49}$$

Strong Coupling Limit

In the strong coupling limit, eq.(8.49) can be expressed as

$$\langle \exp\left(i\sum_P B_{n\mu}\right)\rangle_+ = Z_+^{-1}\sum_k \frac{1}{k!}\left(\frac{1}{2g^2}\right)^k \prod_m \prod_\mu \int_{-\pi}^{\pi} dB_{m\mu} \sum_{\ell_1\pi_1\sigma_1}\cdots\sum_{\ell_k\pi_k\sigma_k}$$

$$\times \exp\left(i\sum_P B_{n\mu} + i\left(f_{\ell_1\pi_1\sigma_1}+\cdots+f_{\ell_k\pi_k\sigma_k}\right)\right). \quad (8.50)$$

Eq.(8.50) has a finite contribution only when the exponent in the integrand is zero. Therefore, the nonzero terms in the integral are those for which satisfy

$$\sum_P B_{n\mu} + f_{\ell_1\pi_1\sigma_1}+\cdots+f_{\ell_k\pi_k\sigma_k} = 0. \quad (8.51)$$

Now, if one specifies the path P which contains K blanquettes, then one should find the number of $f_{\ell_k\pi_k\sigma_k}$ with $k = K$. In this case, one sees that eq.(8.51) is indeed satisfied. Therefore, the k is the number of the blanquettes that are surrounded by the path P.

Area Law

If this area is denoted by A, then one finds that

$$k = K = \frac{A}{a^2}.$$

Therefore, it is easy to find

$$\langle \exp\left(i\sum_P B_{n\mu}\right)\rangle_+ \sim (g^2)^{-A/a^2}, \quad (8.52)$$

where A should be described by some physical quantity like $A = RT$ with T time distance in Euclidean space. Since the expectation value of the Wilson loop is related to the potential in eq.(8.48), one finds the potential as

$$V(R) \simeq \frac{\ln g^2}{a^2} R, \quad (8.53)$$

which is first obtained by Wilson. The potential has a linear rising shape and therefore quarks must be confined.

8.4.4. Critical Review on Wilson's Results

The confinement potential in eq.(8.53) has been commonly accepted as the most important ingredients for the quark confinement physics. However, one notices that there are some problems in the procedure and result of Wilson's evaluations [6].

Proper Dimensions

When one finds the potential of eq.(8.53), then one should have an uneasy feeling on the dimensions of the potential. It is written in terms of the lattice constant a which is units of space and time in this field theory model. The units should not appear in the physical quantities. In this respect, it is clear that eq.(8.53) does not have a proper dimension in view of a basic physics exercise problem.

Potential in Abelian Gauge Fields

The area law obtained above is derived in the case of the $U(1)$ gauge field theory. The result and procedure in the evaluation of the Wilson loop have nothing to do with the non-abelian characters such as the gauge non-invariance of the quark color charge. It is clear that the potential between fermions in the $U(1)$ gauge field theory must be always a Coulomb type

$$V(R) \simeq -\frac{g^2}{4\pi R}. \tag{8.54}$$

This is simply because there is no other dimensional quantity by which one can make the potential $V(R)$ a proper energy dimension together with the dimensionless coupling constant g. Apart from the strength or the sign in front of $1/R$, the Coulomb potential should be always given in eq.(8.54), and there is no exception.

8.4.5. Problems in Wilson's Action

What should be the main reason why such an unphysical result is obtained? This can be easily clarified if one looks into Wilson's action in depth

$$Z_+ = \prod_m \prod_\mu \int_{-\pi}^{\pi} dB_{m\mu} \exp\left(\frac{1}{2g^2}\sum_{n\mu\nu} e^{if_{n\mu\nu}}\right). \tag{8.55}$$

This has a proper continuum limit. That is, when one makes the lattice constant very small $a \to 0$, then one can recover the right Lagrangian density of gauge fields.

Conjugate Path Integral Function

However, if one defines a new path integral function Z_- by

$$Z_- = \prod_m \prod_\mu \int_{-\pi}^{\pi} dB_{m\mu} \exp\left(\frac{1}{2g^2}\sum_{n\mu\nu} e^{-if_{n\mu\nu}}\right), \tag{8.56}$$

where the sign in front of $f_{n\mu\nu}$ is reversed, then one should obtain the same expectation value of the Wilson loop integral. This is simply because the corresponding action

$$S_- = \frac{1}{2g^2}\sum_{n\mu\nu} e^{-if_{n\mu\nu}} \tag{8.57}$$

has also a right continuum limit, that is, in the limit of $a \to 0$

$$S_- = -\frac{1}{4}a^4 \sum_{n\mu\nu} F_{n\mu\nu}^2 + C. \tag{8.58}$$

Employing the new action of S_-, one can easily evaluate the Wilson loop integral just in the same way as in eq.(8.50) in the strong coupling limit

$$\langle \exp\left(i\sum_P B_{n\mu} \right) \rangle_- = Z_-^{-1} \sum_k \frac{1}{k!} \left(\frac{1}{2g^2} \right)^k \prod_m \prod_\mu \int_{-\pi}^{\pi} dB_{m\mu} \sum_{\ell_1 \pi_1 \sigma_1} \cdots \sum_{\ell_k \pi_k \sigma_k}$$

$$\times \exp\left(i\sum_P B_{n\mu} - i(f_{\ell_1\pi_1\sigma_1} + \cdots + i f_{\ell_k\pi_k\sigma_k}) \right). \tag{8.59}$$

No Area Law

In this case, one finds that

$$\langle \exp\left(i\sum_P B_{n\mu} \right) \rangle_- = 0 \tag{8.60}$$

since this time there is no way to satisfy

$$\sum_P B_{n\mu} - (f_{\ell_1\pi_1\sigma_1} + \cdots + f_{\ell_k\pi_k\sigma_k}) = 0. \tag{8.61}$$

Projection Operator of Integers in Integrand

Mathematically, one can understand why eq.(8.50) has a finite value of the integral while eq.(8.59) has no contribution to the Wilson loop integral. The integrand in the path integral function

$$\exp\left[i(f_{\ell_1\pi_1\sigma_1} + \cdots + f_{\ell_k\pi_k\sigma_k}) \right] \tag{8.62}$$

acts as the projection operator to find the number of blanquettes in the area enclosed by the path of the Wilson loop integral

$$\exp\left(i\sum_P B_{n\mu} \right). \tag{8.63}$$

Therefore, the sign in front of $if_{\ell_1\pi_1\sigma_1}$ in eq.(8.59) is crucial since the path has a direction which is connected with the line integral. This simply indicates that the finite value of the expectation value of the Wilson loop in terms of Wilson's action of eq.(8.55) is accidental. Therefore, the potential of eq.(8.53) is the result of the artifacts of Wilson's action. In other words, Wilson's action inevitably picks up unphysical contributions to the Wilson loop integral in the non-perturbative fashion, and therefore it results in producing the area law which is written in terms of unphysical variables.

8.4.6. Confinement of Quarks

In this sense, one should find a right physics concerning the confinement of quarks, and this is a future problem. Here, it should be noted that the confinement of quarks must be related to the non-abelian character of the color charge. The color charges of the quarks are not gauge invariant, and therefore they cannot be observed as a physical observable.

In this respect, the MIT bag model should be the simplest way to confine the quarks inside hadrons since it confines the quarks by the boundary conditions that the quark color currents cannot get out of the bags [21]. For the qualitative description of the confinement mechanism, the MIT bag model may well have a right picture of confining quarks inside hadrons. In shorts, the MIT bag model confines quarks by the kinematical constraints while the linear potential tries to confine quarks in terms of dynamics, and physically the MIT bag picture must be the correct one.

Chapter 9

Quantum Gravity

The quantum field theory of gravitation is constructed in terms of Lagrangian density of Dirac fields which couple to the electromagnetic field A_μ as well as the gravitational field $\mathcal{G}$. The gravity appears in the mass term as $m(1+g\mathcal{G})\bar{\psi}\psi$ with the coupling constant of g. In addition to the gravitational force between fermions, the electromagnetic field A_μ interacts with the gravity as the fourth order effects and its strength amounts to α times the gravitational force. Therefore, the interaction of photon with gravity is not originated from Einstein's general relativity. Further, we present a renormalization scheme for the gravity and show that the graviton stays massless.

9.1. Problems of General Relativity

The motion of the earth is governed by the gravitational force between the earth and the sun, and the Newton equation is written as

$$m\ddot{\boldsymbol{r}} = -G_0 mM \frac{\boldsymbol{r}}{r^3}, \tag{9.1}$$

where G_0, m and M denote the gravitational constant, the mass of the earth and the mass of the sun, respectively. This is the classical mechanics which works quite well.

The gravitational potential that appears in eq.(9.1) is experimentally determined. However, the theoretical derivation of the gravity cannot be achieved in any of the equations such as Newton equation or Maxwell equations. Einstein presented the equation of general relativity which should be some analogous equations to the Maxwell equations in the sense that the gravitational field should be determined by the equation of general relativity. However, since he employed the principle of equivalence which has nothing to do with real experiments, the general relativity became an equation that determines the metric tensor. This does not mean that one can determine the gravitational interaction, and indeed, we have completely lost the correct direction in physics. Therefore, we should find a theoretical frame work to determine the gravitational interaction with fermions in some way or the other.

Before constructing a theory that can describe the field equation under the gravity, we discuss the fundamental problems in the theory of general relativity. Basically, there are two serious problems in the general relativity, the lack of field equation under the gravity and the assumption of the principle of equivalence.

9.1.1. Field Equation of Gravity

When one wishes to write the Dirac equation for a particle under the gravitational interaction, then one faces to the difficulty. Since the Dirac equation for a hydrogen-like atom can be written as

$$\left(-i\nabla\cdot\boldsymbol{\alpha}+m\beta-\frac{Ze^2}{r}\right)\Psi=E\Psi \tag{9.2a}$$

one may write the Dirac equation for the gravitational potential $V(r)=-\frac{G_0mM}{r}$ as

$$\left(-i\nabla\cdot\boldsymbol{\alpha}+m\beta-\frac{G_0mM}{r}\right)\Psi=E\Psi. \tag{9.2b}$$

But there is no foundation for this equation. At least, one cannot write the Lagrangian density which can describe the Dirac equation for the gravitational interaction. This is clear since one does not know whether the interaction can be put into the zero-th component of a vector type or a simple scalar type in the Dirac equation. That is, it may be of the following type

$$\left[-i\nabla\cdot\boldsymbol{\alpha}+\left(m-\frac{G_0mM}{r}\right)\beta\right]\Psi=E\Psi. \tag{9.2c}$$

In fact, this is a right Dirac equation for a particle in the gravitational potential.

9.1.2. Principle of Equivalence

The theory of general relativity is entirely based on the principle of equivalence. Namely, Einstein started from the Gedanken experiment that physics of the two systems (a system under the uniform external gravity and a system that moves with a constant acceleration) must be the same. This looks plausible from the experience on the earth. However, one can easily convince oneself that the system that moves with a constant acceleration cannot be defined properly since there is no such an isolated system in a physical world. The basic problem is that the assumption of the principle of equivalence is concerned with the two systems which specify space and time, not just the numbers in connection with the acceleration of a particle. Note that the acceleration of a particle is indeed connected to the gravitational acceleration, $\ddot{z}=-g$, but this is, of course, just the Newton equation. Therefore, the principle of equivalence inevitably leads Einstein to the space deformation. It is clear that physics must be the same between two inertia systems, and any assumption which contradicts this basic principle cannot be justified at all.

Frame or Coordinate Transformation

Besides, this problem can be viewed differently in terms of Lagrangian. For the system under the uniform external gravity, one can write the corresponding Lagrangian. On the other hand, there is no way to construct any Lagrangian for the system that moves with a constant acceleration. One can define a Lagrangian for a particle that moves with a constant acceleration, but one cannot write the system (or space and time) that moves with a constant acceleration.

Physics in one inertia frame must be equivalent to that of another inertia frame, and this requirement is very severe. It is not only a coordinate change of space and time with Lorentz transformation, but also physical observables must be the same between two systems. In this respect, the principle of equivalence violates this important condition, and therefore, it is very hard to accept the assumption of the principle of equivalence even with the most modest physical intuition.

9.1.3. General Relativity

Einstein generalized the Poisson type equation for gravity

$$\nabla^2\phi_g = 4\pi G_0 \rho$$

to the tensor equations which should have some similarities with the Maxwell equations. Therefore, he had to find some tensor quantity like the field strength $F_{\mu\nu}$ of the electromagnetic field, and the metric tensor $g_{\mu\nu}$ is chosen as the basic tensor field since he started from the principle of equivalence. Thus, the general relativity is the equation for the metric tensor $g_{\mu\nu}$ which, he believed, should be connected to the gravitational field ϕ_g. By noting that

$$g_{00} \simeq 1 + \frac{2\phi_g}{c^2}$$

together with

$$T_{00} \simeq \rho$$

with the energy momentum tensor of $T_{\mu\nu}$, he arrived at the equation of the general relativity

$$R^{\mu\nu} - \frac{1}{2} g^{\mu\nu} R = 8\pi G_0 T^{\mu\nu},$$

where $R^{\mu\nu}$ denotes the Ricci tensor which can be described in terms of the metric tensor $g_{\mu\nu}$. However, the physical meaning of the $g_{\mu\nu}$ is unclear, and that is the basic problem of the general relativity.

Mathematics vs Physics

It should be important to note that Einstein's equation is mathematically complicated, but physically it is just simple. First, we should understand the physics of the Poisson type equation, and the equation is to determine the gravitational field ϕ_g when there is a matter field density ρ. Just in the same way, Einstein's equation can describe the behavior of the metric tensor $g_{\mu\nu}$ when there is the energy-momentum tensor $T_{\mu\nu}$ which should be generated by the matter field density ρ. However, the Poisson type equation is unfortunately insufficient to determine the gravitational interaction with the matter field since one has to know in which way the matter field should be influenced by the gravitational field, and this should be determined by the equation for the matter field like the Dirac equation. In the same way, Einstein's equation is insufficient to find the interaction with the matter field. In addition, the energy-momentum tensor $T_{\mu\nu}$ cannot be defined well unless one has a good field theoretical picture of fermions. In this respect, it is most important to find the Lagrangian density which includes the gravitational field interacting with the fermions, and this will be explained in the next section.

Before going to the discussion of the Lagrangian density of the gravity, it should be important to clarify the origin of the coordinate x_μ in the metric tensor $g_{\mu\nu}(x)$ from where it is measured. From Einstein's equation, it is clear that the origin of the coordinate should be found in the matter field center. Therefore, one can see that the metric tensor $g_{\mu\nu}$ should be in contradiction with the principle of special relativity since its space and time become different from the space and time of the other inertia system. This peculiar behavior of the metric tensor $g_{\mu\nu}$ is just the result of the principle of equivalence which is not consistent with the principle of special relativity as discussed above.

9.2. Lagrangian Density for Gravity

We should start from constructing the quantum mechanics of the gravitation. In other words, we should find the Dirac equation for electron when it moves in the gravitational potential. In this chapter, we present a model Lagrangian density which can describe electrons interacting with the electromagnetic field A_μ as well as the gravitational field $\mathcal{G}$.

9.2.1. Lagrangian Density for QED

We first write the Lagrangian density for electrons interacting with the electromagnetic field A_μ as given in eq.(5.1)

$$\mathcal{L}_{el} = i\bar{\psi}\gamma^\mu \partial_\mu \psi - e\bar{\psi}\gamma^\mu A_\mu \psi - m\bar{\psi}\psi - \frac{1}{4}F_{\mu\nu}F^{\mu\nu}, \tag{9.3}$$

where

$$F_{\mu\nu} = \partial_\mu A_\nu - \partial_\nu A_\mu .$$

This Lagrangian density of QED is best studied and is most reliable in many respects. In particular, the renormalization scheme of QED is theoretically well understood and is experimentally well examined, and there is no problem at all in the perturbative treatment of QED. All the physical observables can be described in terms of the free Fock space terminology after the renormalization, and therefore one can compare any prediction of the physical quantities with experiment. However, it should be noted that QED is the only field theory model in four dimensions which works perfectly well without any conceptual difficulties.

9.2.2. Lagrangian Density for QED Plus Gravity

Now, we propose to write the Lagrangian density for electrons interacting with the electromagnetic field as well as the gravitational field $\mathcal{G}$ [36]

$$\mathcal{L} = i\bar{\psi}\gamma^\mu \partial_\mu \psi - e\bar{\psi}\gamma^\mu A_\mu \psi - m(1+g\mathcal{G})\bar{\psi}\psi - \frac{1}{4}F_{\mu\nu}F^{\mu\nu} + \frac{1}{2}\partial_\mu \mathcal{G}\,\partial^\mu \mathcal{G}, \tag{9.4}$$

where the gravitational field $\mathcal{G}$ is assumed to be a massless scalar field. It is easy to prove that the new Lagrangian density is invariant under the local gauge transformation

$$A_\mu \to A_\mu + \partial_\mu \chi, \quad \psi \to e^{-ie\chi}\psi. \tag{9.5}$$

This is, of course, quite important since the introduction of the gravitational field does not change the most important local symmetry.

9.2.3. Dirac Equation with Gravitational Interactions

Now, one can easily obtain the Dirac equation for electrons from the new Lagrangian density

$$i\gamma^\mu \partial_\mu \psi - e\gamma^\mu A_\mu \psi - m(1 + g\mathcal{G})\psi = 0. \tag{9.6}$$

Also, one can write the equation of motion of gravitational field

$$\partial_\mu \partial^\mu \mathcal{G} = -mg\bar{\psi}\psi. \tag{9.7}$$

The symmetry property of the new Lagrangian density can be easily examined, and one can confirm that it has a right symmetry property under the time reversal transformation, parity transformation and the charge conjugation.

9.2.4. Total Hamiltonian for QED Plus Gravity

The Hamiltonian can be constructed from the Lagrangian density in eq.(9.4)

$$H = \int \left\{ \bar{\psi}\left(-i\boldsymbol{\gamma}\cdot\nabla + m(1+g\mathcal{G})\right)\psi - e\boldsymbol{j}\cdot\boldsymbol{A} \right\} d^3r + \frac{e^2}{8\pi}\int \frac{j_0(\boldsymbol{r}')j_0(\boldsymbol{r})\, d^3r\, d^3r'}{|\boldsymbol{r}'-\boldsymbol{r}|}$$

$$+\frac{1}{2}\int \left(\dot{\boldsymbol{A}}^2 + (\nabla\times\boldsymbol{A})^2\right) d^3r + \frac{1}{2}\int \left(\dot{\mathcal{G}}^2 + (\nabla\mathcal{G})^2\right) d^3r, \tag{9.8}$$

where j_μ is defined as $j_\mu = \bar{\psi}\gamma_\mu\psi$. In this expression of the Hamiltonian, the gravitational energy is still written without making use of the equation of motion. In the next section, we will treat the gravitational energy and rewrite it into an expression which should enable us to easily understand the structure of gravitational force between fermions.

9.3. Static-dominance Ansatz for Gravity

In eq.(9.4), the gravitational field $\mathcal{G}$ is introduced as a *real scalar* field, and therefore it cannot be a physical observable as a classical field [48, 80]. In this case, since the real part of the right hand side in eq.(9.7) should be mostly time independent, it may be reasonable to assume that the gravitational field $\mathcal{G}$ can be written as the sum of the static and time-dependent terms and that the static part should carry the information of diagonal term in the external source term. Thus, the gravitational field $\mathcal{G}$ is assumed to be written as

$$\mathcal{G} = \mathcal{G}_0(\boldsymbol{r}) + \bar{\mathcal{G}}(x), \tag{9.9}$$

where $\mathcal{G}_0(\boldsymbol{r})$ does not depend on time. This ansatz is only a sufficient condition, and its validity cannot be verified mathematically, but it can be examined experimentally.

The equations of motion for $\mathcal{G}_0(\boldsymbol{r})$ and $\overline{\mathcal{G}}(x)$ become

$$\boldsymbol{\nabla}^2 \mathcal{G}_0 = mg\rho_g, \tag{9.10}$$

$$\partial_\mu \partial^\mu \overline{\mathcal{G}}(x) = -mg\Big\{ (\bar{\psi}\psi)_{[\text{non-diagonal}]} + (\bar{\psi}\psi)_{[\text{diagonal rest}]} \Big\}, \tag{9.11}$$

where ρ_g is defined as

$$\rho_g \equiv (\bar{\psi}\psi)_{[\text{diagonal}]}, \tag{9.12}$$

where $(\bar{\psi}\psi)_{[\text{diagonal}]}$ denotes the diagonal part of the $\bar{\psi}\psi$, that is, the terms proportional to $[a_{\boldsymbol{k}}^{\dagger(s)} a_{\boldsymbol{k}'}^{(s)} - b_{\boldsymbol{k}}^{\dagger(s)} b_{\boldsymbol{k}'}^{(s)}]$ of the fermion operators which will be defined in eq.(9.19). Further, $(\bar{\psi}\psi)_{[\text{non-diagonal}]}$ term is a non-diagonal part which is connected to the creation and annihilation of fermion pairs, that is, $[a_{\boldsymbol{k}}^{\dagger(s)} b_{-\boldsymbol{k}'}^{\dagger(s)} + b_{-\boldsymbol{k}'}^{(s)} a_{\boldsymbol{k}}^{(s)}]$ of the fermion operators. In addition, the term $(\bar{\psi}\psi)_{[\text{diagonal rest}]}$ denotes time dependent parts of the diagonal term in the fermion density, and this may also have some effects when the gravity is quantized.

In this case, we can solve eq.(9.10) exactly and find a solution

$$\mathcal{G}_0(\boldsymbol{r}) = -\frac{mg}{4\pi} \int \frac{\rho_g(\boldsymbol{r}')}{|\boldsymbol{r}' - \boldsymbol{r}|} d^3r', \tag{9.13}$$

which is a special solution that satisfies eq.(9.7), but not the general solution. Clearly as long as the solution can satisfy the equation of motion of eq.(9.7), it is physically sufficient. The solution of eq.(9.13) is quite important for the gravitational interaction since this is practically a dominant gravitational force in nature.

Here, we assume that the diagonal term of $(\bar{\psi}\psi)_{[\text{diagonal}]}$ is mostly time independent, and in this case, the static gravitational energy which we call H_G^S can be written as

$$H_G^S = mg \int \rho_g \mathcal{G}_0 \, d^3r + \frac{1}{2} \int (\boldsymbol{\nabla} \mathcal{G}_0)^2 d^3r = -\frac{m^2 G_0}{2} \int \frac{\rho_g(\boldsymbol{r}')\rho_g(\boldsymbol{r})}{|\boldsymbol{r}' - \boldsymbol{r}|} d^3r \, d^3r', \tag{9.14}$$

where the gravitational constant G_0 is related to the coupling constant g as

$$G_0 = \frac{g^2}{4\pi}. \tag{9.15}$$

Eq.(9.14) is just the gravitational interaction energy for the matter fields, and one sees that the gravitational interaction between electrons is always attractive. This is clear since the gravitational field is assumed to be a massless scalar. It may also be important to note that the H_G^S of eq.(9.14) is obtained without making use of the perturbation theory, and it is indeed exact, apart from the static ansatz of the field $\mathcal{G}_0(\boldsymbol{r})$.

9.4. Quantization of Gravitational Field

In quantum field theory, we should quantize fields. For fermion fields, we should quantize the Dirac field by the anti-commutation relations of fermion operators. This is required from the experiment in terms of the Pauli principle, that is, a fermion can occupy only one quantum state. In order to accommodate this experimental fact, we should always quantize the fermion fields with the anti-commutation relations. On the other hand, for gauge fields, we must quantize the vector field in terms of the commutation relation which is also required from the experimental observation that one photon is emitted by the transition between $2p$-state and $1s$-state in hydrogen atoms. That is, a photon is created from the vacuum of the electromagnetic field, and therefore the field quantization is an absolutely necessary procedure. However, it is not very clear whether the gravitational field $\mathcal{G}$ should be quantized according to the bosonic commutation relation or not. In fact, there must be two choices concerning the quantization of the gravitational field $\mathcal{G}$.

9.4.1. No Quantization of Gravitational Field

As the first choice, we may take a standpoint that the gravitational field $\mathcal{G}$ should not be quantized since there is no requirement from experiments. In this sense, there is no definite reason that we have to quantize the scalar field and therefore the gravitational field $\mathcal{G}$ should remain to be a classical field. In this case, we do not have to worry about the renormalization of the graviton propagator, and we obtain the gravitational interaction between fermions as we saw it in eq.(9.14) which is always attractive, and this is consistent with the experimental requirement.

9.4.2. Quantization Procedure

Now, we take the second choice and should quantize the gravitational field $\bar{\mathcal{G}}$. This can be done just in the same way as usual scalar fields

$$\bar{\mathcal{G}}(x) = \sum_{\boldsymbol{k}} \frac{1}{\sqrt{2V\omega_k}} \left[d_k e^{-i\omega_k t + i\boldsymbol{k}\cdot\boldsymbol{r}} + d_k^\dagger e^{i\omega_k t - i\boldsymbol{k}\cdot\boldsymbol{r}} \right], \tag{9.16}$$

where $\omega_k = |\boldsymbol{k}|$. The annihilation and creation operators d_k and $d_k^\dagger$ are assumed to satisfy the following commutation relations

$$\left[d_{\boldsymbol{k}}, d_{\boldsymbol{k}'}^\dagger \right] = \delta_{\boldsymbol{k},\boldsymbol{k}'} \tag{9.17}$$

and all other commutation relations should vanish. Since the graviton can couple to the time dependent external field which is connected to the creation or annihilation of the fermion pairs, the graviton propagator should be affected from the vacuum polarization of fermions. Therefore, we should carry out the renormalization procedure of the graviton propagator such that it can stay massless. We will discuss the renormalization procedure in the later section.

9.4.3. Graviton

Once the gravitational field $\mathcal{G}$ is quantized, then the graviton should appear. From eq.(9.16), one can see that the graviton can indeed propagate as a free massless particle after it is quantized, and this situation is just the same as the gauge field case in QED, namely, photon after the quantization becomes a physical observable. However, it should be noted that the gauge field has a special feature in the sense that the classical gauge field ($\boldsymbol{A}$) is gauge dependent and therefore it is not a physical observable. After the gauge fixing, the gauge field can be quantized since one can uniquely determine the gauge field from the equation of motion, and therefore its quantization is possible.

On the other hand, the gravitational field is assumed to be a real scalar field, and therefore it cannot be a physical observable as a classical field [48, 80]. Only after the quantization, it becomes a physical observable as a graviton, and this can be seen from eq.(9.16) since the creation of the graviton should be made through the second term of eq.(9.16). In this case, the graviton field is a complex field which is an eigenstate of the momentum and thus it is a free graviton state, which can propagate as a free particle.

9.5. Interaction of Photon with Gravity

From the Lagrangian density of eq.(9.4), one sees that photon should interact with the gravity in the fourth order Feynman diagrams as shown in Fig.9.1. The interaction Hamiltonian H_I can be written as

$$H_I = \int (mg\mathcal{G}\,\bar{\psi}\psi - e\bar{\psi}\boldsymbol{\gamma}\psi\cdot\boldsymbol{A})\,d^3r, \tag{9.18}$$

where the fermion field ψ is quantized in the normal way

$$\psi(\boldsymbol{r},t) = \sum_{\boldsymbol{p},s} \frac{1}{\sqrt{L^3}} \left(a_{\boldsymbol{p}}^{(s)} u_{\boldsymbol{p}}^{(s)} e^{i\boldsymbol{p}\cdot\boldsymbol{r}-iE_{\boldsymbol{p}}t} + b_{\boldsymbol{p}}^{\dagger(s)} v_{\boldsymbol{p}}^{(s)} e^{-i\boldsymbol{p}\cdot\boldsymbol{r}+iE_{\boldsymbol{p}}t} \right), \tag{9.19}$$

where $u_{\boldsymbol{p}}^{(s)}$ and $v_{\boldsymbol{p}}^{(s)}$ denote the spinor part of the plane wave solutions of the free Dirac equation. $a_{\boldsymbol{p}}^{(s)}$ and $b_{\boldsymbol{p}}^{(s)}$ are annihilation operators for particle and anti-particle states, and they should satisfy the following anti-commutation relations,

$$\{a_{\boldsymbol{p}}^{(s)}, a^{\dagger}{}_{\boldsymbol{p}'}^{(s')}\} = \delta_{s,s'}\delta_{\boldsymbol{p},\boldsymbol{p}'}, \quad \{b_{\boldsymbol{p}}^{(s)}, b^{\dagger}{}_{\boldsymbol{p}'}^{(s')}\} = \delta_{s,s'}\delta_{\boldsymbol{p},\boldsymbol{p}'} \tag{9.20}$$

and all other anticommutation relations should vanish. The gauge field $\boldsymbol{A}$ can be quantized as given in eq.(3.24) in Chapter 3

$$\boldsymbol{A}(x) = \sum_{\boldsymbol{k}} \sum_{\lambda=1}^{2} \frac{1}{\sqrt{2V\omega_{\boldsymbol{k}}}} \boldsymbol{\varepsilon}^{\lambda}(\boldsymbol{k}) \left[c_{\boldsymbol{k},\lambda} e^{-ikx} + c_{\boldsymbol{k},\lambda}^{\dagger} e^{ikx} \right], \tag{9.21}$$

where $\omega_{\boldsymbol{k}} = |\boldsymbol{k}|$. The polarization vector $\boldsymbol{\varepsilon}^{\lambda}(\boldsymbol{k})$ should satisfy the following relations

$$\boldsymbol{\varepsilon}^{\lambda}(\boldsymbol{k})\cdot\boldsymbol{k} = 0, \quad \boldsymbol{\varepsilon}^{\lambda}(\boldsymbol{k})\cdot\boldsymbol{\varepsilon}^{\lambda'}(\boldsymbol{k}) = \delta_{\lambda,\lambda'}. \tag{9.22}$$

The annihilation and creation operators $c_{\boldsymbol{k},\lambda}$, $c_{\boldsymbol{k},\lambda}^{\dagger}$ should satisfy the following commutation relations

$$[c_{\boldsymbol{k},\lambda}, c_{\boldsymbol{k}',\lambda'}^{\dagger}] = \delta_{\boldsymbol{k},\boldsymbol{k}'}\delta_{\lambda,\lambda'} \tag{9.23}$$

and all other commutation relations should vanish.

The calculation of the S-matrix can be carried out in a straightforward way [14, 89, 94], and we can write

$$S = (ie)^2 \varepsilon_{\mu}^{\lambda}(k)\varepsilon_{\nu}^{\lambda'}(k') \left(\frac{mm'g^2}{q^2} \right) \bar{u}(p')u(p)$$

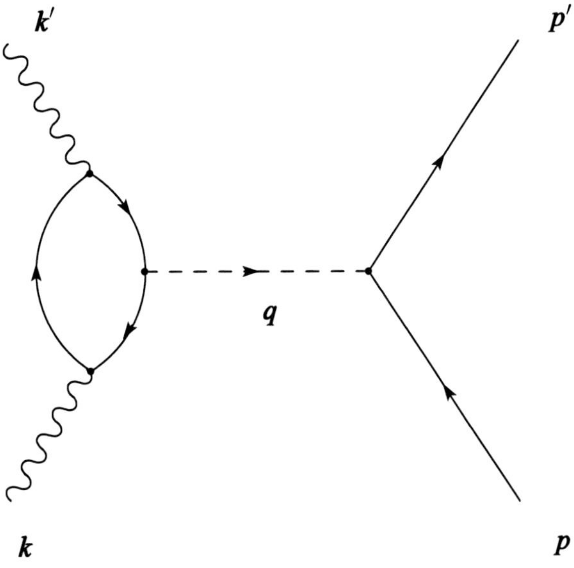

Figure 9.1. The fourth order Feynman diagram.

$$\times \int \frac{d^4a}{(2\pi)^4} \mathrm{Tr}\left[\gamma_\mu \frac{i}{\not{a}-m+i\varepsilon}\gamma_\nu \frac{i}{\not{b}-m+i\varepsilon}\frac{i}{\not{c}-m+i\varepsilon}\right], \tag{9.24}$$

where k and k' denote the four momenta of the initial and final photons while p and p' denote the four momenta of the initial and final fermions, respectively. m and m' denote the mass of the fermion for the vacuum polarization and the mass of the external fermion. a, b, c and q can be written in terms of k and p as

$$q = p' - p,\ k = a - b,\ k' = a - c,\ q = k - k'.$$

Therefore, the S-matrix can be written as

$$S = ie^2 mm' g^2 \varepsilon_\mu^\lambda(k)\varepsilon_\nu^{\lambda'}(k')$$

$$\times \frac{1}{q^2}\bar{u}(p')u(p)\int \frac{d^4a}{(2\pi)^4}\frac{1}{a^2-m^2}\frac{1}{(a-k)^2-m^2}\frac{1}{(a-k')^2-m^2}$$

$$\times \mathrm{Tr}\left[\gamma_\mu(\not{a}+m)\gamma_\nu(\not{a}-\not{k}+m)(\not{a}-\not{k}'+m)\right]. \tag{9.25}$$

Since the term proportional to q does not contribute to the interaction, we can safely approximate in the evaluation of the trace and the integration of a as

$$k' \approx k.$$

Now, we define the trace part as

$$N_{\mu\nu} = \mathrm{Tr}\Big[\gamma_\mu(\not{a}+m)\gamma_\nu(\not{a}-\not{k}+m)(\not{a}-\not{k}'+m)\Big], \tag{9.26a}$$

which can be evaluated as

$$N_{\mu\nu} = 4m\Big[(k^2-a^2+m^2)g_{\mu\nu} + 4a_\mu a_\nu - 2a_\mu k_\nu - 2a_\nu k_\mu\Big]. \tag{9.26b}$$

Defining the integral by

$$I_{\mu\nu} \equiv \int \frac{d^4a}{(2\pi)^4} \frac{N_{\mu\nu}}{(a^2-m^2)[(a-k)^2-m^2][(a-k')^2-m^2]} \tag{9.27}$$

we can rewrite it using Feynman integral

$$I_{\mu\nu} = 2\int \frac{d^4a}{(2\pi)^4} \int_0^1 z\,dz \frac{N_{\mu\nu}}{[(a-kz)^2-m^2+z(1-z)k^2]^3}. \tag{9.28}$$

Therefore, introducing the variable $w = a - kz$ we obtain the S-matrix as

$$S = 8ie^2m^2m'g^2\varepsilon_\mu^\lambda(k)\varepsilon_\nu^{\lambda'}(k')\frac{1}{q^2}\bar{u}(p')u(p)\int_0^1 z\,dz$$

$$\times \int \frac{d^4w}{(2\pi)^4}\left[\frac{(-w^2g_{\mu\nu}+4w_\mu w_\nu)}{[w^2-m^2+z(1-z)k^2]^3} + \frac{\{m^2+k^2(1-z^2)\}g_{\mu\nu}-4k_\mu k_\nu z(1-z)}{[w^2-m^2+z(1-z)k^2]^3}\right]. \tag{9.29}$$

The first part of the integration can be carried out in a straightforward way and we find

$$\int \frac{d^4w}{(2\pi)^4}\frac{(-w^2g_{\mu\nu}+4w_\mu w_\nu)}{[w^2-m^2+z(1-z)k^2]^3} = 0.$$

Thus, the two divergent parts just cancel with each other, and the cancellation here is not due to the regularization as employed in the self-energy diagrams in QED, but it is a kinematical and thus rigorous result. This situation is quite similar to the Feynman diagram of $\pi^0 \to 2\gamma$ decay process [94] and the calculated result of the Feynman diagram is indeed finite and is consistent with the experiment.

The finite part can be easily evaluated [36], and therefore we obtain the S-matrix as

$$S = \frac{e^2}{8\pi^2}m^2m'g^2(\varepsilon^\lambda\varepsilon^{\lambda'})\frac{1}{q^2}\bar{u}(p')u(p), \tag{9.30}$$

where we made use of the relation $k^2 = 0$ for free photon at the end of the calculation.

9.6. Renormalization Scheme for Gravity

At the present stage, it is difficult to judge whether we should quantize the gravitational field or not. At least, there is no experiment which shows any necessity of the quantization of the gravity. Nevertheless, it should be worth checking whether the gravitational interaction with fermions can be renormalizable or not. We know that the interaction of the gravity with fermions is extremely small, but we need to examine whether the graviton can stay massless or not within the perturbation scheme.

Here, we present a renormalization scheme for the scalar field theory which couples to fermion fields. The renormalization scheme for scalar fields is formulated just in the same way as the QED scheme since QED is most successful.

9.6.1. Self-Energy of Graviton

First, we discuss the self-energy of graviton in gravitational interaction. As shown in Appendix J, the self-energy diagram of photon should not be considered for the renormalization procedure in QED. In the same manner, we see that there is no renormalization procedure necessary for the self-energy of graviton. Intuitively, this can be easily understood from eq.(9.7)

$$\partial_\mu \partial^\mu \mathcal{G} = -mg\bar{\psi}\psi. \tag{9.7}$$

As can be seen, the gravitational field $\mathcal{G}$ does not appear in the right hand side of eq.(9.7). This means that the change of the gravitational field from the second order perturbative calculations should be described by the fermion fields and cannot be written in terms of the gravitational field $\mathcal{G}$.

Therefore, the gravitational interaction is not affected from the graviton self-energy diagram. In the S-matrix evaluation, one sometimes finds that the calculated Feynman diagrams do not have any corresponding physical processes and the self-energy diagram of the graviton is just the case. Thus the graviton stays always massless.

Even though there is no practically interesting observables in higher order perturbation diagrams in the gravitational interaction, it is quite nice and transparent that the graviton propagator is not affected by the perturbation theory.

9.6.2. Fermion Self-Energy from Gravity

The fermion self-energy term in QED is calculated to be

$$\begin{aligned}\Sigma_{QED}(p) &= -ie^2 \int \frac{d^4k}{(2\pi)^4} \gamma_\mu \frac{1}{\not{p} - \not{k} - m} \gamma^\mu \frac{1}{k^2} \\ &= \frac{e^2}{8\pi^2} \ln\left(\frac{\Lambda}{m}\right)(-\not{p} + 4m) + \text{finite terms.}\end{aligned} \tag{9.31}$$

In the same way, we can calculate the fermion self-energy due to the gravity

$$\begin{aligned}\Sigma_G(p) &= im^2g^2 \int \frac{d^4k}{(2\pi)^4} \frac{1}{\not{p} - \not{k} - m} \frac{1}{k^2} \\ &= -\frac{m^2g^2}{8\pi^2} \ln\left(\frac{\Lambda}{m}\right)(-\not{p} + 4m) + \text{finite terms},\end{aligned} \tag{9.32}$$

which is just the same as the QED case, apart from the factor in front. Therefore, the renormalization procedure can be carried out just in the same way as the QED case since the total fermion self-energy term within the present model becomes

$$\Sigma(p) = \frac{1}{8\pi^2} \ln\left(\frac{\Lambda}{m}\right)(e^2 - m^2g^2)(-\not{p} + 4m) + \text{finite terms.} \tag{9.33}$$

9.6.3. Vertex Correction from Gravity

Concerning the vertex corrections which arise from the gravitational interaction and electromagnetic interaction with fermions, it may well be that the vertex corrections do not become physically very important. It is obviously too small to measure any effects of the higher order terms from the gravity and electromagnetic interactions. However, we should examine the renormalizability of the vertex corrections and can show that they are indeed well renormalized into the wave function. The vertex corrections from the electromagnetic interaction and the gravity can be evaluated as

$$\Lambda_{QED}(k,q) = imge^2 \int \frac{d^4p}{(2\pi)^4} \left[\gamma_\mu \frac{1}{(\not{k}-\not{p}-m)(\not{k}-\not{p}-\not{q}-m)p^2}\gamma^\mu\right], \tag{9.34a}$$

$$\Lambda_G(k,q) = -im^3g^3 \int \frac{d^4p}{(2\pi)^4} \left[\frac{1}{(\not{k}-\not{p}-m)(\not{k}-\not{p}-\not{q}-m)p^2}\right]. \tag{9.34b}$$

We can easily calculate the integrations and obtain the total vertex corrections for the zero momentum case of $q = 0$ as

$$\Lambda(k,0) = \Lambda_{QED}(k,0) + \Lambda_G(k,0) = \frac{mg}{\pi^2} \ln\left(\frac{\Lambda}{m}\right)(e^2 - m^2g^2) + \text{finite terms}, \tag{9.35}$$

which is logarithmic divergence and is indeed renormalizable just in the same way as the QED case.

9.6.4. Renormalization Procedure

Since the infinite contributions to the fermion self-energy and to the vertex corrections in the second order diagrams are just the same as the QED case, one can carry out the renormalization procedure just in the same way as the QED case. In this way, we can achieve a successful renormalization scheme for the gravity, even though we do not know any occasions in which the higher order contributions may become physically important.

9.7. Gravitational Interaction of Photon with Matter

From eq.(9.30), one finds that the gravitational potential $V(r)$ for photon with matter field can be written as

$$V(r) = -\frac{G_0\alpha m_t^2 M}{2\pi}\frac{1}{r}, \tag{9.36}$$

where m_t and M denote the sum of all the fermion masses and the mass of matter field, respectively. α denotes the fine structure constant $\alpha = \frac{1}{137}$. In this case, the equation of motion for photon $\boldsymbol{A}_\lambda$ under the gravitational field becomes

$$\left(\frac{\partial^2}{\partial t^2} - \nabla^2 - \frac{G_0\alpha m_t^2 M}{2\pi}\frac{1}{r}\right)\boldsymbol{A}_\lambda = 0. \tag{9.37}$$

Assuming the time dependence of the photon field $\boldsymbol{A}_\lambda$ as

$$\boldsymbol{A}_\lambda = \boldsymbol{\varepsilon}_\lambda e^{-i\omega t} A_0(\boldsymbol{r}) \tag{9.38}$$

we obtain

$$\left(-\nabla^2 - \frac{G_0\alpha m_t^2 M}{2\pi}\frac{1}{r}\right)A_0(\boldsymbol{r}) = \omega^2 A_0(\boldsymbol{r}). \tag{9.39}$$

This equation shows that there is no bound state for photon even for the strong coupling limit of $G_0 \to \infty$.

9.7.1. Photon-Gravity Scattering Process

Eq.(9.39) can be written with $|\boldsymbol{k}| = \omega$ as

$$\left(\nabla^2 + \boldsymbol{k}^2\right) A_0(\boldsymbol{r}) = -\frac{G_0 \alpha m_t^2 M}{2\pi} \frac{1}{r} A_0(\boldsymbol{r}), \tag{9.40}$$

which is just the same as the scattering process of a particle under the Coulomb interaction. When the non-relativistic particle with its mass m_0 and momentum $\boldsymbol{k}$ scatters elastically with a point nucleus with its charge Z, the Schrödinger equation becomes

$$\left(\nabla^2 + \boldsymbol{k}^2\right) \psi(\boldsymbol{r}) = -\frac{2m_0 Z e^2}{r} \psi(\boldsymbol{r}). \tag{9.41}$$

The solution of eq.(9.41) is well studied and therefore we can make use of this equation to solve eq.(9.39). In this case, we can obtain the differential cross section of the photon-gravity scattering process

$$\frac{d\sigma}{d\Omega} = \frac{\alpha_g^2}{16\omega^4 \sin^4 \frac{\theta}{2}}, \tag{9.42}$$

where α_g is defined as

$$\alpha_g = \frac{G_0 \alpha m_t^2 M}{2\pi}. \tag{9.43}$$

This differential cross section is just the same as the Rutherford cross section.

9.8. Cosmology

What should be a possible picture of our universe in the new quantum theory of gravity? By now we have sufficient knowledge concerning the cosmology how the present universe is created and what should be a fate of the present universe. Below is a simple picture one can easily draw, even though it is almost a story. In order to make it into physics, hard works may be required, though it must be a doable task.

9.8.1. Cosmic Fireball Formation

Since the gravity is always attractive, it is clear that all of the galaxies should eventually get together. A question may arise in which way these galaxies would collapse into a Cosmic Fireball. It is most likely true that, after the end of the expansion of the present universe, a few galaxies should coalesce into a larger galaxy, and this coalescence should take place repeatedly until two or three giant clusters of galaxies should be formed. Finally, these giant clusters would eventually collide into a Cosmic Fireball which should be quite similar to the initial stage of the big bang. After the Cosmic Fireball is created, it should rapidly expand, and during the expansion, light nuclei should be created. In this picture, galaxies should be naturally formed since the expansion after the explosion should not be very uniform. This is in contrast to the big bang cosmology in which the galaxy formation must be quite difficult since the big bang should be extremely uniform.

In this respect, the universe should repeatedly make the same formation of galaxies. The universe should have existed from the infinite time of past, and it should make the galaxy formation and collision in the infinite time of future.

Here, it should be noted that the concept of the infinite time of past or future is beyond the understanding of human being. Also, the whole universe should have infinite space, but again the infinite space should not be a target of physics research.

9.8.2. Relics of Preceding Universe

According to the present picture of the universe, there may well be some relics of the preceding universe before the Cosmic Fireball.

Large Scale Structure of Universe

In the present universe, there is a large scale structure of the universe among cluster of galaxies such as the Great Attractor. This should be related to the remnants of the Cosmic Fireball formation when the preceding universe got together into the Cosmic Fireball.

Photon Baryon Ratio

Another possible relic must be the large number of photons compared to the number of baryons in the present universe. This photon-baryon ratio may well be understood in terms of the relic of photons in the preceding universe since photon has some interactions with strong gravitational fields and therefore some of photons may be trapped during the Cosmic Fireball formation. On the other hand, neutrino should not be trapped due to the lack of the interactions with baryons. Therefore, the number of neutrinos must be much smaller than the number of photons in the present universe.

9.8.3. Remarks

The gravitational interaction appears always as the mass term and induces always the attractive force between fermions. In addition, there is an interaction between photon and the gravity as the fourth order Feynman diagrams. The behavior of photon under the gravitational field may have some similarity with the result of the general relativity.

The renormalization procedure of the gravitational interaction is carried out in the same way as the QED case, and therefore the propagator of the graviton stays massless, which is just the same as the QED case in which photon stays always massless. This can be easily understood from the observation that the self-energy of photon as well as graviton should not be considered for the renormalization scheme.

Here, it is still an open question whether the gravitational field should be quantized or not. This is basically because there is no definite requirement from experiment for the quantization. For the quantized theory of gravitational field, one may ask as to whether there is any method to observe a graviton or not. The graviton should be created through the fermion pair annihilation. Since this graviton can propagate as a free graviton like a photon, one may certainly have some chance to observe it through the creation of the fermion pair. But this probability must be extremely small since the coupling constant is very small, and there is no enhancement in this process unless a strong gravitational field like a neutron star may rapidly change as a function of time.

9.9. Time Shifts of Mercury and Earth Motions

The new gravity model is applied to the description of the observed advance shifts of the Mercury perihelion, the earth rotation and the GPS satellite motion. First, we obtain the gravitational potential which can be calculated from the non-relativistic reduction of the Dirac equation in terms of the Foldy-Wouthuysen transformation. Then, we should make the classical limit of the Hamiltonian so that we can obtain the classical potential for the gravity.

9.9.1. Non-relativistic Gravitational Potential

The Hamiltonian of the Dirac equation in the gravitational field can be written as

$$H = -i\boldsymbol{\nabla}\cdot\boldsymbol{\alpha} + \left(m - \frac{GmM}{r}\right)\beta, \tag{9.44}$$

where M denotes the mass of the gravity center. This Hamiltonian can be easily reduced to the non-relativistic equation of motion by making use of the Foldy-Wouthuysen transformation [14]. Here, we only write the result in terms of the Hamiltonian H [46]

$$H = m + \frac{\boldsymbol{p}^2}{2m} - \frac{GmM}{r} + \frac{1}{2m^2}\frac{GmM}{r}\boldsymbol{p}^2 - \frac{1}{2m^2}\frac{GMm}{r^3}(\boldsymbol{s}\cdot\boldsymbol{L}), \tag{9.45}$$

where the last term denotes the spin-orbit force, but we do not consider it here. Now, we make the classical limit to derive the Newton equation. In this case, it is safe to assume the factorization ansatz for the third term, that is,

$$\left\langle \frac{1}{2m^2}\frac{GmM}{r}\boldsymbol{p}^2 \right\rangle = \left\langle \frac{1}{2m^2}\frac{GmM}{r} \right\rangle \langle \boldsymbol{p}^2 \rangle . \tag{9.46}$$

By making use of the Virial theorem for the gravitational potential

$$\left\langle \frac{\boldsymbol{p}^2}{m} \right\rangle = \left\langle \frac{GmM}{r} \right\rangle \tag{9.47}$$

we obtain the new gravitational potential for the Newton equation

$$V(r) = -\frac{GmM}{r} + \frac{1}{2mc^2}\left(\frac{GmM}{r}\right)^2 , \tag{9.48}$$

where we explicitly write the light velocity c in the last term of the equation.

9.9.2. Time Shifts of Mercury, GPS Satellite and Earth

The Newton equation with the new gravitational potential can be written as

$$m\ddot{r} = -\frac{GmM}{r^2} + \frac{\ell^2}{mr^3} + \frac{G^2M^2m}{c^2r^3}. \tag{9.49}$$

Therefore, we can introduce a new angular momentum L as

$$L^2 \equiv \ell^2 + \frac{G^2M^2m^2}{c^2}. \tag{9.50}$$

Further, we define the angular velocity ω and radius R by

$$\omega \equiv \frac{\ell}{mR^2}, \quad R \equiv \frac{\ell^2}{GMm^2(1-\varepsilon^2)^{\frac{3}{4}}}, \tag{9.51a}$$

where ε denotes the eccentricity. Correspondingly, we can define a new angular velocity Ω associated with ω as

$$\Omega^2 \equiv \omega^2 + \frac{G^2M^2}{c^2R^4} = \omega^2(1+\eta), \tag{9.51b}$$

where η is defined as

$$\eta = \frac{G^2M^2}{c^2R^4\omega^2}. \tag{9.52}$$

The equation (9.49) can be immediately solved, and one finds the solution of the orbit

$$r = \frac{A}{1+\varepsilon\cos\left(\frac{L}{\ell}\varphi\right)}, \tag{9.53}$$

where A and ε are given as

$$A = \frac{L^2}{GMm^2}, \quad \varepsilon = \sqrt{1+\frac{2L^2E}{m(GmM)^2}}. \tag{9.54}$$

Physical observables can be obtained by integrating $\dot{\varphi} = \frac{\ell}{mr^2}$ over the period T

$$\frac{\ell}{m}\int_0^T dt = \int_0^{2\pi} r^2 d\varphi = A^2\int_0^{2\pi} \frac{1}{\left(1+\varepsilon\cos\left(\frac{L}{\ell}\varphi\right)\right)^2} d\varphi. \tag{9.55}$$

This can be easily calculated to be

$$\omega T = 2\pi(1+2\eta)(1-\varepsilon\eta) \simeq 2\pi\{1+(2-\varepsilon)\eta\}, \tag{9.56}$$

where ε is assumed to be small. Therefore, the new gravity potential gives rise to the advance shift of the time shift, and it can be written as

$$\left(\frac{\Delta T}{T}\right)_{th} \simeq (2-\varepsilon)\eta. \tag{9.57}$$

This is a physical observable which indeed can be compared to experiment.

9.9.3. Mercury Perihelion Shift

The Mercury perihelion advance shift $\Delta\theta$ is well known to be [92]

$$\Delta\theta \simeq 43\,'' \text{ per } 100 \text{ year.} \tag{9.58}$$

Since Mercury has the 0.24 year period, it can amount to the shift ratio $\delta\theta$

$$\delta\theta_{obs} \equiv \left(\frac{\Delta T}{T}\right)_{obs} \simeq 8.0 \times 10^{-8}. \tag{9.59}$$

The theoretical calculation of the new gravity model shows

$$\eta = \frac{G^2 M^2}{c^2 R^4 \omega^2} \simeq 2.65 \times 10^{-8} \tag{9.60}$$

where the following values are used for the Mercury case

$$R = 5.73 \times 10^{10} \text{ m}, \quad M = 1.989 \times 10^{30} \text{ kg}, \quad \omega = 8.30 \times 10^{-7}.$$

Therefore, the theoretical shift ratio $\delta\theta_{th}$ becomes

$$\delta\theta_{th} \equiv \left(\frac{\Delta T}{T}\right)_{th} \simeq 4.8 \times 10^{-8} \tag{9.61}$$

which should be compared to the observed value in eq.(9.59). As can be seen, this agreement is indeed remarkable since there is no free parameter in the theoretical calculation.

9.9.4. GPS Satellite Advance Shift

Many GPS satellites which are orbiting around the earth should be influenced rather heavily by the new gravitational potential. The GPS satellite advance shift can be estimated just in the same way as above, and we obtain

$$\eta = \frac{G^2 M^2}{c^2 R^4 \omega^2} \simeq 1.69 \times 10^{-10} \tag{9.62}$$

where we employ the following values for the GPS satellite [8, 100]

$$R = 2.6561 \times 10^{7} \text{ m}, \quad M = 5.974 \times 10^{24} \text{ kg}, \quad \omega = 1.4544 \times 10^{-4} \tag{9.63}$$

since the satellite circulates twice per day. Therefore, the advance shift of the GPS satellite becomes

$$\left(\frac{\Delta T}{T}\right)_{th} \simeq 3.4 \times 10^{-10}. \tag{9.64}$$

This should be compared to the observed value of

$$\left(\frac{\Delta T}{T}\right)_{exp} \simeq 4.5 \times 10^{-10}. \tag{9.65}$$

As seen from the comparison between the calculation and the observed value, the new gravity theory can indeed achieve a remarkable agreement with experiment.

9.9.5. Time Shift of Earth Rotation – Leap Second

Here, we calculate the time shift of the earth rotation around the sun [47]. First, we evaluate the η

$$\eta = \frac{G^2M^2}{c^2R^4\omega^2} \simeq 0.992 \times 10^{-8} \tag{9.66}$$

where we employ the following values for R, M and ω

$$R = 1.496 \times 10^{11} \text{ m}, \; M = 1.989 \times 10^{30} \text{ kg}, \; \omega = 1.991 \times 10^{-7}. \tag{9.67}$$

In this case, we find the time shift for one year

$$(\Delta T)_{th} \simeq 0.621 \;\; s/year \tag{9.68}$$

where $\varepsilon = 0.0167$ is taken. In fact, people have been making corrections for the leap second, and according to the data, they made the first leap second correction in June of 1972. After that, they have made the leap second corrections from December 1972 to December 2008. The total corrections amount to 23 seconds for 36.5 years since we should start from June 1972. This corresponds to the time shift per year

$$(\Delta T)_{exp} \simeq 0.63 \pm 0.02 \;\; s/year \tag{9.69}$$

where the errors are supposed to come from one year shift of the observation. This agrees surprisingly well with the theoretical time shift of the earth.

9.9.6. Observables from General Relativity

Now, we discuss the calculated results by the general relativity [28, 92]. For the Mercury perihelion shift, the result is quite well known, and it can be written in terms of the angular shift. In fact, the angular variable φ is modified by the general relativity to

$$\cos\varphi \longrightarrow \cos(1-\gamma)\varphi \tag{9.70}$$

where γ is found to be

$$\gamma = \frac{3G^2M^2}{c^2R^4\omega^2}. \tag{9.71}$$

This change of the shift in the angular variable could explain the observed Mercury perihelion shift. However, as can be seen from eq.(9.53), this effect vanishes to zero in the case of $\varepsilon = 0$, that is, for the circular orbit. This is, of course, unphysical in that the effect of the general relativity is valid only for the elliptic orbit case. In Newton dynamics, the angular momentum ℓ is the only quantity which can be affected from the external effects like the general relativity or the additional potential.

9.9.7. Prediction from General Relativity

Now, we should calculate the physical observables as to how the general relativity can induce the perihelion shift. In this case, one finds that the change appears in eq.(9.53) as

$$r = \frac{A}{1+\varepsilon\cos\left(\frac{L}{\ell}\varphi\right)} \quad \Rightarrow \quad r = \frac{A}{1+\varepsilon\cos\left((1-\gamma)\varphi\right)} \tag{9.72}$$

and thus the physical observable becomes

$$\omega T \simeq 2\pi(1+2\varepsilon\gamma). \tag{9.73}$$

Therefore, the advance shift of the Mercury perihelion becomes

$$\delta\theta_{th} \equiv \left(\frac{\Delta T}{T}\right)_{th} \simeq 3.3\times 10^{-8} \tag{9.74}$$

which is a factor of 2.5 smaller than the observed value of the Mercury perihelion shift. It should be noted that the predicted shift in eq.(9.72) is indeed the advance shift of the Mercury perihelion as given in eq.(9.73).

In addition, the GPS satellite shift predicted by the general relativity becomes

$$\left(\frac{\Delta T}{T}\right)_{th} \simeq 0.10\times 10^{-10} \tag{9.75}$$

which is very small. This is because the GPS satellite motion has almost the circular orbit around the earth.

Further, the time shift of the earth rotation around the sun predicted by the general relativity becomes

$$(\Delta T)_{th} \simeq 0.031 \ \ s/year. \tag{9.76}$$

This shows that it is much too small compared to the observed time shift of the earth rotation around the sun.

In reality, if the angular momentum is affected from the external potential as given in eq.(9.50), then not only the angular variable but also A in eq.(9.54) should be changed, and therefore as the total effects of the physical observables in the general relativity, eq.(9.73) is modified to

$$\omega T \simeq 2\pi\{1-2(2-\varepsilon)\gamma\} \tag{9.77}$$

which is, unfortunately, a retreat shift since ε is smaller than unity.

9.9.8. Summary of Comparisons between Calculations and Data

We summarize the calculated results of the Mercury perihelion shift, GPS satellite advance shift and Leap Second corrections due to the new gravity model as well as the general relativity. Here, the observed data are compared with the predictions of the model calculations in Table 9.1.

Table 9.1

	Mercury $(\Delta T/T)$	GPS $(\Delta T/T)$	Leap Second ΔT
Observed data	8.0×10^{-8}	4.5×10^{-10}	0.63 ± 0.02 $s/year$
New Gravity	4.8×10^{-8}	3.4×10^{-10}	0.62 $s/year$
General Relativity	3.3×10^{-8}	0.10×10^{-10}	0.031 $s/year$

Table 9.1 shows the calculated results of the Mercury perihelion shift, GPS satellite advance shift and Leap Second corrections together with the observed data. The New Gravity shows the prediction of the new gravity model calculations which are discussed in this paper. The General Relativity is the calculation in which we only consider the angular shift following Einstein. From this table, one sees that the general relativity cannot describe the observed data.

9.9.9. Intuitive Picture of Time Shifts

It may be interesting to note that the velocity of the Mercury or the earth around the sun is one of the fastest objects we can observe as a classical motion. This velocity v is around $v \sim 1.0 \times 10^{-4}\ c$, which leads to the correction of the relativistic effects in physical observables as

$$\left(\frac{v}{c}\right)^2 \sim 1.0 \times 10^{-8}$$

which is just the same magnitude as the values observed in the Mercury perihelion shift ($\Delta T/T \sim 5 \times 10^{-8}$) and the leap second corrections ($\Delta T/T \sim 2 \times 10^{-8}$). Therefore, it should not be surprising at all that the new additional gravitational potential which is obtained as the relativistic effects of the gravity potential in Dirac equation can account for the advance shifts of the planets orbiting around the sun.

In this sense, the physical effect of the earth rotation velocity on the perihelion shift can be compared to the Michelson-Morley experiment. The interesting point is that the Michelson-Morley experiment is essentially to examine the kinematical effect of the relativity that the light velocity is not influenced by the earth rotation velocity, even though the classical mechanics indicates it should be affected. On the other hand, the leap second correction is the relativistic effect of the dynamical motion of the earth rotation, and it is a deviation from the Newton mechanics. Both of the observed facts can be understood by the relativistic effects of the earth motion around the sun, and in fact, the Michelson-Morley experiment proves that the light velocity is independent of the speed of the earth rotation, which leads to the concept of the special relativity, while the perihelion shift of the planets confirms the existence of the new additional gravity potential which is derived from the non-relativistic reduction of the Dirac equation with the gravitational potential.

9.9.10. Leap Second Dating

Since we know quite accurately the time shift of the earth rotation around the sun by now, we may apply this time shift to the dating of some archaeological objects such as pyramids or Stonehenge. For example, the time shift of 1000 years amounts to 10.3 minutes, and some of the archaeological objects may well possess a special part of the building which can be pointed to the sun at the equinox. In this case, one may be able to find out the date when this object was constructed. This new dating procedure is basically useful for the stone-made archaeological objects in contrast to the dating of the wooden buildings which can be determined from the Carbon dating. It should be noted that the new dating method has an important assumption that there should be no major earthquake in the region of the archaeological objects.

It should be worthwhile noting that one should be careful for the Leap Second Dating method in the realistic application. This is clear since the earth is also rotating in its own axis when it is rotating around the sun. Therefore, the advance time shift of the earth rotation around the sun should correspond to the retreat time shift of the earth's own rotation if one measures it at one fixed point of the earth surface.

Appendix A

Introduction to Field Theory

This Appendix is intended for readers who may not be very familiar with the field theory terminology. In particular, the basic notations which are used in this textbook are explained in detail. The notation in physics is important since it is just like the language with which all the communications become possible. Therefore, the notation must be defined well, but readers should remember them all.

In quantum field theory, the basic concept is, for sure, based on quantum mechanics. Therefore, we explain some of the important ingredients in quantum mechanics. The most difficult part of quantum mechanics is the quantization itself, and the quantization procedure which is often called the *first quantization* is consistent with all the experiments even though there is no fundamental principle that may lead to the concept of the quantization. Also, the Schrödinger field is described in terms of the non-relativistic field theory, and apart from the kinematics, the behavior of the Schrödinger field from the field theoretical point of view should be similar to the relativistic field.

The relativistic quantum mechanics of fermions is described such that readers may understand and remember it all. Apart from the creation and annihilation of particles, the Dirac theory can describe physics properly. Some properties of the hydrogen atom are explained.

The relativistic boson fields are also discussed here. However, we present questions rather than descriptions of the boson field properties since unfortunately these questions are not answered well in this textbook. On the other hand, it is by now quite possible that the Klein–Gordon equation may not be derived from the fundamental principle any more [82, 62]. These conceptual problems are still not very well organized in this textbook.

Maxwell equation is reviewed here rather in detail from the point of view of the gauge field theory. In this textbook, the Maxwell equation is considered to be most fundamental for all.

Regularization and renormalization are briefly discussed in terms of physical observables. In addition, we describe the path integral formulation since it is an interesting tool. However, it is not very useful for practical calculations, apart from the derivation of Feynman diagrams. In this regard, the path integral formulation should be taken as an alternative tool which is expressed in terms of many dimensional integrals rather than the differential equations in order to obtain any physical observables.

In Appendix H, we present a new concept of the first quantization itself. Mathematically, there is nothing new in the new picture, but the understanding of quantum mechanics may well have a conceptual change in future physics. At least, it should be easy to understand why one can replace the energy and momentum by the differential operators in deriving the Schrödinger equation while there is a good reason to believe that the Klein–Gordon equation for elementary fields cannot be derived from the fundamental principle.

Finally, we review briefly the renormalization scheme in QED. This is well explained in the standard field theory textbooks, and there is no special need for the presentation of the renormalization in QED. Here, we stress that the renormalization in QED itself is well constructed since the Fock space of the unperturbed Hamiltonian is prepared in advance.

Notations in Field Theory

In field theory, one often employs special notations which are by now commonly used. In this Appendix, we explain some of the notations which are particularly useful in field theory.

A.1. Natural Units

In this text, we employ the natural units because of its simplicity

$$c = 1, \quad \hbar = 1. \tag{A.1.1}$$

If one wishes to get the right dimensions out, one should use

$$\hbar c = 197.33 \ \text{MeV} \cdot \text{fm}. \tag{A.1.2}$$

For example, pion mass is $m_\pi \simeq 140$ MeV/c^2. Its Compton wave length is

$$\frac{1}{m_\pi} = \frac{\hbar c}{m_\pi c^2} = \frac{197 \text{ MeV} \cdot \text{ fm}}{140 \text{ MeV}} \simeq 1.4 \text{ fm}.$$

The fine structure constant α is expressed by the coupling constant e which is defined in some different ways

$$\alpha = e^2 = \frac{e^2}{\hbar c} = \frac{e^2}{4\pi} = \frac{e^2}{4\pi\hbar c} = \frac{1}{137.036}.$$

Some constants:

Electron mass: $m_e = 0.511 \ \text{MeV}/c^2$

Muon mass: $m_\mu = 105.66 \ \text{MeV}/c^2$

Proton mass: $M_p = 938.28 \ \text{MeV}/c^2$

Bohr radius: $a_0 = \dfrac{1}{m_e e^2} = 0.529 \times 10^{-8}$ cm

Magnetic moments: $\begin{pmatrix} \text{Electron}: & \mu_e = 1.00115965219 & \dfrac{e\hbar}{2m_e c} \\ \text{Theory}: & \mu_e = 1.0011596524 & \dfrac{e\hbar}{2m_e c} \\ \text{Muon}: & \mu_\mu = 1.001165920 & \dfrac{e\hbar}{2m_\mu c} \end{pmatrix}.$

A.2. Hermite Conjugate and Complex Conjugate

For a complex c-number A

$$A = a + bi \ \ (a, b: \text{ real}). \tag{A.2.1}$$

its complex conjugate A^* is defined as

$$A^* = a - bi. \tag{A.2.2}$$

Matrix A

If A is a matrix, one defines the hermite conjugate $A^\dagger$

$$(A^\dagger)_{ij} = A^*_{ji}. \tag{A.2.3}$$

Differential Operator $\hat{A}$

If $\hat{A}$ is a differential operator, then the hermite conjugate can be defined only when the Hilbert space and its scalar product are defined. For example, suppose $\hat{A}$ is written as

$$\hat{A} = i\frac{\partial}{\partial x}. \tag{A.2.4}$$

In this case, its hermite conjugate $\hat{A}^\dagger$ becomes

$$\hat{A}^\dagger = -i\left(\frac{\partial}{\partial x}\right)^T = i\frac{\partial}{\partial x} = \hat{A} \tag{A.2.5}$$

which means $\hat{A}$ is Hermitian. This can be easily seen in a concrete fashion since

$$\langle\psi|\hat{A}\psi\rangle = \int_{-\infty}^{\infty} \psi^\dagger(x) i\frac{\partial}{\partial x}\psi(x)\,dx = -i\int_{-\infty}^{\infty}\left(\frac{\partial}{\partial x}\psi^\dagger(x)\right)\psi(x)\,dx = \langle\hat{A}\psi|\psi\rangle, \tag{A.2.6}$$

where $\psi(\pm\infty) = 0$ is assumed. The complex conjugate of $\hat{A}$ is simply

$$\hat{A}^* = -i\frac{\partial}{\partial x} \neq \hat{A}. \tag{A.2.7}$$

Field ψ

If the $\psi(x)$ is a c-number field, then the hermite conjugate $\psi^\dagger(x)$ is just the same as the complex conjugate $\psi^*(x)$. However, when the field $\psi(x)$ is quantized, then one should always take the hermite conjugate $\psi^\dagger(x)$. When one takes the complex conjugate of the field as $\psi^*(x)$, one may examine the time reversal invariance as discussed in Chapter 2.

A.3. Scalar and Vector Products (Three Dimensions) :

Scalar Product

For two vectors in three dimensions

$$\boldsymbol{r} = (x,y,z) \equiv (x_1,x_2,x_3), \quad \boldsymbol{p} = (p_x,p_y,p_z) \equiv (p_1,p_2,p_3) \tag{A.3.1}$$

the scalar product is defined

$$\boldsymbol{r}\cdot\boldsymbol{p} = \sum_{k=1}^{3} x_k p_k \equiv x_k p_k, \tag{A.3.2}$$

where, in the last step, we omit the summation notation if the index k is repeated twice.

Vector Product

The vector product is defined as

$$\boldsymbol{r}\times\boldsymbol{p} \equiv (x_2p_3 - x_3p_2, x_3p_1 - x_1p_3, x_1p_2 - x_2p_1). \tag{A.3.3}$$

This can be rewritten in terms of components,

$$(\boldsymbol{r}\times\boldsymbol{p})_i = \varepsilon_{ijk} x_j p_k, \tag{A.3.4}$$

where ε_{ijk} denotes anti-symmetric symbol with

$$\varepsilon_{123} = \varepsilon_{231} = \varepsilon_{312} = 1, \quad \varepsilon_{132} = \varepsilon_{213} = \varepsilon_{321} = -1, \quad \text{otherwise} = 0.$$

A.4. Scalar Product (Four Dimensions)

For two vectors in four dimensions,

$$x^\mu \equiv (t,x,y,z) = (x_0,\boldsymbol{r}), \quad p^\mu \equiv (E,p_x,p_y,p_z) = (p_0,\boldsymbol{p}) \tag{A.4.1}$$

the scalar product is defined

$$x\cdot p \equiv Et - \boldsymbol{r}\cdot\boldsymbol{p} = x_0p_0 - x_kp_k. \tag{A.4.2}$$

This can be also written as

$$x_\mu p^\mu \equiv x_0p^0 + x_1p^1 + x_2p^2 + x_3p^3 = Et - \boldsymbol{r}\cdot\boldsymbol{p} = x\cdot p, \tag{A.4.3}$$

where x_μ and p_μ are defined as

$$x_\mu \equiv (x_0, -\boldsymbol{r}), \quad p_\mu \equiv (p_0, -\boldsymbol{p}). \tag{A.4.4}$$

Here, the repeated indices of the Greek letters mean the four dimensional summation $\mu = 0,1,2,3$. The repeated indices of the roman letters always denote the three dimensional summation throughout the text.

A.4.1. Metric Tensor

It is sometimes convenient to introduce the metric tensor $g^{\mu\nu}$ which has the following properties

$$g^{\mu\nu} = g_{\mu\nu} = \begin{pmatrix} 1 & 0 & 0 & 0 \\ 0 & -1 & 0 & 0 \\ 0 & 0 & -1 & 0 \\ 0 & 0 & 0 & -1 \end{pmatrix}. \qquad (A.4.5)$$

In this case, the scalar product can be rewritten as

$$x \cdot p = x^\mu p^\nu g_{\mu\nu} = Et - \boldsymbol{r} \cdot \boldsymbol{p}. \qquad (A.4.6)$$

A.5. Four Dimensional Derivatives ∂_μ

The derivative ∂_μ is introduced for convenience

$$\partial_\mu \equiv \frac{\partial}{\partial x^\mu} = \left(\frac{\partial}{\partial x^0}, \frac{\partial}{\partial x^1}, \frac{\partial}{\partial x^2}, \frac{\partial}{\partial x^3} \right) = \left(\frac{\partial}{\partial t}, \frac{\partial}{\partial x}, \frac{\partial}{\partial y}, \frac{\partial}{\partial z} \right) = \left(\frac{\partial}{\partial t}, \boldsymbol{\nabla} \right), \qquad (A.5.1)$$

where the lower index has the positive space part. Therefore, the derivative ∂^μ becomes

$$\partial^\mu \equiv \frac{\partial}{\partial x_\mu} = \left(\frac{\partial}{\partial t}, -\frac{\partial}{\partial x}, -\frac{\partial}{\partial y}, -\frac{\partial}{\partial z} \right) = \left(\frac{\partial}{\partial t}, -\boldsymbol{\nabla} \right). \qquad (A.5.2)$$

A.5.1. $\hat{p}^\mu$ and Differential Operator

Since the operator $\hat{p}^\mu$ becomes a differential operator as

$$\hat{p}^\mu = (\hat{E}, \hat{\boldsymbol{p}}) = \left(i\frac{\partial}{\partial t}, -i\boldsymbol{\nabla} \right) = i\partial^\mu$$

the negative sign, therefore, appears in the space part. For example, if one defines the current j^μ in four dimension as

$$j^\mu = (\rho, \boldsymbol{j}),$$

then the current conservation is written as

$$\partial_\mu j^\mu = \frac{\partial \rho}{\partial t} + \boldsymbol{\nabla} \cdot \boldsymbol{j} = \frac{1}{i} \hat{p}_\mu j^\mu = 0. \qquad (A.5.3)$$

A.5.2. Laplacian and d'Alembertian Operators

The Laplacian and d'Alembertian operators, Δ and $\Box$ are defined as

$$\Delta \equiv \boldsymbol{\nabla} \cdot \boldsymbol{\nabla} = \frac{\partial^2}{\partial x^2} + \frac{\partial^2}{\partial y^2} + \frac{\partial^2}{\partial z^2},$$

$$\Box \equiv \partial_\mu \partial^\mu = \frac{\partial^2}{\partial t^2} - \Delta.$$

A.6. γ-Matrices

Here, we present explicit expressions of the γ-matrices in two and four dimensions. Before presenting the representation of the γ-matrices, we first give the explicit representation of Pauli matrices.

A.6.1. Pauli Matrices

Pauli matrices are given as

$$\sigma_x = \sigma_1 = \begin{pmatrix} 0 & 1 \\ 1 & 0 \end{pmatrix}, \quad \sigma_y = \sigma_2 = \begin{pmatrix} 0 & -i \\ i & 0 \end{pmatrix}, \quad \sigma_z = \sigma_3 = \begin{pmatrix} 1 & 0 \\ 0 & -1 \end{pmatrix}. \qquad (A.6.1)$$

Below we write some properties of the Pauli matrices.

Hermiticity

$$\sigma_1^\dagger = \sigma_1, \quad \sigma_2^\dagger = \sigma_2, \quad \sigma_3^\dagger = \sigma_3.$$

Complex Conjugate

$$\sigma_1^* = \sigma_1, \quad \sigma_2^* = -\sigma_2, \quad \sigma_3^* = \sigma_3.$$

Transposed

$$\sigma_1^T = \sigma_1, \quad \sigma_2^T = -\sigma_2, \quad \sigma_3^T = \sigma_3 \quad (\sigma_k^T = \sigma_k^*).$$

Useful Relations

$$\sigma_i \sigma_j = \delta_{ij} + i\varepsilon_{ijk}\sigma_k, \qquad (A.6.2)$$

$$[\sigma_i, \sigma_j] = 2i\varepsilon_{ijk}\sigma_k. \qquad (A.6.3)$$

A.6.2. Representation of γ-matrices

(a) Two dimensional representations of γ-matrices

$$\text{Dirac}: \ \gamma_0 = \begin{pmatrix} 1 & 0 \\ 0 & -1 \end{pmatrix}, \quad \gamma_1 = \begin{pmatrix} 0 & 1 \\ -1 & 0 \end{pmatrix}, \quad \gamma_5 = \gamma_0\gamma_1 = \begin{pmatrix} 0 & 1 \\ 1 & 0 \end{pmatrix},$$

$$\text{Chiral}: \ \gamma_0 = \begin{pmatrix} 0 & 1 \\ 1 & 0 \end{pmatrix}, \quad \gamma_1 = \begin{pmatrix} 0 & -1 \\ 1 & 0 \end{pmatrix}, \quad \gamma_5 = \gamma_0\gamma_1 = \begin{pmatrix} 1 & 0 \\ 0 & -1 \end{pmatrix}.$$

(b) Four dimensional representations of gamma matrices

$$\text{Dirac}: \gamma_0 = \beta = \begin{pmatrix} \mathbf{1} & \mathbf{0} \\ \mathbf{0} & -\mathbf{1} \end{pmatrix}, \quad \boldsymbol{\gamma} = \begin{pmatrix} \mathbf{0} & \boldsymbol{\sigma} \\ -\boldsymbol{\sigma} & \mathbf{0} \end{pmatrix},$$

$$\gamma_5 = i\gamma_0\gamma_1\gamma_2\gamma_3 = \begin{pmatrix} \mathbf{0} & \mathbf{1} \\ \mathbf{1} & \mathbf{0} \end{pmatrix}, \quad \boldsymbol{\alpha} = \begin{pmatrix} \mathbf{0} & \boldsymbol{\sigma} \\ \boldsymbol{\sigma} & \mathbf{0} \end{pmatrix},$$

$$\text{Chiral}: \gamma_0 = \beta = \begin{pmatrix} \mathbf{0} & \mathbf{1} \\ \mathbf{1} & \mathbf{0} \end{pmatrix}, \quad \boldsymbol{\gamma} = \begin{pmatrix} \mathbf{0} & -\boldsymbol{\sigma} \\ \boldsymbol{\sigma} & \mathbf{0} \end{pmatrix},$$

$$\gamma_5 = i\gamma_0\gamma_1\gamma_2\gamma_3 = \begin{pmatrix} \mathbf{1} & \mathbf{0} \\ \mathbf{0} & -\mathbf{1} \end{pmatrix}, \quad \boldsymbol{\alpha} = \begin{pmatrix} \boldsymbol{\sigma} & \mathbf{0} \\ \mathbf{0} & -\boldsymbol{\sigma} \end{pmatrix}.$$

$$\text{where} \quad \mathbf{0} \equiv \begin{pmatrix} 0 & 0 \\ 0 & 0 \end{pmatrix}, \quad \mathbf{1} \equiv \begin{pmatrix} 1 & 0 \\ 0 & 1 \end{pmatrix}.$$

A.6.3. Useful Relations of γ-Matrices

Here, we summarize some useful relations of the γ-matrices.

Anti-commutation relations

$$\{\gamma^\mu, \gamma^\nu\} = 2g^{\mu\nu}, \quad \{\gamma^5, \gamma^\nu\} = 0. \tag{A.6.4}$$

Hermiticity

$$\gamma_\mu^\dagger = \gamma_0\gamma_\mu\gamma_0 \ (\gamma_0^\dagger = \gamma_0, \ \gamma_k^\dagger = -\gamma_k), \quad \gamma_5^\dagger = \gamma_5. \tag{A.6.5}$$

Complex Conjugate

$$\gamma_0^* = \gamma_0, \quad \gamma_1^* = \gamma_1, \quad \gamma_2^* = -\gamma_2, \quad \gamma_3^* = \gamma_3, \quad \gamma_5^* = \gamma_5. \tag{A.6.6}$$

Transposed

$$\gamma_\mu^T = \gamma_0\gamma_\mu^*\gamma_0, \quad \gamma_5^T = \gamma_5. \tag{A.6.7}$$

A.7. Transformation of State and Operator

When one transforms a quantum state $|\psi\rangle$ by a unitary transformation U which satisfies

$$U^\dagger U = 1$$

one writes the transformed state as

$$|\psi'\rangle = U|\psi\rangle. \tag{A.7.1}$$

The unitarity is important since the norm must be conserved, that is,

$$\langle \psi' | \psi' \rangle = \langle \psi | U^\dagger U | \psi \rangle = 1.$$

In this case, an arbitrary operator O is transformed as

$$O' = U O U^{-1}. \qquad (A.7.2)$$

This can be obtained since the expectation value of the operator O must be the same between two systems, that is,

$$\langle \psi | O | \psi \rangle = \langle \psi' | O' | \psi' \rangle. \qquad (A.7.3)$$

Since

$$\langle \psi' | O' | \psi' \rangle = \langle \psi | U^\dagger O' U | \psi \rangle = \langle \psi | O | \psi \rangle$$

one finds

$$U^\dagger O' U = O$$

which is just eq.(A.7.2).

A.8. Fermion Current

We summarize the fermion currents and their properties of the Lorentz transformation. We also give their nonrelativistic expressions since the basic behaviors must be kept in the nonrelativistic expressions. Here, the approximate expressions are obtained by making use of the plane wave solutions for the Dirac wave function.

$$\text{Fermioncurrents}: \begin{pmatrix} \text{Scalar}: & \bar{\psi}\psi \simeq 1 \\ \text{Pseudoscalar}: & \bar{\psi}\gamma_5\psi \simeq \frac{\boldsymbol{\sigma}\cdot\boldsymbol{p}}{m} \\ \text{Vector}: & \bar{\psi}\gamma_\mu\psi \simeq \left(1, \frac{\boldsymbol{p}}{m}\right) \\ \text{Axialvector}: & \bar{\psi}\gamma_\mu\gamma_5\psi \simeq \left(\frac{\boldsymbol{\sigma}\cdot\boldsymbol{p}}{m}, \boldsymbol{\sigma}\right) \end{pmatrix}. \qquad (A.8.1)$$

Therefore, under the parity $\hat{P}$ and time reversal $\hat{T}$ transformation, the above currents behave as

$$\text{Parity } \hat{P}: \begin{pmatrix} \bar{\psi}'\psi' = \bar{\psi}\hat{P}^{-1}\hat{P}\psi = \bar{\psi}\psi \\ \bar{\psi}'\gamma_5\psi' = \bar{\psi}\hat{P}^{-1}\gamma_5\hat{P}\psi = -\bar{\psi}\gamma_5\psi \\ \bar{\psi}'\gamma_k\psi' = \bar{\psi}\hat{P}^{-1}\gamma_k\hat{P}\psi = -\bar{\psi}\gamma_k\psi \\ \bar{\psi}'\gamma_k\gamma_5\psi' = \bar{\psi}\hat{P}^{-1}\gamma_k\gamma_5\hat{P}\psi = \bar{\psi}\gamma_k\gamma_5\psi \end{pmatrix}, \qquad (A.8.2)$$

$$\text{Timereversal } \hat{T}: \begin{pmatrix} \bar{\psi}'\psi' = \bar{\psi}\hat{T}^{-1}\hat{T}\psi = \bar{\psi}\psi \\ \bar{\psi}'\gamma_5\psi' = \bar{\psi}\hat{T}^{-1}\gamma_5\hat{T}\psi = \bar{\psi}\gamma_5\psi \\ \bar{\psi}'\gamma_k\psi' = \bar{\psi}\hat{T}^{-1}\gamma_k\hat{T}\psi = -\bar{\psi}\gamma_k\psi \\ \bar{\psi}'\gamma_k\gamma_5\psi' = \bar{\psi}\hat{T}^{-1}\gamma_k\gamma_5\hat{T}\psi = -\bar{\psi}\gamma_k\gamma_5\psi \end{pmatrix}. \qquad (A.8.3)$$

A.9. Trace in Physics

A.9.1. Definition

The trace of $N \times N$ matrix A is defined as

$$\text{Tr}\{A\} = \sum_{i=1}^{N} A_{ii}. \tag{A.9.1}$$

This is simply the summation of the diagonal elements of the matrix A. It is easy to prove

$$\text{Tr}\{AB\} = \text{Tr}\{BA\}. \tag{A.9.2}$$

A.9.2. Trace in Quantum Mechanics

In quantum mechanics, the trace of the Hamiltonian H becomes

$$\text{Tr}\{H\} = \text{Tr}\{UHU^{-1}\} = \sum_{n=1} E_n, \tag{A.9.3}$$

where U is a unitary operator that diagonalizes the Hamiltonian, and E_n denotes the energy eigenvalue of the Hamiltonian. Therefore, the trace of the Hamiltonian has the meaning of the sum of all the eigenvalues of the Hamiltonian.

A.9.3. Trace in $SU(N)$

In the special unitary group $SU(N)$, one often describes the element U^a in terms of the generator T^a as

$$U^a = e^{iT^a}. \tag{A.9.4}$$

In this case, the generator must be hermitian and traceless since

$$\det U^a = \exp\left(\text{Tr}\{\ln U^a\}\right) = \exp\left(i\,\text{Tr}\{T^a\}\right) = 1 \tag{A.9.5}$$

and thus

$$\text{Tr}\{T^a\} = 0. \tag{A.9.6}$$

The generators of $SU(N)$ group satisfy the following commutation relations

$$[T^a, T^b] = iC^{abc}T^c, \tag{A.9.7}$$

where C^{abc} denotes a structure constant in the Lie algebra. The generators are normalized in this textbook such that

$$\text{Tr}\{T^a T^b\} = \frac{1}{2}\delta^{ab}. \tag{A.9.8}$$

A.9.4. Trace of γ-Matrices and $\not{p}$

The Trace of the γ-matrices is also important. First, we have

$$\mathrm{Tr}\{1\} = 4, \quad \mathrm{Tr}\{\gamma_\mu\} = 0, \quad \mathrm{Tr}\{\gamma_5\} = 0. \qquad (A.9.9)$$

In field theory, one often defines a symbol of $\not{p}$ just for convenience

$$\not{p} \equiv p_\mu \gamma^\mu.$$

In this case, the following relation holds

$$\not{p}\not{q} = pq - i\sigma_{\mu\nu} p^\mu q^\nu. \qquad (A.9.10)$$

The following relations may also be useful

$$\mathrm{Tr}\{\not{p}\not{q}\} = 4pq, \qquad (A.9.11)$$

$$\mathrm{Tr}\{\gamma_5\not{p}\not{q}\} = 0, \qquad (A.9.12)$$

$$\mathrm{Tr}\{\not{p}_1\not{p}_2\not{p}_3\not{p}_4\} = 4\Big\{(p_1p_2)(p_3p_4) - (p_1p_3)(p_2p_4) + (p_1p_4)(p_2p_3)\Big\}, \qquad (A.9.13)$$

$$\mathrm{Tr}\{\gamma_5\not{p}_1\not{p}_2\not{p}_3\not{p}_4\} = 4i\varepsilon_{\alpha\beta\gamma\delta}\, p_1^\alpha\, p_2^\beta\, p_3^\gamma\, p_4^\delta. \qquad (A.9.14)$$

Basic Equations and Principles

A.10. Lagrange Equation

In classical field theory, the equation of motion is most important, and it is derived from the Lagrange equation. Therefore, we review briefly how we can obtain the equation of motion from the Lagrangian density.

A.10.1. Lagrange Equation in Classical Mechanics

Before going to the field theory treatment, we first discuss the Lagrange equation (Newton equation) in classical mechanics. In order to obtain the Lagrange equation by the variational principle in classical mechanics, one starts from the action S as defined

$$S = \int L(q,\dot{q})\, dt, \qquad (A.10.1)$$

where the Lagrangian $L(q,\dot{q})$ depends on the general coordinate q and its velocity $\dot{q}$. At the time of deriving equation of motion by the variational principle, q and $\dot{q}$ are independent as the function of t. This is clear since, in the action S, the functional dependence of $q(t)$ is unknown and therefore one cannot make any derivative of $q(t)$ with respect to time t. Once the equation of motion is established, then one can obtain $\dot{q}$ by time differentiation of $q(t)$ which is a solution of the equation of motion.

The Lagrange equation can be obtained by requiring that the action S should be a minimum with respect to the variation of q and $\dot{q}$.

$$\delta S = \int \delta L(q,\dot{q})\,dt = \int \left(\frac{\partial L}{\partial q}\delta q + \frac{\partial L}{\partial \dot{q}}\delta\dot{q}\right) dt$$

$$= \int \left(\frac{\partial L}{\partial q} - \frac{d}{dt}\frac{\partial L}{\partial \dot{q}}\right)\delta q\,dt = 0, \qquad (A.10.2)$$

where the surface terms are assumed to vanish. Therefore, one obtains the Lagrange equation

$$\frac{\partial L}{\partial q} - \frac{d}{dt}\frac{\partial L}{\partial \dot{q}} = 0. \qquad (A.10.3)$$

A.10.2. Hamiltonian in Classical Mechanics

The Lagrangian $L(q,\dot{q})$ must be invariant under the infinitesimal time displacement ε of $q(t)$ as

$$q(t+\varepsilon) \to q(t) + \dot{q}\varepsilon, \quad \dot{q}(t+\varepsilon) \to \dot{q}(t) + \ddot{q}\varepsilon + \dot{q}\frac{d\varepsilon}{dt}. \qquad (A.10.4)$$

Therefore, one finds

$$\delta L(q,\dot{q}) = L(q(t+\varepsilon),\dot{q}(t+\varepsilon)) - L(q,\dot{q}) = \frac{\partial L}{\partial q}\dot{q}\varepsilon + \frac{\partial L}{\partial \dot{q}}\ddot{q}\varepsilon + \frac{\partial L}{\partial \dot{q}}\dot{q}\frac{d\varepsilon}{dt} = 0. \qquad (A.10.5)$$

Neglecting the surface term, one obtains

$$\delta L(q,\dot{q}) = \left[\frac{\partial L}{\partial q}\dot{q} + \frac{\partial L}{\partial \dot{q}}\ddot{q} - \frac{d}{dt}\left(\frac{\partial L}{\partial \dot{q}}\dot{q}\right)\right]\varepsilon = \left[\frac{d}{dt}\left(L - \frac{\partial L}{\partial \dot{q}}\dot{q}\right)\right]\varepsilon = 0. \qquad (A.10.6)$$

Thus, if one defines the Hamiltonian H as

$$H \equiv \frac{\partial L}{\partial \dot{q}}\dot{q} - L \qquad (A.10.7)$$

then it is a conserved quantity.

A.10.3. Lagrange Equation for Fields

The Lagrange equation for fields can be obtained almost in the same way as the particle case. For fields, we should start from the Lagrangian density $\mathcal{L}$ and the action is written as

$$S = \int \mathcal{L}\left(\psi, \dot{\psi}, \frac{\partial \psi}{\partial x_k}\right) d^3r\,dt, \qquad (A.10.8)$$

where $\psi(x)$, $\dot{\psi}(x)$ and $\frac{\partial \psi}{\partial x_k}$ are independent functional variables.

The Lagrange equation can be obtained by requiring that the action S should be a minimum with respect to the variation of ψ, $\dot{\psi}$ and $\frac{\partial \psi}{\partial x_k}$,

$$\delta S = \int \delta\mathcal{L}\left(\psi, \dot{\psi}, \frac{\partial \psi}{\partial x_k}\right) d^3r\,dt = \int \left(\frac{\partial \mathcal{L}}{\partial \psi}\delta\psi + \frac{\partial \mathcal{L}}{\partial \dot{\psi}}\delta\dot{\psi} + \frac{\partial \mathcal{L}}{\partial(\frac{\partial \psi}{\partial x_k})}\delta\left(\frac{\partial \psi}{\partial x_k}\right)\right) d^3r\,dt$$

$$= \int \left(\frac{\partial \mathcal{L}}{\partial \psi} - \frac{\partial}{\partial t}\frac{\partial \mathcal{L}}{\partial \dot{\psi}} - \frac{\partial}{\partial x_k}\frac{\partial \mathcal{L}}{\partial (\frac{\partial \psi}{\partial x_k})} \right) \delta\psi \, d^3r \, dt = 0, \qquad (A.10.9)$$

where the surface terms are assumed to vanish. Therefore, one obtains

$$\frac{\partial \mathcal{L}}{\partial \psi} = \frac{\partial}{\partial t}\frac{\partial \mathcal{L}}{\partial \dot{\psi}} + \frac{\partial}{\partial x_k}\frac{\partial \mathcal{L}}{\partial (\frac{\partial \psi}{\partial x_k})}, \qquad (A.10.10)$$

which can be expressed in the relativistic covariant way as

$$\frac{\partial \mathcal{L}}{\partial \psi} = \partial_\mu \left(\frac{\partial \mathcal{L}}{\partial (\partial_\mu \psi)} \right). \qquad (A.10.11)$$

This is the Lagrange equation for field ψ, which should hold for any independent field ψ.

A.11. Noether Current

If the Lagrangian density is invariant under the transformation of the field with a continuous variable, then there is always a conserved current associated with this symmetry. This is called *Noether current* and can be derived from the invariance of the Lagrangian density and the Lagrange equation.

A.11.1. Global Gauge Symmetry

The Lagrangian density which is discussed in this textbook should have the following functional dependence in general

$$\mathcal{L} = i\bar{\psi}\gamma_\mu\partial^\mu\psi - m\bar{\psi}\psi + \mathcal{L}_I\left[\bar{\psi}\psi, \bar{\psi}\gamma_5\psi, \bar{\psi}\gamma_\mu\psi\right].$$

This Lagrangian density is obviously invariant under the global gauge transformation

$$\psi' = e^{i\alpha}\psi, \quad \psi'^\dagger = e^{-i\alpha}\psi^\dagger, \qquad (A.11.1)$$

where α ia a real constant. Therefore, the Noether current is conserved in this system. To derive the Noether current conservation for the global gauge transformation, one can consider the infinitesimal global transformation, that is, $|\alpha| \ll 1$. In this case, the transformation becomes

$$\psi' = \psi + \delta\psi, \quad \delta\psi = i\alpha\psi. \qquad (A.11.2a)$$

$$\psi'^\dagger = \psi^\dagger + \delta\psi^\dagger, \quad \delta\psi^\dagger = -i\alpha\psi^\dagger. \qquad (A.11.2b)$$

Invariance of Lagrangian Density

Now, it is easy to find

$$\delta\mathcal{L} = \mathcal{L}(\psi', \psi'^\dagger, \partial_\mu\psi', \partial_\mu\psi'^\dagger) - \mathcal{L}(\psi, \psi^\dagger, \partial_\mu\psi, \partial_\mu\psi^\dagger) = 0. \qquad (A.11.3a)$$

At the same time, one can easily evaluate $\delta\mathcal{L}$

$$\begin{aligned}\delta\mathcal{L} &= \frac{\partial\mathcal{L}}{\partial\psi}\delta\psi + \frac{\partial\mathcal{L}}{\partial(\partial_\mu\psi)}\delta(\partial_\mu\psi) + \frac{\partial\mathcal{L}}{\partial\psi^\dagger}\delta\psi^\dagger + \frac{\partial\mathcal{L}}{\partial(\partial_\mu\psi^\dagger)}\delta(\partial_\mu\psi^\dagger) \\ &= i\alpha\left[\left(\partial_\mu\frac{\partial\mathcal{L}}{\partial(\partial_\mu\psi)}\right)\psi + \frac{\partial\mathcal{L}}{\partial(\partial_\mu\psi)}\partial_\mu\psi - \left(\partial_\mu\frac{\partial\mathcal{L}}{\partial(\partial_\mu\psi^\dagger)}\right)\psi^\dagger - \frac{\partial\mathcal{L}}{\partial(\partial_\mu\psi^\dagger)}\partial_\mu\psi^\dagger\right] \\ &= i\alpha\partial_\mu\left[\frac{\partial\mathcal{L}}{\partial(\partial_\mu\psi)}\psi - \frac{\partial\mathcal{L}}{\partial(\partial_\mu\psi^\dagger)}\psi^\dagger\right] = 0, \qquad (A.11.3b)\end{aligned}$$

where the equation of motion for ψ is employed.

Current Conservation

Therefore, if one defines the current j_μ as

$$j^\mu \equiv -i\left[\frac{\partial\mathcal{L}}{\partial(\partial_\mu\psi)}\psi - \frac{\partial\mathcal{L}}{\partial(\partial_\mu\psi^\dagger)}\psi^\dagger\right] \qquad (A.11.4)$$

then one has

$$\partial_\mu j^\mu = 0. \qquad (A.11.5)$$

For Dirac fields with electromagnetic interactions or self-interactions, one can obtain as a conserved current

$$j^\mu = \bar{\psi}\gamma^\mu\psi. \qquad (A.11.6)$$

A.11.2. Chiral Symmetry

When the Lagrangian density is invariant under the chiral transformation,

$$\psi' = e^{i\alpha\gamma_5}\psi \qquad (A.11.7)$$

then there is another Noether current. Here, $\delta\psi$ as defined in eq.(A.11.2) becomes

$$\delta\psi = i\alpha\gamma_5\psi. \qquad (A.11.8)$$

Therefore, a corresponding conserved current for massless Dirac fields with electromagnetic interactions or self-interactions can be obtained

$$j_5^\mu = -i\frac{\partial\mathcal{L}}{\partial(\partial_\mu\psi)}\gamma_5\psi = \bar{\psi}\gamma^\mu\gamma_5\psi. \qquad (A.11.9)$$

In this case, we have

$$\partial_\mu j_5^\mu = 0 \qquad (A.11.10)$$

which is the conservation of the axial vector current. The conservation of the axial vector current is realized for field theory models with massless fermions.

A.12. Hamiltonian Density

The Hamiltonian density $\mathcal{H}$ is constructed from the Lagrangian density $\mathcal{L}$. The field theory models which we consider should possess the translational invariance. If the Lagrangian density is invariant under the translation a^μ, then there is a conserved quantity which is the energy momentum tensor $\mathcal{T}^{\mu\nu}$. The Hamiltonian density is constructed from the energy momentum tensor of $\mathcal{T}^{00}$.

A.12.1. Hamiltonian Density from Energy Momentum Tensor

Now, the Lagrangian density is given as $\mathcal{L}\left(\psi_i, \dot{\psi}_i, \frac{\partial \psi_i}{\partial x_k}\right)$. If one considers the following infinitesimal translation a^μ of the field ψ_i and $\psi_i^\dagger$

$$\psi_i' = \psi_i + \delta\psi_i, \quad \delta\psi_i = (\partial_\nu \psi_i)a^\nu,$$

$$\psi_i^{\dagger\prime} = \psi_i^\dagger + \delta\psi_i^\dagger, \quad \delta\psi_i^\dagger = (\partial_\nu \psi_i^\dagger)a^\nu,$$

then the Lagrangian density should be invariant

$$\delta\mathcal{L} \equiv \mathcal{L}(\psi_i', \partial_\mu \psi_i') - \mathcal{L}(\psi_i, \partial_\mu \psi_i)$$

$$= \sum_i \left[\frac{\partial \mathcal{L}}{\partial \psi_i}\delta\psi_i + \frac{\partial \mathcal{L}}{\partial(\partial_\mu \psi_i)}\delta(\partial_\mu \psi_i) + \frac{\partial \mathcal{L}}{\partial \psi_i^\dagger}\delta\psi_i^\dagger + \frac{\partial \mathcal{L}}{\partial(\partial_\mu \psi_i^\dagger)}\delta(\partial_\mu \psi_i^\dagger)\right] = 0. \qquad (A.12.1)$$

Making use of the Lagrange equation, one obtains

$$\delta\mathcal{L} = \sum_i \left[\frac{\partial \mathcal{L}}{\partial \psi_i}(\partial_\nu \psi_i) + \frac{\partial \mathcal{L}}{\partial(\partial_\mu \psi_i)}(\partial_\mu \partial_\nu \psi_i) - \partial_\mu \left(\frac{\partial \mathcal{L}}{\partial(\partial_\mu \psi_i)}\partial_\nu \psi_i \right)\right] a^\nu$$

$$+ \sum_i \left[\frac{\partial \mathcal{L}}{\partial \psi_i^\dagger}(\partial_\nu \psi_i^\dagger) + \frac{\partial \mathcal{L}}{\partial(\partial_\mu \psi_i^\dagger)}(\partial_\mu \partial_\nu \psi_i^\dagger) - \partial_\mu \left(\frac{\partial \mathcal{L}}{\partial(\partial_\mu \psi_i^\dagger)}\partial_\nu \psi_i^\dagger \right)\right] a^\nu$$

$$= \partial_\mu \left[\mathcal{L} g^{\mu\nu} - \sum_i \left(\frac{\partial \mathcal{L}}{\partial(\partial_\mu \psi_i)}\partial^\nu \psi_i + \frac{\partial \mathcal{L}}{\partial(\partial_\mu \psi_i^\dagger)}\partial^\nu \psi_i^\dagger \right)\right] a_\nu = 0. \qquad (A.12.2)$$

Energy Momentum Tensor $\mathcal{T}^{\mu\nu}$

Therefore, if one defines the energy momentum tensor $\mathcal{T}^{\mu\nu}$ by

$$\mathcal{T}^{\mu\nu} \equiv \sum_i \left(\frac{\partial \mathcal{L}}{\partial(\partial_\mu \psi_i)}\partial^\nu \psi_i + \frac{\partial \mathcal{L}}{\partial(\partial_\mu \psi_i^\dagger)}\partial^\nu \psi_i^\dagger \right) - \mathcal{L} g^{\mu\nu} \qquad (A.12.3)$$

then $\mathcal{T}^{\mu\nu}$ is a conserved quantity, that is

$$\partial_\mu \mathcal{T}^{\mu\nu} = 0.$$

This leads to the definition of the Hamltonian density $\mathcal{H}$ in terms of $\mathcal{T}^{00}$

$$\mathcal{H} \equiv \mathcal{T}^{00} = \sum_i \left(\frac{\partial \mathcal{L}}{\partial(\partial_0 \psi_i)}\partial^0 \psi_i + \frac{\partial \mathcal{L}}{\partial(\partial_0 \psi_i^\dagger)}\partial^0 \psi_i^\dagger \right) - \mathcal{L}. \qquad (A.12.4)$$

A.12.2. Hamiltonian Density from Conjugate Fields

When the Lagrangian density is given as $\mathcal{L}(\psi_i, \dot{\psi}_i, \frac{\partial \psi_i}{\partial x_k})$, one can define the conjugate fields Π_{ψ_i} and $\Pi_{\psi_i^\dagger}$ as

$$\Pi_{\psi_i} \equiv \frac{\partial \mathcal{L}}{\partial \dot{\psi}_i}, \quad \Pi_{\psi_i^\dagger} \equiv \frac{\partial \mathcal{L}}{\partial \dot{\psi}_i^\dagger}.$$

In this case, the Hamiltonian density can be written as being consistent with eq.(A.12.4)

$$\mathcal{H} = \sum_i \left(\Pi_{\psi_i} \dot{\psi}_i + \Pi_{\psi_i^\dagger} \dot{\psi}_i^\dagger \right) - \mathcal{L}. \tag{A.12.5}$$

It should be noted that this way of making the Hamiltonian density is indeed easier to remember than the construction starting from the energy momentum tensor.

Hamiltonian

The Hamiltonian is defined by integrating the Hamiltoian density over all space

$$H = \int \mathcal{H}\, d^3r = \int \left[\sum_i (\Pi_{\psi_i} \dot{\psi}_i + \Pi_{\psi_i^\dagger} \dot{\psi}_i^\dagger) - \mathcal{L} \right] d^3r.$$

A.12.3. Hamiltonian Density for Free Dirac Fields

For a free Dirac field with its mass m, the Lagrangian density becomes

$$\mathcal{L} = \psi_i^\dagger \dot{\psi}_i + \psi_i^\dagger \left[i\gamma_0 \boldsymbol{\gamma} \cdot \boldsymbol{\nabla} - m\gamma_0 \right]_{ij} \psi_j.$$

Therefore, the conjugate fields Π_{ψ_i} and $\Pi_{\psi_i^\dagger}$ are obtained

$$\Pi_{\psi_i} \equiv \frac{\partial \mathcal{L}}{\partial \dot{\psi}_i} = \psi_i^\dagger, \quad \Pi_{\psi_i^\dagger} = 0.$$

Thus, the Hamiltonian density becomes

$$\mathcal{H} = \sum_i \left(\Pi_{\psi_i} \dot{\psi}_i + \Pi_{\psi_i^\dagger} \dot{\psi}_i^\dagger \right) - \mathcal{L} = \bar{\psi}_i \left[-i\gamma_k \partial_k + m \right]_{ij} \psi_j = \bar{\psi} \left[-i\boldsymbol{\gamma} \cdot \boldsymbol{\nabla} + m \right] \psi. \tag{A.12.6}$$

A.12.4. Hamiltonian for Free Dirac Fields

The Hamiltonian H is obtained by integrating the Hamiltonian density over all space and thus can be written as

$$H = \int \mathcal{H}\, d^3r = \int \bar{\psi} [-i\boldsymbol{\gamma} \cdot \boldsymbol{\nabla} + m] \psi \, d^3r. \tag{A.12.7}$$

In classical field theory, this Hamiltonian is not an operator but is just the field energy itself. However, this field energy cannot be evaluated unless one knows the shape of the field $\psi(x)$ itself. Therefore, one should determine the shape of the field $\psi(x)$ by the equation of motion in the classical field theory.

A.12.5. Role of Hamiltonian

We should comment on the usefulness of the classical field Hamiltonian itself for field theory models. In fact, the Hamiltonian alone is not useful. This is similar to the classical mechanics case in which the Hamiltonian of a point particle itself does not tell a lot. Instead, one has to derive the Hamilton equations in order to calculate some physical properties of the system and the Hamilton equations are equivalent to the Lagrange equations in classical mechanics.

Classical Field Theory

In classical field theory, the situation is just the same as the classical mechanics case. If one stays in the classical field theory, then one should derive the field equation from the Hamiltonian by the functional variational principle as will be discussed in the next section.

Quantized Field Theory

The Hamiltonian of the field theory becomes important when the fields are quantized. In this case, the Hamiltonian becomes an operator, and thus one has to solve the eigenvalue problem for the quantized Hamiltonian $\hat{H}$

$$\hat{H}|\Psi\rangle = E|\Psi\rangle, \qquad (A.12.8)$$

where $|\Psi\rangle$ is called *Fock state* and should be written in terms of the creation and annihilation operators of fermion and anti-fermion. The space spanned by the Fock states is called *Fock space*.

In normal circumstances of the field theory models such as QED and QCD, it is practically impossible to find the eigenstate of the quantized Hamiltonian. The difficulty of the quantized field theory comes mainly from two reasons. Firstly, one has to construct the vacuum state which is composed of infinite many negative energy particles interacting with each other. The vacuum state should be the eigenstate of the Hamiltonian

$$\hat{H}|\Omega\rangle = E_\Omega|\Omega\rangle,$$

where E_Ω denotes the energy of the vacuum and it is in general infinity with the negative sign. The vacuum state $|\Omega\rangle$ is composed of infinitely many negative energy particles

$$|\Omega\rangle = \prod_{p,s} b^{\dagger(s)}_{p}|0\rangle\rangle,$$

where $|0\rangle\rangle$ denotes the null vacuum state. In the realistic calculations, the number of the negative energy particles must be set to a finite value, and this should be reasonable since physical observables should not depend on the properties of the deep negative energy particles. However, it is most likely that the number of the negative energy particles should be, at least, larger than a few thousand for two dimensional field theory models.

The second difficulty arises from the operators in the Hamiltonian which can change the fermion and anti-fermion numbers and therefore can induce infinite series of the transitions among the Fock states. Since the spectrum of bosons and baryons can be obtained by operating the fermion and anti-fermion creation operators on the vacuum state, the Fock space which is spanned by the creation and annihilation operators becomes infinite. In the realistic calculations, the truncation of the Fock space becomes most important, even though it is difficult to find any reasonable truncation scheme.

In this respect, the Thirring model is an exceptional case where the exact eigenstate of the quantized Hamiltonian is found. This is, however, understandable since the Thirring model Hamiltonian does not contain the operators which can change the fermion and anti-fermion numbers.

A.13. Variational Principle in Hamiltonian

When one has the Hamiltonian, then one can derive the equation of motion by requiring that the Hamiltonian should be minimized with respect to the functional variation of the state $\psi(\boldsymbol{r})$.

A.13.1. Schrödinger Field

When one minimizes the Hamiltonian

$$H = \int \left[-\frac{1}{2m} \psi^\dagger \nabla^2 \psi + \psi^\dagger U \psi \right] d^3 r \qquad (A.13.1)$$

with respect to $\psi(\boldsymbol{r})$, then one can obtain the static Schrödinger equation.

Functional Derivative

First, one defines the functional derivative for an arbitrary function $\psi_i(\boldsymbol{r})$ by

$$\frac{\delta \psi_i(\boldsymbol{r}')}{\delta \psi_j(\boldsymbol{r})} = \delta_{ij} \delta(\boldsymbol{r} - \boldsymbol{r}'). \qquad (A.13.2)$$

This is the most important equation for the functional derivative, and once one accepts this definition of the functional derivative, then one can evaluate the functional variation just in the same way as normal derivative of the function $\psi_i(\boldsymbol{r})$.

Functional Variation of Hamiltonian

For the condition on $\psi(\boldsymbol{r})$, one requires that it should be normalized according to

$$\int \psi^\dagger(\boldsymbol{r}) \psi(\boldsymbol{r})\, d^3 r = 1. \qquad (A.13.3)$$

In order to minimize the Hamiltonian with the above condition, one can make use of the Lagrange multiplier and make a functional derivative of the following quantity with respect to $\psi^\dagger(\boldsymbol{r})$

$$H[\psi] = \int \left[-\frac{1}{2m} \psi^\dagger(\boldsymbol{r}') \nabla'^2 \psi(\boldsymbol{r}') + \psi^\dagger(\boldsymbol{r}') U \psi(\boldsymbol{r}') \right] d^3 r'$$

$$-E\left(\int \psi^\dagger(\boldsymbol{r}')\psi(\boldsymbol{r}')\,d^3r' - 1\right), \qquad (A.13.4)$$

where E denotes a Lagrange multiplier and just a constant. In this case, one obtains

$$\frac{\delta H[\psi]}{\delta\psi^\dagger(\boldsymbol{r})} = \int \delta(\boldsymbol{r}-\boldsymbol{r}')\left[-\frac{1}{2m}\boldsymbol{\nabla}'^2\psi(\boldsymbol{r}') + U\psi(\boldsymbol{r}') - E\psi(\boldsymbol{r}')\right] d^3r' = 0. \qquad (A.13.5)$$

Therefore, one finds

$$-\frac{1}{2m}\boldsymbol{\nabla}^2\psi(\boldsymbol{r}) + U\psi(\boldsymbol{r}) = E\psi(\boldsymbol{r}) \qquad (A.13.6)$$

which is just the static Schrödinger equation.

A.13.2. Dirac Field

The Dirac equation for free field can be obtained by the variational principle of the Hamiltonian eq.(A.12.7). Below, we derive the static Dirac equation in a concrete fashion by the functional variation of the Hamiltonian.

Functional Variation of Hamiltonian

For the condition on $\psi_i(\boldsymbol{r})$, one requires that it should be normalized according to

$$\int \psi_i^\dagger(\boldsymbol{r})(\gamma^0)_{ij}\psi_j(\boldsymbol{r})\,d^3r = 1. \qquad (A.13.7)$$

Now, the Hamiltonian should be minimized with the condition of eq.(A.13.7)

$$H[\psi_i] = \int \psi_i^\dagger(\boldsymbol{r})\left[-i(\gamma^0\boldsymbol{\gamma}\cdot\boldsymbol{\nabla})_{ij} + m(\gamma^0)_{ij}\right]\psi_j(\boldsymbol{r})\,d^3r$$

$$-E\left(\int \psi_i^\dagger(\boldsymbol{r})(\gamma^0)_{ij}\psi_j(\boldsymbol{r})\,d^3r - 1\right), \qquad (A.13.8)$$

where E is just a constant of the Lagrange multiplier. By minimizing the Hamiltonian with respect to $\psi_i^\dagger(\boldsymbol{r})$, one obtains

$$(-i\boldsymbol{\gamma}\cdot\boldsymbol{\nabla} + m)\,\psi(\boldsymbol{r}) - E\psi(\boldsymbol{r}) = 0 \qquad (A.13.9)$$

which is just the static Dirac equation for free field.

Appendix B

Non-relativistic Quantum Mechanics

The quantization has two kinds of the procedure, the first quantization and the second quantization. By the first quantization, we mean that the coordinate $\boldsymbol{r}$ and the momentum $\boldsymbol{p}$ of a point particle do not commute with each other. That is,

$$[x_i, p_j] = \hbar\delta_{ij}.$$

In terms of this quantization procedure, we can obtain the Schrödinger equation by requiring that the particle Hamiltonian should be an operator and therefore the state ψ should be introduced.

There is another quantization procedure, the second quantization, which is the quantization of fields. From the experimental observations of creations and annihilations of particle pairs or photons, one needs to quantize fields. The field quantization is closely connected to the relativistic field equations which inevitably includes anti-particle states (negative energy states in fermion field case). In this respect, one does not have to quantize the Schrödinger field in the non-relativistic quantum mechanics. Therefore, we discuss only the first quantization procedure and problems related to the quantization.

B.1. Procedure of First Quantization

In the standard procedure of the first quantization, the energy E and momentum $\boldsymbol{p}$ are regarded as operators, and the simplest expressions of $\hat{E}$ and $\hat{\boldsymbol{p}}$ are given as

$$\hat{E} \to i\frac{\partial}{\partial t}, \quad \hat{\boldsymbol{p}} \to -i\boldsymbol{\nabla}. \qquad (B.1.1)$$

For a free point particle with its mass m, the dispersion relation can be written as

$$E = \frac{\boldsymbol{p}^2}{2m}. \qquad (B.1.2)$$

If one employs the quantization procedure of eq.(B.1.1), then one should prepare some state which receives the operation of eq.(B.1.1). This state is called wave function and is often denoted as $\psi(\boldsymbol{r},t)$. In this case, eq.(B.1.2) becomes

$$i\frac{\partial}{\partial t}\psi(\boldsymbol{r},t) = -\frac{1}{2m}\boldsymbol{\nabla}^2\psi(\boldsymbol{r},t) \qquad (B.1.3)$$

which is the Schrödinger equation.

New Picture of First Quantization

At a glance, one may feel that the procedure of eqs.$(B.1.1)$ and $(B.1.2)$ are more fundamental than eq.$(B.1.3)$ itself. However, this is not so trivial. If one looks into the Maxwell equation, then one realizes that the Maxwell equation is already a quantized equation for classical electromagnetic fields. In this respect, the Maxwell equation does not have any corresponding classical equation of motions like the Newton equation. In this sense, eq.$(B.1.3)$ can be regarded as a fundamental equation for quantum mechanics as well, even though one can derive eq.$(B.1.3)$ from eqs.$(B.1.1)$ and $(B.1.2)$.

In fact, in Appendix H, we treat the derivation of the Dirac equation from the Maxwell equation and the local gauge invariance, and there we see that the first quantization of eq.(B.1.1) is not needed and therefore it is not the fundamental principle any more. Instead, the Schrödinger equation is obtained from the non-relativistic reduction of the Dirac equation. In this respect, the derivation of the Schrödinger equation does not involve the first quantization.

B.2. Mystery of Quantization or Hermiticity Problem?

Here, we present a problem related to the quantization in box with the periodic boundary conditions. We restrict ourselves to the one dimensional case, but the result is easily generalized to three dimensions.

B.2.1. Free Particle in Box

By denoting the wave function as

$$\psi(x) = e^{-iEt}u(x)$$

the static Schrödinger equation without interactions is written

$$-\frac{1}{2m}\frac{\partial^2 u(x)}{\partial x^2} = Eu(x). \qquad (B.2.1)$$

The solution can be obtained as

$$u(x) = \left\{ \frac{1}{\sqrt{L}}e^{ikx}, \ \frac{1}{\sqrt{L}}e^{-ikx} \right\}, \ \text{with } E = \frac{k^2}{2m}, \qquad (B.2.2)$$

where one puts the particle into a box with its length L. Now, one requires that the wave function $u(x)$ should satisfy the periodic boundary conditions and should be the eigenstate of the momentum. In this case, one has

$$\hat{p}u_k(x) = ku_k(x), \ \ k = \frac{2\pi n}{L}, \ \ n = 0, \pm 1, \dots .$$

Therefore, one can write the eigenstate wave function as

$$u_n(x) = \frac{1}{\sqrt{L}}e^{i\frac{2\pi n}{L}x}, \ \ n = 0, \pm 1, \dots .$$

B.2.2. Hermiticity Problem

Now, the quantization relation is written

$$\hat{p}x - x\hat{p} = -i \tag{B.2.3}$$

and one takes the expectation value of the quantization relation with the wave function $u_n(x)$ and obtains

$$\langle u_n|\hat{p}x - x\hat{p}|u_m\rangle = -i\delta_{nm}. \tag{B.2.4}$$

If one makes use of the hermiticity of $\hat{p}$, then one obtains

$$(n-m)\frac{2\pi}{L}\langle u_n|x|u_m\rangle = -i\delta_{nm}. \tag{B.2.5}$$

However, the above equation does not hold for $n = m$ since the left hand side is zero while the right hand side is $-i$.

What is wrong with the calculation? The answer is simple, and one should not make use of the hermiticity of the momentum $\hat{p}$ because the surface term at the boundary does not vanish for the periodic boundary condition. In fact, in the above evaluation, the surface term just gives the missing constant of $-i$ for $n = m$. In other words, one can easily show the following equation

$$\langle u_n|\hat{p}x|u_n\rangle = -i + \langle \hat{p}u_n|x|u_n\rangle. \tag{B.2.6}$$

It should be noted that the hermiticity of the momentum in the following sense is valid

$$\langle u_n|\hat{p}|u_m\rangle = \langle \hat{p}u_n|u_m\rangle. \tag{B.2.7}$$

From this exercise, one learns that the quantization condition of eq.(B.2.3) should be all right, but the hermiticity of the momentum operator cannot necessarily be justified as long as one employs the periodic boundary conditions for the wave functions. Also, the periodic boundary conditions must be physically acceptable. Therefore, one should be careful for treating the momentum operator and it should be operated always on the right hand side as it is originally meant. In this case, one does not make any mistakes.

This argument must be valid even for a very large L as long as one keeps the periodic boundary conditions. This may look slightly odd, but the free particle should be present anywhere in the physical space, and therefore one should give up the vanishing of the surface term in the plane wave case. The criteria of right physics must be given from the observation that physical observables should not depend on L. In other words, one should take the value of L much larger than any other scales in the model, and this is called *thermodynamic limit.* In order to obtain any physical observables, one should always take the thermodynamic limit.

B.3. Schrödinger Fields

The Schrödinger equation with a potential $U(\boldsymbol{r})$ is written as

$$i\frac{\partial\psi(\boldsymbol{r},t)}{\partial t} = \left(-\frac{1}{2m}\nabla^2 + U(\boldsymbol{r})\right)\psi(\boldsymbol{r},t). \tag{B.3.1}$$

From this equation, one can derive the vector current conservation

$$\frac{\partial \rho}{\partial t} + \nabla \cdot \boldsymbol{j} = 0, \qquad (B.3.2)$$

where ρ and $\boldsymbol{j}$ are defined as

$$\rho = \psi^\dagger \psi, \quad \boldsymbol{j} = -\frac{i}{2m}\left(\psi^\dagger \nabla \psi - (\nabla \psi^\dagger)\psi\right). \qquad (B.3.3)$$

B.3.1. Currents of Bound State

Now, it is interesting to observe how the currents from this Schrödinger field behave in the realistic physical situations. Since the time dependence of the Schrödinger field ψ is factorized as

$$\psi(\boldsymbol{r}, t) = e^{-iEt} u(\boldsymbol{r})$$

the basic properties of the field are represented by the field $u(\boldsymbol{r})$. When the field $u(\boldsymbol{r})$ represents a bound state, then $u(\boldsymbol{r})$ becomes a real field. In this case, the current density $\boldsymbol{j}$ vanishes to zero,

$$\boldsymbol{j} = -\frac{i}{2m}\Big(u(\boldsymbol{r})\nabla u(\boldsymbol{r}) - (\nabla u(\boldsymbol{r}))u(\boldsymbol{r})\Big) = 0. \qquad (B.3.4)$$

On the other hand, the probability density of $\rho \equiv |u(\boldsymbol{r})|^2$ is always time-independent, and since the bound state wave function is confined within a limited area of space, the ρ is also limited within some area of space.

B.3.2. Free Fields (Static)

When there is no potential, that is

$$U(\boldsymbol{r}) = 0$$

then the field can be described as a free particle solution. This solution is obtained when the theory is put into a box with its volume V,

$$\psi(\boldsymbol{r}, t) = \frac{1}{\sqrt{V}} e^{-iEt} e^{i\boldsymbol{k}\cdot\boldsymbol{r}}, \quad \frac{1}{\sqrt{V}} e^{-iEt} e^{-i\boldsymbol{k}\cdot\boldsymbol{r}}, \qquad (B.3.5)$$

where E is the energy of the particle and $\boldsymbol{k}$ denotes a quantum number which should correspond to the momentum of a particle. In this case, the probability density is finite and constant

$$\rho = \frac{1}{V} \qquad (B.3.6)$$

while the current $\boldsymbol{j}$ is also a constant and can be written as

$$\boldsymbol{j} = \frac{\boldsymbol{k}}{m} \qquad (B.3.7)$$

which just corresponds to the velocity of a particle.

Real Field Condition

It is important to note that the Schrödinger field ψ should be a complex function. If one imposes the condition that the field should be real,

$$\psi(\boldsymbol{r},t) = \psi^{\dagger}(\boldsymbol{r},t)$$

then one obtains from the Schrödinger equation eq.(B.1.3)

$$\frac{\partial \psi(\boldsymbol{r},t)}{\partial t} = 0.$$

Therefore, the Schrödinger field ψ should be time independent. In this case, one sees immediately that the energy E must be zero since

$$\left(-\frac{1}{2m}\nabla^2 + U(\boldsymbol{r})\right)\psi(\boldsymbol{r},t) = 0.$$

Therefore, the real field condition of ψ is too strong and it should not be imposed before solving the Schrödinger equation. This concept should always hold in the Schrödinger field, and therefore it is most likely true that the same concept should hold for relativistic boson fields as well. However, this statement may not be justified if the Klein–Gordon field should not have any correspondence with the Schrödinger field in the non-relativistic limit.

B.3.3. Degree of Freedom of Schrödinger Field

The Schrödinger field is a complex field. However, the Schrödinger field ψ itself should correspond to one particle. It is clear that one cannot make the following separation of the field into real and imaginary parts

$$\psi(\boldsymbol{r},t) = \rho(\boldsymbol{r},t)e^{i\xi(\boldsymbol{r},t)} \tag{B.3.8}$$

and claim that $\rho(\boldsymbol{r},t)$ and $\xi(\boldsymbol{r},t)$ describe two independent fields (particles). When ψ is a complex field, it has right properties as a field, and the current density of the ψ field has a finite value.

B.4. Hydrogen-like Atoms

When the potential $U(\boldsymbol{r})$ is a Coulomb type,

$$U(\boldsymbol{r}) = -\frac{Ze^2}{r} \tag{B.4.1}$$

then the Schrödinger equation can be solved exactly and the Schrödinger field ψ together with the energy eigenvalue E can be obtained as

$$\psi(\boldsymbol{r}) = R_{n\ell}(r)Y_{\ell m}(\theta,\varphi), \tag{B.4.2}$$

$$E_n = -\frac{mZ^2e^4}{2n^2} = -\frac{m}{2n^2}\left(\frac{Z}{137}\right)^2, \tag{B.4.3}$$

where $R_{n\ell}(r)$ and $Y_{\ell m}$ denote the radial wave function and the spherical harmonics, respectively. The principal quantum number n runs as $n = 1, 2, \ldots, \infty$, and ℓ runs as $\ell = 0, 1, 2, \ldots, \infty$, satisfying the condition

$$\ell \leq n - 1. \tag{B.4.4}$$

It should be worth writing the explicit shape of the wave functions for a few lowest states of $1s$, $2p$ and $2s$ with the Bohr radius $a_0 = \frac{1}{me^2}$.

$$1s\text{−statee} : \; R_{1s}(r) = \left(\frac{Z}{a_0}\right)^{\frac{3}{2}} 2e^{-\frac{Zr}{a_0}}, \quad Y_{00}(\theta, \varphi) = \frac{1}{\sqrt{4\pi}},$$

$$2p\text{−state} : \; R_{2p}(r) = \left(\frac{Z}{2a_0}\right)^{\frac{3}{2}} \frac{Zr}{\sqrt{3}\,a_0} e^{-\frac{Zr}{2a_0}},$$

$$\begin{cases} Y_{11}(\theta, \varphi) = -\sqrt{\dfrac{3}{8\pi}} \sin\theta e^{i\varphi} \\ Y_{10}(\theta, \varphi) = \sqrt{\dfrac{3}{4\pi}} \cos\theta \\ Y_{1-1}(\theta, \varphi) = \sqrt{\dfrac{3}{8\pi}} \sin\theta e^{-i\varphi} \end{cases},$$

$$2s\text{−state} : \; R_{2s}(r) = \frac{1}{2\sqrt{2}} \left(\frac{Z}{a_0}\right)^{\frac{3}{2}} \left(2 - \frac{Zr}{a_0}\right) e^{-\frac{Zr}{2a_0}}, \quad Y_{00}(\theta, \varphi) = \frac{1}{\sqrt{4\pi}}.$$

B.5. Harmonic Oscillator Potential

The harmonic oscillator potential $U(x)$

$$U(x) = \frac{1}{2} m\omega^2 x^2$$

is a very special potential which is often used in quantum mechanics exercise problems since the Schrödinger equation with the harmonic oscillator potential can be solved exactly. However, the harmonic oscillator potential is not realistic since it does not have a free field-like solution. The Schrödinger field in the harmonic oscillator potential is always confined and there is no scattering state solution.

Nevertheless, it should be worth writing solutions of the Schrödinger field ψ with its mass m in the one dimensional harmonic oscillator potential. The Schrödinger equation can be written as

$$\left(-\frac{1}{2m}\frac{\partial^2}{\partial x^2} + \frac{1}{2} m\omega^2 x^2\right) \psi(x) = E\psi(x). \tag{B.5.1}$$

In this case, the solution of the Schrödinger field ψ together with the energy eigenvalue E can be obtained as

$$\psi_n(x) = \left(\frac{\alpha^2}{4^n \pi (n!)^2}\right)^{\frac{1}{4}} H_n(\alpha x) e^{-\frac{1}{2}\alpha^2 x^2}, \tag{B.5.2a}$$

$$E_n = \omega\left(n + \frac{1}{2}\right), \quad n = 0, 1, 2, \ldots, \tag{B.5.2b}$$

where α is given as

$$\alpha = \sqrt{m\omega}. \qquad (B.5.3)$$

$H_n(\xi)$ denotes the hermite polynomial and is given as

$$H_n(\xi) = (-)^n e^{\xi^2} \frac{d^n}{d\xi^n} e^{-\xi^2}$$

since $H_n(\xi)$ can be expressed in terms of the generating function as

$$e^{-x^2+2\xi x} = e^{-(x-\xi)^2+\xi^2} = \sum_{n=0}^{\infty} \frac{1}{n!} H_n(\xi) x^n.$$

Some of them are given below

$$H_0(\xi) = 1, \quad H_1(\xi) = 2\xi, \quad H_2(\xi) = 4\xi^2 - 2, \qquad (B.5.4a)$$

$$H_3(\xi) = 8\xi^3 - 12\xi, \quad H_4(\xi) = 16\xi^4 - 48\xi^2 + 12. \qquad (B.5.4b)$$

B.5.1. Creation and Annihilation Operators

The Hamiltonian of the one dimensional harmonic oscillator potential

$$\hat{H} = \frac{\hat{p}^2}{2m} + \frac{1}{2} m\omega^2 x^2 \qquad (B.5.5)$$

can be rewritten in terms of creation $a^\dagger$ and annihilation a operators

$$a^\dagger = \sqrt{\frac{m\omega}{2}} x - \frac{i}{\sqrt{2m\omega}} \hat{p}, \quad a = \sqrt{\frac{m\omega}{2}} x + \frac{i}{\sqrt{2m\omega}} \hat{p} \qquad (B.5.6)$$

as

$$\hat{H} = \omega \left(a^\dagger a + \frac{1}{2} \right). \qquad (B.5.7)$$

$a^\dagger$ and a satisfies the following commutation relation

$$[a, a^\dagger] = 1 \qquad (B.5.8)$$

because of the definition of $a^\dagger$ and a in eq.(B.5.6).

Number Operator $\hat{N}$

By introducing the number operator $\hat{N}$ as

$$\hat{N} = a^\dagger a \qquad (B.5.9)$$

one finds

$$[\hat{N}, a] = -a, \quad [\hat{N}, a^\dagger] = a^\dagger. \qquad (B.5.10)$$

With the eigenstate $|\phi_n\rangle$ of the number operator $\hat{N}$ and its eigenvalue n

$$\hat{N}|\phi_n\rangle = n|\phi_n\rangle \qquad (B.5.11)$$

one can easily prove the following equations

$$a^\dagger|\phi_n\rangle = \sqrt{n+1}\,|\phi_{n+1}\rangle, \quad a|\phi_n\rangle = \sqrt{n}\,|\phi_{n-1}\rangle. \qquad (B.5.12)$$

This indicates that $a^\dagger$ operator increases the quantum number n by one unit while a decreases it in the same way. Therefore, $a^\dagger$ and a are called *creation* and *annihilation* operators, respectively. In addition, one can evaluate the expectation value of the number operator $\hat{N}$ with the state $|\phi_n\rangle$ as

$$n = \langle\phi_n|a^\dagger a|\phi_n\rangle = \|a\phi_n\|^2 \geq 0$$

which shows that the n must be a non-negative value. Therefore, one finds from eq.(B.5.12)

$$a|\phi_0\rangle = 0. \qquad (B.5.13)$$

Therefore, one sees that the smallest value of n is $n = 0$. This leads to the constraint for n as

$$n = 0, 1, 2, \ldots. \qquad (B.5.14)$$

Operating the Hamiltonian $\hat{H}$ on $|\phi_n\rangle$, one finds

$$\hat{H}|\phi_n\rangle = \omega\left(\hat{N} + \frac{1}{2}\right)|\phi_n\rangle = \omega\left(n + \frac{1}{2}\right)|\phi_n\rangle = E_n|\phi_n\rangle. \qquad (B.5.15)$$

Thus, the energy E_n can be written as

$$E_n = \omega\left(n + \frac{1}{2}\right), \quad n = 0, 1, 2, \ldots$$

which agrees with the result given in eq.(B.5.2). The state $|\phi_n\rangle$ can be easily constructed by operating $a^\dagger$ operator onto the $|\phi_0\rangle$

$$|\phi_n\rangle = \frac{1}{\sqrt{n!}}\left(a^\dagger\right)^n|\phi_0\rangle. \qquad (B.5.16)$$

It should be noted that the state $|\phi_n\rangle$ is specified by the quantum number n. If one wishes to obtain an explicit expression of the wave function, then one should project the state $|\phi_n\rangle$ onto the $|x\rangle$ or $|p\rangle$ representation as given below.

Explicit Wave Function in x-representation

The wave function $\psi_n(x) \equiv \langle x|\phi_n\rangle$ in the x-representation can be obtained in the following way. First, one solves the differential equation from eq.(B.5.13)

$$\langle x|a|x\rangle\langle x|\phi_0\rangle = \left(\sqrt{\frac{m\omega}{2}}x + \frac{1}{\sqrt{2m\omega}}\frac{\partial}{\partial x}\right)\psi_0(x) = 0 \qquad (B.5.17)$$

which leads to the ground state wave function

$$\psi_0(x) = \left(\frac{\alpha^2}{\pi}\right)^{\frac{1}{4}} e^{-\frac{1}{2}\alpha^2x^2}, \text{ with } \alpha = \sqrt{m\omega}. \qquad (B.5.18)$$

From eq.(B.5.16), one obtains the wave function for an arbitrary state $\psi_n(x)$

$$\psi_n(x) = \frac{1}{\sqrt{n!}} \langle x| \left(a^\dagger\right)^n |x\rangle \langle x|\phi_0\rangle$$
$$= \frac{1}{\sqrt{n!}} \left(\frac{\alpha^2}{\pi}\right)^{\frac{1}{4}} \left(\sqrt{\frac{m\omega}{2}}x - \frac{1}{\sqrt{2m\omega}} \frac{\partial}{\partial x}\right)^n e^{-\frac{1}{2}\alpha^2 x^2}$$

which can be shown to be just identical to eq.(B.5.2a).

Appendix C

Relativistic Quantum Mechanics of Bosons

When a particle motion becomes comparable to the velocity of light, then one has to consider the relativistic effects. The dispersion relation between the energy and momentum of the particle with its mass m becomes

$$E = \sqrt{\boldsymbol{p}^2 + m^2}\,. \tag{C.0.1}$$

If one employs the quantization procedure of eq.(B.1.1),

$$\hat{E} \to i\frac{\partial}{\partial t}, \qquad \hat{\boldsymbol{p}} \to -i\nabla$$

then one obtains the following equation from eq.(C.0.1)

$$i\frac{\partial}{\partial t}\phi = \sqrt{-\nabla^2 + m^2}\,\phi. \tag{C.0.2}$$

However, it is easy to realize that the differential operator in square root cannot be defined well.

C.1. Klein–Gordon Equation

Therefore, it is essential to rewrite eq.(C.0.1) for the quantization procedure

$$E^2 = \boldsymbol{p}^2 + m^2. \tag{C.1.1}$$

From this dispersion relation, one obtains

$$\left(\frac{\partial^2}{\partial t^2} - \nabla^2 + m^2\right)\phi = 0 \tag{C.1.2}$$

which is the Klein–Gordon equation. Eq.(C.1.2) can describe bosons which must be a spinless particle. This is clear since the field ϕ has only one component, and therefore it should correspond to one component particle which is the spin zero state.

It should be interesting to compare eq.(C.1.2) with the equation for the electromagnetic field eq.(E.5.1) which is the resulting equation from the Maxwell equation. One sees that the Maxwell equation has just the same structure as the Klein–Gordon equation. In this respect, the Maxwell equation does correspond already to the relativistic quantum mechanics. Indeed, in Appendix H, we will see that the Dirac equation can be obtained from the assumption that the Maxwell equation and the local gauge invariance are the most fundamental principles. In this case, one sees that the Klein–Gordon equation cannot be derived from the new principle which shows that the first quantization procedure is not the fundamental principle but is only the result of the Dirac equation as a consequence. Therefore, if one employs this standpoint, then it may well be difficult to justify that the Klein–Gordon equation can be derived as the fundamental equation.

C.2. Scalar Field

Here, we examine whether a real scalar field with a finite mass can exist as a physical observable or not in the Klein–Gordon equation. Normally, one finds that pion with the positive charge is an anti-particle of pion with the negative charge. This can be easily understood if we look into the structure of the pion in terms of quarks. $\pi^{\pm}$ are indeed anti-particle to each other by changing quarks into anti-quarks. Since pion is not an elementary particle, their dynamics must be governed by the complicated quark dynamics. Under some drastic approximations, the motion of pion may be governed by the Klein–Gordon like equation.

C.2.1. Physical Scalar Field

It looks that eq.(C.1.2) contains the negative energy state. However, one sees that eq.(C.1.2) is only one component equation and, therefore the eigenvalue of E^2 can be obtained as a physical observable. There is no information from the Klein–Gordon equation for the energy E itself, but only E^2 as we see it below,

$$\left(-\boldsymbol{\nabla}^2+m^2\right)\phi=E^2\phi. \tag{C.2.1}$$

Therefore, if one obtains the eigenvalue of E^2 as

$$E^2=\alpha, \quad \alpha>0$$

then one cannot say which one of $E=\sqrt{\alpha}$ or $E=-\sqrt{\alpha}$ should be taken. Both solutions can be all right, but one should take only one of the solutions. Here, we take a positive value of E. In this case, the solution of eq.(C.2.1) should be described just in the same way as the Schrödinger field

$$\phi_{\boldsymbol{k}}(x)=A(t)e^{i\boldsymbol{k}\cdot\boldsymbol{r}} \tag{C.2.2}$$

which should be an eigenfunction of the momentum operator $\hat{\boldsymbol{p}}=-i\boldsymbol{\nabla}$. The coefficient $A(t)$ can be determined by putting eq.(C.2.2) into eq.(C.1.1). One can easily find that $A(t)$ should be written as

$$A(t)=\frac{1}{\sqrt{V\omega_{\boldsymbol{k}}}}a_{\boldsymbol{k}}e^{-i\omega_{\boldsymbol{k}}t}, \tag{C.2.3}$$

where $a_{\boldsymbol{k}}$ is a constant, and $\omega_{\boldsymbol{k}}$ is given as

$$\omega_{\boldsymbol{k}} = \sqrt{\boldsymbol{k}^2 + m^2}\,. \tag{C.2.4}$$

The shape of $A(t)$ in eq.(C.2.3) can be also determined from the Lorentz invariance. Now, one sees that eq.(C.2.2) has a right non-relativistic limit. This is indeed a physical scalar field solution of the Klein–Gordon equation.

C.2.2. Current Density

Now, we discuss the current density of the Klein–Gordon field which is defined as

$$\rho(x) = i\left(\phi^\dagger(x)\frac{\partial\phi(x)}{\partial t} - \frac{\partial\phi^\dagger(x)}{\partial t}\phi(x)\right), \tag{C.2.5a}$$

$$\boldsymbol{j}(x) = -i\left(\phi^\dagger(x)(\boldsymbol{\nabla}\phi(x)) - (\boldsymbol{\nabla}\phi^\dagger(x))\phi(x)\right). \tag{C.2.5b}$$

It should be noted that the current density must be hermitian and therefore the shape of eqs.(C.2.5) is uniquely determined. One cannot change the order between

$$\frac{\partial\phi^\dagger(x)}{\partial t} \quad \text{and} \quad \phi(x)$$

in the second term of eq.(C.2.5a).

Classical Real Scalar Field

Now, we come to an important observation that a real scalar field should have a serious problem. The real scalar field $\phi(x)$ can be written as

$$\phi(x) = \sum_{\boldsymbol{k}} \frac{1}{\sqrt{2V\omega_k}}\left[a_k e^{-i\omega_k t + i\boldsymbol{k}\cdot\boldsymbol{r}} + a_k^* e^{i\omega_k t - i\boldsymbol{k}\cdot\boldsymbol{r}}\right], \tag{C.2.6}$$

where V denotes the box. We assume that the $\phi(x)$ is still a classical field, that is, a_k^* and a_k are not operators, but just the c-number.

In this case, it is easy to prove that the current density of $(\rho(x), \boldsymbol{j}(x))$ which is constructed from the real scalar field $\phi(x)$ vanishes to zero

$$\rho(x) = 0, \quad \boldsymbol{j}(x) = 0.$$

This means that the real scalar field cannot propagate classically. This is clear since a real wave function even in the Schrödinger equation cannot propagate. That is, this real state should have the energy $E = 0$.

The basic problem is that one cannot impose the condition that the scalar field should be a real, and this is too strong as a physical condition. Within the classical field theory, the scalar Klein–Gordon field should be reduced to the Schrödinger field when it moves much more slowly than the velocity of light. Therefore, it is quite difficult to make any physical pictures of the real scalar Klein–Gordon field since it is not compatible with the Schrödinger field in the non-relativistic limit.

Quantized Real Scalar Field

Now, we come to the current density when the field is quantized. Below it is shown that the current density of the real scalar field has some problem even if quantized, contrary to a common belief [99].

When one quantizes the boson field of ϕ, then $a_k^\dagger$ and a_k become creation and annihilation operators

$$\hat{\phi}(x) = \sum_k \frac{1}{\sqrt{2V\omega_k}} \left[a_k e^{-i\omega_k t + i\boldsymbol{k}\cdot\boldsymbol{r}} + a_k^\dagger e^{i\omega_k t - i\boldsymbol{k}\cdot\boldsymbol{r}} \right]. \qquad (C.2.7)$$

In this case, the current density of eq.(C.2.4) becomes

$$\hat{\rho} = i\left(\hat{\phi}(x)\hat{\Pi}(x) - \hat{\Pi}(x)\hat{\phi}(x)\right) = i\left[\hat{\phi}(x), \hat{\Pi}(x)\right], \qquad (C.2.8)$$

where $\hat{\Pi}(x)$ is a conjugate field of $\hat{\phi}(x)$, that is, $\Pi(x) = \dot{\phi}(x)$. However, the quantization condition of the boson fields with eq.(C.2.7) becomes

$$\left[\hat{\phi}(\boldsymbol{r}), \hat{\Pi}(\boldsymbol{r}')\right]_{t=t'} = i\delta(\boldsymbol{r} - \boldsymbol{r}'). \qquad (C.2.9)$$

Therefore, eq.(C.2.8) becomes

$$\hat{\rho} = i\left[\hat{\phi}(x), \hat{\Pi}(x)\right] = -\delta(\mathbf{0}) \qquad (C.2.10)$$

which is infinity. Thus, the current density of the quantized real boson field is divergent after the quantization! Therefore, it is by now obvious that the current density of the real scalar field has an improper physical meaning. This should be related to the fact that the quantized field ϕ is assumed to be a real field which must be a wrong condition in nature.

Physical Scalar Field

The physical scalar field $\phi(x)$ can be written as in eqs.(C.2.2) and (C.2.3)

$$\phi(x) = \sum_k \frac{1}{\sqrt{V\omega_{\boldsymbol{k}}}} a_{\boldsymbol{k}} e^{i\boldsymbol{k}\cdot\boldsymbol{r} - i\omega_{\boldsymbol{k}} t}.$$

In this case, we can calculate the current density for the scalar field with the fixed momentum of $\boldsymbol{k}$ and obtain

$$\rho = \frac{|a_{\boldsymbol{k}}|^2}{V}$$

which is positive definite and finite. Therefore, the physical scalar field does not have any basic problems.

C.2.3. Complex Scalar Field

Since a real scalar field has a difficulty not only in the classical case but also in the quantized case, it should be worth considering a complex scalar field. In this case, the complex scalar field can be written as

$$\hat{\phi}(x) = \sum_k \frac{1}{\sqrt{2V\omega_k}} \left[a_k e^{-i\omega_k t + i\boldsymbol{k}\cdot\boldsymbol{r}} + b_k^\dagger e^{i\omega_k t - i\boldsymbol{k}\cdot\boldsymbol{r}} \right]. \qquad (C.2.11)$$

In this case, the current density is well defined and has no singularity when quantized. It is commonly believed that the complex scalar field should describe charged bosons, one which has a positive charge and the other which has a negative charge. However, a question may arise as to where this degree of freedom comes from? By now, one realizes that there is no negative energy solution in the Klein–Gordon equation. If one took into account the negative energy solution, then one should have had field equations of two components. On the other hand, eq.(C.2.11) assumes a scalar field with two components, and not the result of the field equations. It is therefore most important to seek for the two component Klein–Gordon like equation which should be somewhat similar to the Dirac equation.

Boson Number

Since the vector current of the boson field is conserved, one can define the boson number N_B

$$N_B = \int \rho(x)\, d^3r \qquad (C.2.12)$$

which is a conserved quantity. In the Schrödinger field, the N_B is positive definite, but in the Klein–Gordon field of eq.(C.2.11), the N_B is not necessarily positive definite due to the negative energy solutions of the Klein–Gordon equation.

Charge vs. Boson Number

One cannot interpret the boson number as the charge. This is clear since the charge is associated with the coupling constant g of the gauge field, and the negative charge of the positron in QED is associated with the quantum number of the negative energy degree of freedom. Therefore, unless the Klein–Gordon field could take into account the negative energy degree of freedom in a proper manner, there is no way to interpret the negative value of the boson number N_B in terms of proper physics terminology. This cannot be remedied even if one quantizes the boson field. The quantum number has nothing to do with the field quantization, but it is determined from the equation of motion as well as the symmetry of the Lagrangian density.

C.2.4. Composite Bosons

Pions and ρ-mesons are composed of quark and anti-quark fields. Suppressing the isospin variables, we can describe the boson fields in terms of the Dirac field $\psi_q(x)\chi_{\frac{1}{2}}$ only with the large components, for simplicity

$$\Psi^{(B)} = \psi_q(x_1)\psi_{\bar{q}}(x_2)\left(\chi^{(1)}_{\frac{1}{2}} \otimes \chi^{(2)}_{\frac{1}{2}}\right) = \Phi^{(Rel)}(x)\Phi^{(CM)}(X)\xi_{s,s_z}, \qquad (C.2.13)$$

where $x = x_1 - x_2$, $X = \frac{1}{2}$. Here, $\psi_{\bar{q}}(x)\chi_{\frac{1}{2}}$ denotes the anti-particle field and ξ_{s,s_z} is the spin wave function of the boson. $\Phi^{(Rel)}(x)$ denotes the internal structure of the boson and $\Phi^{(CM)}(X)$ corresponds to the boson field. Now, it is clear that the boson field $\Phi^{(CM)}(X)$ is a complex field.

Neutral Scalar Field

In the field theory textbooks, the real scalar field is interpreted as a boson with zero charge. But this is not the right interpretation. The charge is a property of the field in units of the coupling constant. The positive and negative charges are connected to the flavor of the scalar fields. For example, a chargeless Schrödinger field, of course, has a finite current density of $\rho(\boldsymbol{r})$. The charge Q of the Schrödinger field is given as $Q = e_0 \int \rho(\boldsymbol{r})\, d^3r$ and for the chargeless field, we simply have $e_0 = 0$, which means that it does not interact with the electromagnetic field due to the absence of the coupling constant.

Schwinger Boson

There is a good example of the physical composite boson which is described in terms of the fermion and anti-fermion operators, and it is a boson in the Schwinger model. The Hamiltonian of the Schwinger model can be bosonized, and it is given in eq.(5.74)

$$H = \frac{1}{2} \sum_p \{ \tilde{\Pi}^\dagger(p)\tilde{\Pi}(p) + p^2 \tilde{\Phi}^\dagger(p)\tilde{\Phi}(p) + \mathcal{M}^2 \tilde{\Phi}^\dagger(p)\tilde{\Phi}(p) \} . \tag{5.74}$$

The boson fields $\tilde{\Phi}(p)$ and its conjugate field $\tilde{\Pi}(p)$ satisfy the bosonic commutation relation

$$\left[\tilde{\Phi}(p), \tilde{\Pi}^\dagger(p')\right] = i\delta_{p,p'} .$$

One can see that the boson field $\tilde{\Phi}(p)$ in the Schwinger model is indeed a complex field.

C.2.5. Gauge Field

The electromagnetic field A_μ is a real vector field which is required from the Maxwell equation, and therefore it has zero current density. However, the gauge field itself is gauge dependent and therefore it is not directly a physical observable. In this case, the energy flow in terms of the Poynting vector becomes a physical quantity. After the gauge fixing and the field quantization, the vector field $\hat{\boldsymbol{A}}(x)$ can be written as

$$\hat{\boldsymbol{A}}(x) = \sum_{\boldsymbol{k}} \sum_{\lambda=1}^{2} \frac{1}{\sqrt{2V\omega_{\boldsymbol{k}}}} \boldsymbol{\varepsilon}(\boldsymbol{k},\lambda) \left[c_{\boldsymbol{k},\lambda} e^{-ikx} + c^\dagger_{\boldsymbol{k},\lambda} e^{ikx} \right] , \tag{C.2.14}$$

where $\omega_{\boldsymbol{k}} = |\boldsymbol{k}|$. Here, $\boldsymbol{\varepsilon}(\boldsymbol{k},\lambda)$ denotes the polarization vector. In this case, one-photon state with $(\boldsymbol{k},\lambda)$ becomes

$$\boldsymbol{A}_{\boldsymbol{k},\lambda}(x) = \langle \boldsymbol{k},\lambda | \hat{\boldsymbol{A}}(x) | 0 \rangle = \frac{1}{\sqrt{2V\omega_{\boldsymbol{k}}}} \boldsymbol{\varepsilon}(\boldsymbol{k},\lambda) e^{-i\boldsymbol{k}\cdot\boldsymbol{r} + i\omega_{\boldsymbol{k}} t} \tag{C.2.15}$$

which is the eigenstate of the momentum operator $\hat{\boldsymbol{p}} = -i\boldsymbol{\nabla}$. In this respect, the gauge field A_μ is completely different from the Klein–Gordon scalar field. Naturally, the gauge field does not have any corresponding non-relativistic field.

C.3. Degree of Freedom of Boson Fields

The Klein–Gordon equations are obtained by just replacing the momentum and energy by the differential operators. Now, we should examine the number of the degrees of freedom. If the positive and negative energy states should be taken into account, there should be two degrees of freedom. Therefore, the states should be described by a spinor with two components. However, the Klein–Gordon equation has only one component for a real scalar field. This should not be sufficient for describing the boson state if it should contain the negative energy state.

Below we discuss the Klein–Gordon equation with a spinor of two components and present some attempt to obtain a Hamiltonian in the two by two matrix form. One example of the Hamiltonian for a free boson is given in the textbook by Gross [63]

$$H = \begin{pmatrix} m + \dfrac{\boldsymbol{p}^2}{2m} & \dfrac{\boldsymbol{p}^2}{2m} \\ -\dfrac{\boldsymbol{p}^2}{2m} & -\left(m + \dfrac{\boldsymbol{p}^2}{2m}\right) \end{pmatrix}. \qquad (C.3.1)$$

In this case, the wave equation becomes

$$i\frac{\partial \phi}{\partial t} = \begin{pmatrix} m + \dfrac{\boldsymbol{p}^2}{2m} & \dfrac{\boldsymbol{p}^2}{2m} \\ -\dfrac{\boldsymbol{p}^2}{2m} & -\left(m + \dfrac{\boldsymbol{p}^2}{2m}\right) \end{pmatrix} \phi. \qquad (C.3.2)$$

If one writes ϕ as

$$\phi^{\pm} = \begin{pmatrix} \xi_1 \\ \xi_2 \end{pmatrix} e^{i\boldsymbol{k}\cdot\boldsymbol{r} \pm iEt}$$

then one has for the energy eigenvalue as $E^2 = m^2 + \boldsymbol{k}^2$ which is a right dispersion relation for a boson. However, one notices that the Hamiltonian is not hermite, and in principle this should give rise to some problems when one wishes to extend it to interacting systems.

There are other attempts to write Klein–Gordon equation in the spinor form with two components. However, one can easily prove that, if one has E, $\boldsymbol{p}$, and m with which one wishes to make a Lorentz scalar as

$$H = E\beta_0 + \boldsymbol{p}\cdot\boldsymbol{\alpha} + m\beta_1 \qquad (C.3.3)$$

then one needs to have 5 independent elements of the hermite matrices. If the hermite matrices are 2×2, then there are only 4 independent components which in fact correspond to the Pauli matrices. Thus, the two dimensional representation of 5 independent hermite matrices are not possible, and this strongly suggests that the Klein–Gordon equation cannot be rewritten in a physically plausible form.

Appendix D

Relativistic Quantum Mechanics of Fermions

Electron has a spin and its magnitude is $1/2$. The relativistic equation of eq.(C.1.2) has only one component of the field ϕ, and therefore it cannot describe the spin one half particle such as electron. Therefore, one has to consider some other equations for electron, and this equation is discovered by Dirac as we describe below.

D.1. Derivation of Dirac Equation

The procedure of obtaining Dirac equation is rather simple. One starts from eq.(C.1.1) and tries to factorize it into a linear equation for E and $\boldsymbol{p}$. This can be realized as

$$E^2 - \boldsymbol{p}^2 - m^2 = (E - \boldsymbol{p}\cdot\boldsymbol{\alpha} - m\beta)(E + \boldsymbol{p}\cdot\boldsymbol{\alpha} + m\beta) = 0, \qquad (D.1.1)$$

where $\boldsymbol{\alpha}$ and β are four by four matrices, and they satisfy the following anti-commutation relations

$$\{\alpha_i, \alpha_j\} = 2\delta_{ij}, \quad \{\alpha_i, \beta\} = 0, \quad \beta^2 = 1,$$

where i and j run $i, j = x, y, z$. Some of the representations are given in Appendix A.6. The relativistic quantum mechanical equation for a free electron discovered by Dirac is given as

$$\left(i\frac{\partial}{\partial t} + i\boldsymbol{\nabla}\cdot\boldsymbol{\alpha} - m\beta \right) \psi = 0, \qquad (D.1.2)$$

where ψ denotes the wave function and should have four components since $\boldsymbol{\alpha}$ and β are four by four matrices,

$$\psi = \begin{pmatrix} \psi_1 \\ \psi_2 \\ \psi_3 \\ \psi_4 \end{pmatrix}. \qquad (D.1.3)$$

In the four components of the wave function, two of them correspond to the spin degrees of freedom and the other two components represent the positive and negative energy states. The static Dirac equation becomes

$$(-i\boldsymbol{\nabla}\cdot\boldsymbol{\alpha} + m\beta)\,\psi = E\psi. \qquad (D.1.4)$$

If the energy eigenvalue contains the negative energy states, which is indeed the case, then the negative energy states must be physical, in contrast to the boson case. For fermions, the negative energy states play a very important role in quantum field theory.

D.2. Negative Energy States

The negative energy states in the Dirac equation are essential for describing the vacuum of quantum field theory, and Dirac interpreted that the vacuum of quantum field theory must be occupied completely by the negative energy states. Because of the Pauli principle, the vacuum is stable as long as the states are full. If one creates a hole in the vacuum, then this corresponds to a new particle with the same mass of the particle which one considers. This can be easily seen since, from the vacuum energy

$$E_v = -\sum_n \sqrt{\mathbf{p}_n^2 + m^2}$$

one extracts the state n_0 which is one hole state

$$E_v^{1hole} = -\sum_{n \neq n_0} \sqrt{\mathbf{p}_n^2 + m^2}.$$

Therefore one obtains a hole state energy

$$E^h \equiv E_v^{1hole} - E_v = \sqrt{\mathbf{p}_{n_0}^2 + m^2} \qquad (D.2.1)$$

which means that the hole state has the same mass as the original fermion. Further, if the fermion has a charge e, then the hole state must have an opposite charge which can be seen in the same way as the energy case.

$$\text{hole charge } e_h \equiv \left(\sum_{n \neq n_0} e - \sum_n e \right) = -e. \qquad (D.2.2)$$

D.3. Hydrogen Atom

The Dirac equation is most successful for describing the spectrum of hydrogen atom. One writes the Dirac equation for the hydrogen-like atoms as

$$\left(-i\nabla \cdot \boldsymbol{\alpha} + m\beta - \frac{Ze^2}{r} \right) \psi = E\psi, \qquad (D.3.1)$$

where Z denotes the charge of nucleus. Before presenting the solution of the above equation, we should make some comments on this equation. It is of course clear that there is no system which is composed of one electron, except a free electron state. Therefore, even though eq.(D.3.1) shows a simple one body problem for an electron with the potential which is measured from the coordinate center, the realistic problem must be at least two body problem, a system composed of electron and proton. In addition, there must be some contributions from photons and virtual pairs of electron and positron in the intermediate states. If one wishes to discuss the problem of hydrogen atom in the field theoretical treatment, it becomes extremely difficult. This is clear since, in this case, one has to evaluate the system as a many body problem. Even a reliable treatment of the center of mass effects is a non-trivial issue if one starts from the Dirac equation.

D.3.1. Conserved Quantities

The Dirac Hamiltonian for electron in hydrogen atom is written as

$$H = -i\nabla \cdot \boldsymbol{\alpha} + m\beta - \frac{Ze^2}{r}. \qquad (D.3.2)$$

Now, one defines the total angular momentum $\boldsymbol{J}$ and an operator K by

$$\boldsymbol{J} = \boldsymbol{L} + \frac{1}{2}\boldsymbol{\Sigma}, \qquad (D.3.3a)$$

$$K = \beta(\boldsymbol{\Sigma} \cdot \boldsymbol{L} + 1), \qquad (D.3.3b)$$

where $\boldsymbol{\Sigma}$ is extended to 4×4 matrix of $\boldsymbol{\sigma}$ and is defined as

$$\boldsymbol{\Sigma} = \begin{pmatrix} \boldsymbol{\sigma} & 0 \\ 0 & \boldsymbol{\sigma} \end{pmatrix}.$$

In this case, it is easy to prove that $\boldsymbol{J}$ and K commute with the Hamiltonian H

$$[H, \boldsymbol{J}] = 0, \quad [H, K] = 0. \qquad (D.3.4)$$

Therefore, the energy eigenvalues can be specified by the eigenvalues of $\boldsymbol{J}^2$ and K

$$\boldsymbol{J}^2 \psi^{\kappa}_{j,j_m} = j(j+1)\psi^{\kappa}_{j,j_m}, \qquad (D.3.5a)$$

$$K\psi^{\kappa}_{j,j_m} = \kappa\psi^{\kappa}_{j,j_m}, \qquad (D.3.5b)$$

where κ takes values according to

$$\kappa = \mp\left(j + \frac{1}{2}\right) \quad \text{for } j = \ell \pm \frac{1}{2}. \qquad (D.3.6)$$

D.3.2. Energy Spectrum

In this case, the energy eigenvalue of the Dirac Hamiltonian in hydrogen atom is given as

$$E_{n,j} = m\left[1 - \frac{(Z\alpha)^2}{n^2 + 2\left(n - (j+\frac{1}{2})\right)\left[\sqrt{(j+\frac{1}{2})^2 - (Z\alpha)^2} - (j+\frac{1}{2})\right]}\right]^{\frac{1}{2}}, \qquad (D.3.7)$$

where α denotes the fine structure constant and is given as

$$\alpha = \frac{1}{137}.$$

The quantum number n runs as $n = 1, 2, \ldots$. The energy $E_{n,j}$ can be expanded to order α^4

$$E_{n,j} - m = -\frac{m(Z\alpha)^2}{2n^2} - \frac{m(Z\alpha)^4}{2n^4}\left(\frac{n}{j+\frac{1}{2}} - \frac{3}{4}\right) + o\left((Z\alpha)^6\right). \qquad (D.3.8)$$

The first term in the energy eigenvalue is the familiar energy spectrum of the hydrogen atom in the non-relativistic quantum mechanics.

D.3.3. Ground State Wave Function ($1s_{\frac{1}{2}}$ – state)

It may be worth to write the Dirac wave function of the lowest state in a hydrogen-like atom. We denote the ground state wave function by

$$\psi^{(-1)}_{\frac{1}{2},m_s}(\boldsymbol{r}) \ \text{for} \ 1s_{\frac{1}{2}}-\text{state}$$

since $\kappa = -1$ and $s = \frac{1}{2}$. The energy eigenvalue $E_{1s_{\frac{1}{2}}}$ is simply written as

$$E_{1s_{\frac{1}{2}}} = m\sqrt{1-(Z\alpha)^2}.$$

The ground state wave function is explicitly given as

$$\psi^{(-1)}_{\frac{1}{2},m_s}(\boldsymbol{r}) = \begin{pmatrix} f^{(-1)}(r) \\ -i\boldsymbol{\sigma}\cdot\hat{\boldsymbol{r}}g^{(-1)}(r) \end{pmatrix} \frac{\chi_{m_s}}{\sqrt{4\pi}}, \qquad (D.3.9)$$

where χ_{m_s} denotes the two component spinor and is given as

$$\chi_{\frac{1}{2}} = \begin{pmatrix} 1 \\ 0 \end{pmatrix}, \quad \chi_{-\frac{1}{2}} = \begin{pmatrix} 0 \\ 1 \end{pmatrix}.$$

$\hat{\boldsymbol{r}}$ is defined as

$$\hat{\boldsymbol{r}} = \frac{\boldsymbol{r}}{r}.$$

The radial wave functions $f^{(-1)}(r)$ and $g^{(-1)}(r)$ can be analytically written

$$f^{(-1)}(r) = N\rho^{\sqrt{1-(Z\alpha)^2}}\frac{e^{-\rho}}{\rho}, \qquad (D.3.10a)$$

$$g^{(-1)}(r) = N\rho^{\sqrt{1-(Z\alpha)^2}}\frac{e^{-\rho}}{\rho}\left[\frac{\varepsilon\left(\sqrt{1-(Z\alpha)^2}-1\right)-Z\alpha}{\sqrt{1-(Z\alpha)^2}+1-Z\alpha\varepsilon}\right], \qquad (D.3.10b)$$

where ρ and ε are defined as

$$\rho = \sqrt{m^2 - E^2_{1s_{\frac{1}{2}}}}\, r = Z\alpha m r,$$

$$\varepsilon = \sqrt{\frac{m - E_{1s_{\frac{1}{2}}}}{m + E_{1s_{\frac{1}{2}}}}} = \frac{1-\sqrt{1-(Z\alpha)^2}}{Z\alpha}.$$

N is a normalization constant and should be determined from

$$\int_0^\infty \left[\left(f^{(-1)}(r)\right)^2 + \left(g^{(-1)}(r)\right)^2\right] r^2\, dr = 1.$$

D.4. Lamb Shifts

The important consequence of the Dirac equation in hydrogen atom is that the energy eigenvalues are specified by the total angular momentum j apart from the principal quantum number n, and indeed this is consistent with experimental observations.

There is one important deviation of the experiment from the Dirac prediction, that is, the degeneracy of $2p_{\frac{1}{2}}$-state and $2s_{\frac{1}{2}}$-state is resolved. The $2p_{\frac{1}{2}}$-state is lower than the $2s_{\frac{1}{2}}$-state in hydrogen atom. This splitting is originated from the second order effect of the vector field $\boldsymbol{A}$ which affects only on the $2s_{\frac{1}{2}}$-state. Intuitively, the second order effects must be always attractive. However, in the calculation of the Lamb shift, one has to consider the renormalization of the mass term, and due to this renormalization effect, the second order contribution of the vector field $\boldsymbol{A}$ becomes repulsive, and therefore the $2s_{\frac{1}{2}}$-state becomes higher than the $2p_{\frac{1}{2}}$-state.

D.4.1. Quantized Vector Field

Here, we briefly explain how to evaluate the Lamb shift in the non-relativistic kinematics. In this calculation, the quantized electromagnetic field $\boldsymbol{A}$ should be employed

$$\hat{\boldsymbol{A}}(x) = \sum_{\boldsymbol{k}} \sum_{\lambda=1}^{2} \frac{1}{\sqrt{2V\omega_{\boldsymbol{k}}}} \boldsymbol{\varepsilon}(\boldsymbol{k},\lambda) \left[c_{\boldsymbol{k},\lambda} e^{-ikx} + c_{\boldsymbol{k},\lambda}^{\dagger} e^{ikx} \right], \tag{D.4.1}$$

where $c_{\boldsymbol{k},\lambda}$ and $c_{\boldsymbol{k},\lambda}^{\dagger}$ denote the creation and annihilation operators which satisfy the following commutation relations

$$[c_{\boldsymbol{k},\lambda}, c_{\boldsymbol{k}',\lambda'}^{\dagger}] = \delta_{\boldsymbol{k},\boldsymbol{k}'} \delta_{\lambda,\lambda'}$$

and all other commutation relations vanish.

D.4.2. Non-relativistic Hamiltonian

We start from the Hamiltonian for electron in the hydrogen atom with the electromagnetic interaction

$$H = \frac{\hat{\boldsymbol{p}}^2}{2m_0} - \frac{Ze^2}{r} - \frac{e}{m_0} \hat{\boldsymbol{p}} \cdot \hat{\boldsymbol{A}}, \tag{D.4.2}$$

where the $\hat{\boldsymbol{A}}^2$ term is ignored in the Hamiltonian.

D.4.3. Second Order Perturbation Energy

Now, the second order perturbation energy due to the electromagnetic interaction for a free electron state can be written as

$$\delta E = -\sum_{\lambda} \sum_{\boldsymbol{k}} \sum_{\boldsymbol{p}'} \left(\frac{e}{m_0} \right)^2 \frac{1}{2V\omega_{\boldsymbol{k}}} \frac{|\langle \boldsymbol{p}' | \boldsymbol{\varepsilon}(\boldsymbol{k},\lambda) \cdot \hat{\boldsymbol{p}} | \boldsymbol{p} \rangle|^2}{E_{\boldsymbol{p}'} + k - E_{\boldsymbol{p}}}, \tag{D.4.3}$$

where $|\boldsymbol{p}\rangle$ and $|\boldsymbol{p}'\rangle$ denote the free electron state with its momentum. Since the photon energy ($\omega_{\boldsymbol{k}} = k$) is much larger than the energy difference of the electron states ($E_{\boldsymbol{p}'} - E_{\boldsymbol{p}}$)

$$|E_{\boldsymbol{p}'} - E_{\boldsymbol{p}}| \ll k \tag{D.4.4}$$

one obtains

$$\delta E = -\frac{1}{6\pi^2}\Lambda\left(\frac{e}{m_0}\right)^2 \boldsymbol{p}^2, \tag{D.4.5}$$

where Λ is the cutoff momentum of photon. This divergence is proportional to the cutoff Λ which is not the logarithmic divergence. However, this is essentially due to the non-relativistic treatment, and if one carries out the relativistic calculation of quantum field theory, then the divergence becomes logarithmic.

D.4.4. Mass Renormalization and New Hamiltonian

Defining the effective mass δm as

$$\delta m = \frac{1}{3\pi^2}\Lambda e^2 \tag{D.4.6}$$

the free energy of electron can be written as

$$E_F = \frac{\boldsymbol{p}^2}{2m_0} - \frac{\boldsymbol{p}^2}{2m_0^2}\delta m \simeq \frac{\boldsymbol{p}^2}{2(m_0+\delta m)}, \tag{D.4.7}$$

where one should keep only the term up to order of e^2 because of the perturbative expansion. Now, one defines the renormalized (physical) electron mass m by

$$m = m_0 + \delta m \tag{D.4.8}$$

and rewrites the Hamiltonian H in terms of the renormalized electron mass m

$$H = \frac{\hat{\boldsymbol{p}}^2}{2m} - \frac{Ze^2}{r} + \frac{\hat{\boldsymbol{p}}^2}{2m^2}\delta m - \frac{e}{m}\hat{\boldsymbol{p}}\cdot\hat{\boldsymbol{A}}. \tag{D.4.9}$$

Here, the third term ($\frac{\hat{p}^2}{2m^2}\delta m$) corresponds to the counter term which cancels out the second order perturbation energy [eq.(D.4.3)].

D.4.5. Lamb Shift Energy

Using eq.(D.4.3), one can calculate the second order perturbation energy due to the electro-magnetic interaction for the $2s_{1/2}$ electron state in hydrogen atom [12]

$$\Delta E_{2s_{1/2}} = \frac{1}{6\pi^2}\Lambda\left(\frac{e}{m}\right)^2 \langle 2s_{1/2}|\hat{\boldsymbol{p}}^2|2s_{1/2}\rangle$$
$$-\sum_{\lambda}\sum_{\boldsymbol{k}}\sum_{n\ell}\left(\frac{e}{m}\right)^2 \frac{1}{2V\omega_{\boldsymbol{k}}} \frac{|\langle n\ell|\boldsymbol{\varepsilon}(\boldsymbol{k},\lambda)\cdot\hat{\boldsymbol{p}}|2s_{1/2}\rangle|^2}{E_{n,\ell}+k-E_{2s_{1/2}}}, \tag{D.4.10}$$

where the first term comes from the counter term. This energy can be rewritten as

$$\Delta E_{2s_{1/2}} = \frac{1}{6\pi^2}\left(\frac{e}{m}\right)^2 \sum_{n,\ell}|\langle n,\ell|\hat{\boldsymbol{p}}|2s_{1/2}\rangle|^2 \int_0^{\Lambda} dk \frac{E_{n,\ell}-E_{2s_{1/2}}}{E_{n,\ell}+k-E_{2s_{1/2}}}, \tag{D.4.11}$$

where the following identity equation is employed

$$\sum_{n,\ell} |\langle n,\ell|\hat{\boldsymbol{p}}|2s_{1/2}\rangle|^2 = \langle 2s_{1/2}|\hat{\boldsymbol{p}}^2|2s_{1/2}\rangle. \qquad (D.4.12)$$

Further, neglecting the dependence of $E_{n,\ell}$ in the energy denominator because of eq.(D.4.4), one can carry out the summation of $\sum_{n,\ell}$ in eq.(D.4.11) as

$$\sum_{n,\ell} |\langle n,\ell|\hat{\boldsymbol{p}}|2s_{1/2}\rangle|^2 (E_{n,\ell} - E_{2s_{1/2}})$$

$$= \frac{1}{2}\langle 2s_{1/2}| \left[[\hat{\boldsymbol{p}},\hat{H}_0],\hat{\boldsymbol{p}}\right] |2s_{1/2}\rangle = 2\pi Ze^2 \langle 2s_{1/2}|\delta(\boldsymbol{r})|2s_{1/2}\rangle, \qquad (D.4.13)$$

where $\hat{H}_0$ is the unperturbed Hamiltonian of hydrogen atom

$$\hat{H}_0 = \frac{\hat{\boldsymbol{p}}^2}{2m} - \frac{Ze^2}{r}.$$

Choosing the value of the cutoff Λ, though there is no justification, as

$$\Lambda \sim m \qquad (D.4.14)$$

one obtains the energy shift for the $2s_{1/2}$ electron state

$$\Delta E_{2s_{1/2}} \simeq 1040 \text{ MHz}$$

which is close to a right Lamb shift for the $2s_{1/2}$ state

$$\Delta E^{exp}_{2s_{1/2}} = 1057.862 \ \pm 0.020 \text{ MHz}.$$

It should be noted that there is no energy shift for the $2p_{1/2}$ state.

D.4.6. Lamb Shift in Muonium

The Lamb shift energy of $2s_{\frac{1}{2}}$ state in muonium (μ^+e^- system) presents an important QED test [7, 79, 97, 113]. This Lamb shift energy of muonium $\Delta E^{(\mu)}_{2s_{1/2}}$ can be related to that of hydrogen atom $\Delta E^{(H)}_{2s_{1/2}}$ as [85]

$$\Delta E^{(\mu)}_{2s_{1/2}} = \left(\frac{m_r^{(\mu)}}{m_r^{(H)}}\right)^3 \Delta E^{(H)}_{2s_{1/2}} \qquad (D.4.15)$$

where $m_r^{(\mu)}$ and $m_r^{(H)}$ are given as

$$m_r^{(\mu)} = \frac{m_e}{1+\frac{m_e}{m_\mu}}, \quad m_r^{(H)} = \frac{m_e}{1+\frac{m_e}{M_p}}.$$

Here, m_μ and M_p denote the masses of muon and proton, respectively. Using the experimental value of the hydrogen Lamb shift energy

$$\Delta E^{(H)}_{2s_{1/2}}(exp) = 1057.845 \ \pm 0.009 \ \text{MHz}$$

we can predict the Lamb shift energy of muonium

$$\Delta E^{(\mu)}_{2s_{1/2}}(th) = 1044 \ \ \text{MHz}.$$

This value should be compared to the observed value [113]

$$\Delta E^{(\mu)}_{2s_{1/2}}(exp) = 1042 \ \ \pm 22 \ \ \text{MHz}$$

which perfectly agrees with the prediction. The important point is that the new relation does not depend on the cutoff Λ. If the observed accuracy of the Lamb shift energy in muonium is improved, then the Lamb shift of muonium should present a very good test of the QED renormalization scheme.

D.4.7. Lamb Shift in Anti-hydrogen Atom

The structure of the negative energy state should be examined if one can measure the Lamb shift of anti-hydrogen atom [4]. In this case, positron should feel the effect of the vacuum state which should be somewhat different from the case in which electron may feel in the same situation. The Lamb shift energy of the $2s_{1/2}$ state in anti-hydrogen atom can be written as

$$\Delta E_{2s_{1/2}} = \frac{2m_r^3\alpha^5}{3m_e^2} \ln\left(\frac{\bar{\Lambda}}{< E_{n,\ell} >}\right) \qquad (D.4.16)$$

where $\bar{\Lambda}$ denotes the effective cutoff value of the anti-hydrogen atom. If the observed value of the $\bar{\Lambda}$ differs from the Λ value, then it should mean that there is some chance of understanding the structure of the negative energy state in the interacting field theory model.

D.4.8. Physical Meaning of Cutoff Λ

The calculated result of the Lamb shift energy depends on the cutoff Λ, which is not satisfactory at all. However, there is no way to avoid the presence of the cutoff Λ as long as we treat the Lamb shift in the non-relativistic field equations. The important point is that we should understand the origin of the value of the cutoff Λ. This should, of course, be understood if one treats it relativistically.

In the non-relativistic treatment, the mass counter term is linear divergent. However, if one treats it relativistically, the divergence is logarithmic. This reason of the one rank down of the divergence is originated from the fact that the relativistic treatment considers the negative energy states which in fact reduce the divergence rank due to the cancellation. Now, we consider the renormalization effect in hydrogen atom, and if we calculate the Lamb shift energy in the non-relativistic treatment, then it has the logarithmic divergence as we saw above, and this is the one rank down of the divergence. This is due to the fact that the evaluation of the Lamb shift energy is based on the cancellation between the counter term and the perturbation energy in hydrogen atom. In the same way, if one can calculate the Lamb shift energy relativistically, then one should obtain the one rank down of the divergence, and this means that it should be finite.

Appendix E

Maxwell Equation and Gauge Transformation

Fundamental equations for electromagnetic fields are the Maxwell equation, and they are written for the electric field $\boldsymbol{E}$ and magnetic field $\boldsymbol{B}$ as

$$\nabla \cdot \boldsymbol{E} = \rho, \tag{E.0.1}$$

$$\nabla \cdot \boldsymbol{B} = 0, \tag{E.0.2}$$

$$\nabla \times \boldsymbol{E} = -\frac{\partial \boldsymbol{B}}{\partial t}, \tag{E.0.3}$$

$$\nabla \times \boldsymbol{B} = \boldsymbol{j} + \frac{\partial \boldsymbol{E}}{\partial t}, \tag{E.0.4}$$

where ρ and $\boldsymbol{j}$ denote the charge and current densities, respectively. These are already equations for the fields and therefore they are quantum mechanical equations. In this respect, it is important to realize that the first quantization procedure ($[x_j, p_i] = i\hbar\delta_{ij}$) is already done in the Maxwell equation.

E.1. Gauge Invariance

The Maxwell equation is written in terms of $\boldsymbol{E}$ and $\boldsymbol{B}$. Now, if one introduces the vector potential $\boldsymbol{A}$ as

$$\boldsymbol{B} = \nabla \times \boldsymbol{A} \tag{E.1.1}$$

then eq.(E.0.2) can be always satisfied since

$$\nabla \cdot \boldsymbol{B} = \nabla \cdot \nabla \times \boldsymbol{A} = \nabla \times \nabla \cdot \boldsymbol{A} = 0.$$

Therefore, one often employs the vector potential in order to solve the Maxwell equation. However, one notices in this case that the number of the degrees of freedom is still 3, that is, A_x, A_y, A_z in spite of the fact that we made use of one equation [eq.(E.0.2)]. This means that there must be a redundancy in the vector potential. This is the gauge freedom, that is, if one transforms

$$\boldsymbol{A} = \boldsymbol{A}' + \nabla \chi \tag{E.1.2a}$$

then the magnetic field $\boldsymbol{B}$ does not depend on χ

$$\boldsymbol{B} = \nabla \times \boldsymbol{A} = \nabla \times \boldsymbol{A}' + \nabla \times \nabla\chi = \nabla \times \boldsymbol{A}',$$

where χ is an arbitrary function that depends on $(\boldsymbol{r}, t)$. Now, Faraday's law [eq.(E.0.3)] can be rewritten by using the vector potential,

$$\nabla \times \left(\boldsymbol{E} + \frac{\partial \boldsymbol{A}}{\partial t} \right) = 0. \tag{E.1.3}$$

This means that one can write the electric field $\boldsymbol{E}$ as

$$\boldsymbol{E} = -\nabla A_0 - \frac{\partial \boldsymbol{A}}{\partial t}, \tag{E.1.4}$$

where A_0 is an arbitrary function of $(\boldsymbol{r}, t)$ and is called electrostatic potential. Since $\boldsymbol{E}$ in eq.(E.1.4) must be invariant under the gauge transformation of eq.(E.1.2a), it suggests that A_0 should be transformed under the gauge transformation as

$$A_0 = A_0' - \frac{\partial \chi}{\partial t}. \tag{E.1.2b}$$

In this case, the electric field is invariant under the gauge transformation of eqs.(E.1.2)

$$\boldsymbol{E} = -\nabla A_0 - \frac{\partial \boldsymbol{A}}{\partial t} = -\nabla \left(A_0' - \frac{\partial \chi}{\partial t} \right) - \frac{\partial}{\partial t} (\boldsymbol{A}' + \nabla\chi) = -\nabla A_0' - \frac{\partial \boldsymbol{A}'}{\partial t}$$

and eq.(E.1.4) can automatically reproduce Faraday's law since

$$\nabla \times \boldsymbol{E} = -\nabla \times \nabla A_0 - \nabla \times \frac{\partial \boldsymbol{A}}{\partial t} = -\frac{\partial \boldsymbol{B}}{\partial t}.$$

E.2. Derivation of Lorenz Force in Classical Mechanics

The interaction of electrons with the electromagnetic forces in nonrelativistic kinematics can be determined from the gauge invariance. This is remarkable and therefore we explain the derivation below since it is indeed interesting to learn the basic mechanism of the interaction. First, one starts from a free electron Lagrangian in classical mechanics

$$L = \frac{1}{2} m \dot{\boldsymbol{r}}^2. \tag{E.2.1}$$

When one wishes to add any interaction of electron with $\boldsymbol{A}$ and A_0 to the above Lagrangian, one sees that the Lagrangian must be linear functions of $\boldsymbol{A}$ and A_0. This is clear since the Lagrangian must be gauge invariant under eqs.(E.1.2). From the parity and time reversal invariance, one can write down the new Lagrangian

$$L = \frac{1}{2} m \dot{\boldsymbol{r}}^2 + g(\dot{\boldsymbol{r}} \cdot \boldsymbol{A} - A_0), \tag{E.2.2}$$

where g is a constant which cannot be determined from the gauge condition. When one makes the gauge transformation

$$\boldsymbol{A} = \boldsymbol{A}' + \nabla\chi, \quad A_0 = A_0' - \frac{\partial \chi}{\partial t}$$

one obtains

$$L = \frac{1}{2} m\dot{\boldsymbol{r}}^2 + g(\dot{\boldsymbol{r}} \cdot \boldsymbol{A}' - A_0') + g\frac{d\chi}{dt}. \tag{E.2.3}$$

Since the total derivative in the Lagrangian does not have any effects on the equation of motion, eq.(E.2.2) is invariant under the gauge transformation. It is amazing that the shape of the Lagrangian for electrons interacting with the electromagnetic fields is determined from the gauge invariance.

It is now easy to calculate the equation of motion for electron,

$$m\ddot{\boldsymbol{r}} = g\dot{\boldsymbol{r}} \times \boldsymbol{B} + g\boldsymbol{E}, \tag{E.2.4}$$

where the first term in the right hand side corresponds to the Lorenz force.

E.3. Number of Independent Functional Variables

The Maxwell equations are described in terms of the electric field $\boldsymbol{E}$ and the magnetic field $\boldsymbol{B}$. Once the charge density ρ and the current density $\boldsymbol{j}$ are given, then one can determine the fields $\boldsymbol{E}, \boldsymbol{B}$. It should be important to count the number of the unknown functional variables and the number of equations.

E.3.1. Electric and Magnetic fields $\boldsymbol{E}$ and $\boldsymbol{B}$

In terms of the electric field $\boldsymbol{E}$ and the magnetic field $\boldsymbol{B}$, it is easy to count the number of the functional variables. The number is six since one has

$$E_x, \ E_y, \ E_z, \ B_x, \ B_y, \ B_z. \tag{E.3.1}$$

On the other hand, the number of equations looks eight since the Gauss law [eq.(E.0.1)] and no magnetic monopole [eq.(E.0.2)] give two equations, and Faraday's law [eq.(E.0.3)] and Ampere's law [eq.(E.0.4)] seem to have six equations. However, Faraday's law gives only two equations since there is one constraint because

$$\nabla \times \boldsymbol{E} + \frac{\partial \boldsymbol{B}}{\partial t} = 0 \longrightarrow \nabla \cdot \left(\nabla \times \boldsymbol{E} + \frac{\partial \boldsymbol{B}}{\partial t} \right) = \nabla \times \nabla \cdot \boldsymbol{E} + \frac{\partial (\nabla \cdot \boldsymbol{B})}{\partial t} = 0. \tag{E.3.2}$$

In addition, Ampere's law has two equations since there is one constraint due to the continuity equation because

$$\nabla \times \boldsymbol{B} - \boldsymbol{j} - \frac{\partial \boldsymbol{E}}{\partial t} = 0 \longrightarrow \nabla \cdot \left(\nabla \times \boldsymbol{B} - \boldsymbol{j} - \frac{\partial \boldsymbol{E}}{\partial t} \right)$$

$$= \nabla \times \nabla \cdot \boldsymbol{B} - \nabla \cdot \boldsymbol{j} - \frac{\partial \rho}{\partial t} = 0. \tag{E.3.3}$$

Therefore, the number of the Maxwell equations is six which agrees with the number of the independent functional variables as expected.

Integrated Gauss's Law

In the electro-static exercise problems, one often employs the integrated Gauss law

$$\int_S \boldsymbol{E}\cdot d\boldsymbol{S} = \int_V \boldsymbol{\nabla}\cdot\boldsymbol{E}\, d^3r = \int_V \rho\, d^3r = Q. \qquad (E.3.4)$$

For the spherical charge distribution of ρ, for example, one can determine the electric field E_r in spite of the fact that one has employed only one equation of the Gauss law. This is of course clear because the symmetry makes it possible to adjust the number of the independent functional variable E_r which is one and the number of equation which is also one.

E.3.2. Vector Field A_μ and Gauge Freedom

When one introduces the vector field A_μ as

$$\boldsymbol{B} = \boldsymbol{\nabla}\times\boldsymbol{A}, \quad \boldsymbol{E} = -\boldsymbol{\nabla}A_0 - \frac{\partial \boldsymbol{A}}{\partial t} \qquad (E.3.5)$$

then the number of the independent fields is four since

$$A_0,\ A_x,\ A_y,\ A_z. \qquad (E.3.6)$$

On the other hand, the number of equations is three since the Gauss law [eq.(E.0.1)] gives one equation

$$-\boldsymbol{\nabla}\cdot\left(\boldsymbol{\nabla}A_0 + \frac{\partial \boldsymbol{A}}{\partial t}\right) = \rho \qquad (E.3.7)$$

and Ampere's law gives two equations as discussed above due to the continuity equation

$$\boldsymbol{\nabla}\times(\boldsymbol{\nabla}\times\boldsymbol{A}) = \boldsymbol{j} - \frac{\partial}{\partial t}\left(\boldsymbol{\nabla}A_0 + \frac{\partial \boldsymbol{A}}{\partial t}\right). \qquad (E.3.8)$$

It is of course easy to see that no magnetic monopole

$$\boldsymbol{\nabla}\cdot\boldsymbol{B} = \boldsymbol{\nabla}\cdot\boldsymbol{\nabla}\times\boldsymbol{A} = \boldsymbol{\nabla}\times\boldsymbol{\nabla}\cdot\boldsymbol{A} = 0 \qquad (E.3.9)$$

and Faraday's law

$$\boldsymbol{\nabla}\times\boldsymbol{E} = \boldsymbol{\nabla}\times\left(-\frac{\partial \boldsymbol{A}}{\partial t} - \boldsymbol{\nabla}A_0\right) = -\boldsymbol{\nabla}\times\frac{\partial \boldsymbol{A}}{\partial t} = -\frac{\partial \boldsymbol{B}}{\partial t} \qquad (E.3.10)$$

are automatically satisfied in terms of the vector potential A_μ.

Gauge Freedom

Therefore, the number of the unknown functional variables is four, but the number of equations is three, and they are not the same. This redundancy of the vector field is just related to the gauge freedom, and if one wishes to solve the Maxwell equations in terms of the vector potential A_μ, then one should reduce the number of the functional variables of the vector potential by fixing the gauge freedom.

Electromagnetic Wave

As an example, if there is no source term present [$\rho = 0$ and $\boldsymbol{j} = 0$], then the solution of the Maxwell equations with the Coulomb gauge fixing gives the electromagnetic wave which is composed of the transverse field only

$$A_0 = 0, \quad A_z = 0, \quad (A_x, A_y) \neq 0, \tag{E.3.11}$$

where the direction of $\boldsymbol{k}$ is chosen to be z-direction.

E.4. Lagrangian Density of Electromagnetic Fields

For the electric field $\boldsymbol{E}$ and magnetic field $\boldsymbol{B}$, the total energy of the system becomes

$$\begin{aligned} E &= \frac{1}{2}\int (E_k E_k + B_k B_k)\, d^3r \\ &= \frac{1}{2}\int \left[\left(\dot{A}_k + \frac{\partial A_0}{\partial x_k}\right)^2 + \left(\frac{\partial A_k}{\partial x_j}\frac{\partial A_k}{\partial x_j} - \frac{\partial A_k}{\partial x_j}\frac{\partial A_j}{\partial x_k}\right)\right] d^3r. \end{aligned} \tag{E.4.1}$$

Now, one introduces the field strength $F_{\mu\nu}$ as

$$F_{\mu\nu} = \partial_\mu A_\nu - \partial_\nu A_\mu \tag{E.4.2}$$

which is gauge invariant. In this case, one sees that $F_{\mu\nu}$ just corresponds to the electric field $\boldsymbol{E}$ and magnetic field $\boldsymbol{B}$ as

$$F_{0k} = F^{k0} = -F_{k0} = -F^{0k} = E_k, \quad F_{ij} = F^{ij} = -F_{ji} = -F^{ji} = -\varepsilon_{ijk}B_k.$$

The Lagrangian density can be written as

$$\mathcal{L} = \frac{1}{2}(E_k E_k - B_k B_k) = \frac{1}{2}\left[\left(\dot{A}_k + \frac{\partial A_0}{\partial x_k}\right)^2 - \left(\frac{\partial A_k}{\partial x_j}\frac{\partial A_k}{\partial x_j} - \frac{\partial A_k}{\partial x_j}\frac{\partial A_j}{\partial x_k}\right)\right] \tag{E.4.3}$$

which leads to the following Lagrangian density

$$\mathcal{L} = \frac{1}{4}\left(-F_{0k}F^{0k} - F_{k0}F^{k0} - F_{jk}F^{jk}\right) = -\frac{1}{4}F_{\mu\nu}F^{\mu\nu}. \tag{E.4.4}$$

The Lagrange equation for A_ν is given as

$$\partial_\mu \frac{\partial \mathcal{L}}{\partial(\partial_\mu A_\nu)} \equiv \frac{\partial}{\partial t}\frac{\partial \mathcal{L}}{\partial \dot{A}_0} + \frac{\partial}{\partial x_k}\frac{\partial \mathcal{L}}{\partial(\frac{\partial A_\nu}{\partial x_k})} = \frac{\partial \mathcal{L}}{\partial A_\nu} \tag{E.4.5}$$

which becomes

$$\begin{aligned} [\nu = 0] &\longrightarrow \frac{\partial}{\partial x_k}\left(\dot{A}_k + \frac{\partial A_0}{\partial x_k}\right) = 0 \longrightarrow \nabla \cdot \boldsymbol{E} = 0, \\ [\nu = k] &\longrightarrow \frac{\partial}{\partial t}\left(\dot{A}_k + \frac{\partial A_0}{\partial x_k}\right) + \frac{\partial}{\partial x_j}(\varepsilon_{jki}B_i) = 0 \longrightarrow \frac{\partial \boldsymbol{E}}{\partial t} - \nabla \times \boldsymbol{B} = 0. \end{aligned}$$

They are just the Maxwell equations [eqs.(E.0.1) and (E.0.4)] without any source terms. Since no magnetic monopole and Faraday's law [eqs.(E.0.2) and (E.0.43] can be automatically satisfied in terms of the vector potential A_μ, the Lagrangian density of eq.(E.4.4) is the right one that reproduces the Maxwell equations.

E.5. Boundary Condition for Photon

When there is no source term present ($\rho = 0$, $\boldsymbol{j} = 0$), then eq.(E.3.8) becomes

$$\left(\boldsymbol{\nabla}^2 - \frac{\partial^2}{\partial t^2}\right)\boldsymbol{A} = 0, \qquad (E.5.1)$$

where the Coulomb gauge fixing condition

$$\boldsymbol{\nabla}\cdot\boldsymbol{A} = 0$$

is employed. In this case, one sees that eq.(E.5.1) is a quantum mechanical equation for photon. Yet, one does not discuss the bound state of photon. This is clear since photon cannot be confined. There is no bound state of photon in quantum mechanics and eq.(E.5.1) has always the plane wave solution

$$\boldsymbol{A}(x) = \sum_{\boldsymbol{k}}\sum_{\lambda=1}^{2}\frac{1}{\sqrt{2V\omega_{\boldsymbol{k}}}}\boldsymbol{\varepsilon}_\lambda\left[c_{\boldsymbol{k},\lambda}e^{-ikx} + c^{\dagger}_{\boldsymbol{k},\lambda}e^{ikx}\right], \qquad (E.5.2)$$

where the polarization vectors $\boldsymbol{\varepsilon}_\lambda$ has two components

$$\boldsymbol{\varepsilon}_1 = (1,0,0), \quad \boldsymbol{\varepsilon}_2 = (0,1,0) \qquad (E.5.3)$$

when the direction of $\boldsymbol{k}$ is chosen to be z-direction.

This is basically due to the fact that photon is massless and therefore one cannot specify the system one measures. It always propagates with the speed of light! But still the equation derived from the Maxwell equation is a quantum mechanical equation of motion, though relativistic.

Appendix F

Regularizations and Renormalizations

In quantum field theory, one often faces to the regularization which is not very easy to understand. Mathematically it is straightforward, but the connection of the regularization with physics is not at all simple.

F.1. Euler's Regularization

Here, we first explain Euler's regularization which has nothing to do with physics, at least, at the time of derivation. However, we can learn the essence of the regularization from this mathematical example.

F.1.1. Abelian Summation

Let us define the following abelian summation N_0

$$N_0 = \sum_{n=0}^{\infty} (-)^n. \tag{F.1.1}$$

This quantity has no definite value as long as one makes the summation as it is.

F.1.2. Regularized Abelian Summation

Now, one regularizes this quantity as

$$N_\lambda = \sum_{n=0}^{\infty} (-)^n e^{-n\lambda}, \tag{F.1.2}$$

where λ is a positive value which is eventually set to zero. The summation of eq.(F.1.2) can be carried out in a straightforward way, and one obtains

$$N_0 = \lim_{\lambda \to 0} \frac{1}{1+e^{-\lambda}} = \frac{1}{2}. \tag{F.1.3}$$

In this way, one obtains the finite value for the summation of $(-)^n$ if one regularizes the summation in a proper way. It is no doubt that quantities defined in eqs.$(F.1.1)$ and $(F.1.2)$ are mathematically different from each other. If there is any model in physics which can realize the procedure described above, then one should employ the above procedure of eq.$(F.1.2)$ as the regularization. Therefore, the regularization is closely connected with field theory models in physical world.

F.2. Chiral Anomaly

In the massless QED_2, one finds that the axial vector current is classically conserved due to Noether's theorem in the Lagrangian density

$$\partial_\mu j_5^\mu = 0.$$

However, the conservation of the axial vector current is violated when the vacuum state is regularized in a gauge invariant fashion. Here, we discuss how the anomaly appears when regularized and show it in an explicit calculation.

F.2.1. Charge and Chiral Charge of Vacuum

In the vacuum of the Dirac fields, all the negative energy states are filled with the negative energy particles. In this case, one often asks what is the charge and the chiral charge of the vacuum states.

Here, we consider the Schwinger model which is the two dimensional field theory model. In this case, the charge and the chiral charge of the vacuum states can be defined as

$$Q = \sum_{n=-\infty}^{n_0} + \sum_{n=n_0+1}^{\infty}, \qquad (F.2.1a)$$

$$Q^5 = \sum_{n=-\infty}^{n_0} - \sum_{n=n_0+1}^{\infty}. \qquad (F.2.1b)$$

It is obvious that the summation of the Q and Q^5 does not make sense.

F.2.2. Large Gauge Transformation

Now, one wishes to regularize the above charge and chiral charge. Here, one makes the regularization so that the regularization procedure is consistent with the so called *large gauge transformation*.

$$Q_\lambda = \sum_{n=-\infty}^{n_0} e^{\lambda(n+\frac{Lg}{2\pi}A_1)} + \sum_{n=n_0+1}^{\infty} e^{-\lambda(n+\frac{Lg}{2\pi}A_1)}, \qquad (F.2.2a)$$

$$Q^5_\lambda = \sum_{n=-\infty}^{n_0} e^{\lambda(n+\frac{Lg}{2\pi}A_1)} - \sum_{n=n_0+1}^{\infty} e^{-\lambda(n+\frac{Lg}{2\pi}A_1)}, \qquad (F.2.2a)$$

where L, g and A_1 denote the box length, the gauge coupling constant and the vector potential, respectively. Here, the vector potential A_1 depends only on time. The reason why the term $(n + \frac{Lg}{2\pi}A_1)$ appears is because the Hamiltonian of the Schwinger model has the invariance under the large gauge transformation

$$A_1 \to A_1 + \frac{2\pi}{Lg}N, \quad N \text{ integer.}$$

Therefore, it is natural to assume that the regularization should keep this invariance.

F.2.3. Regularized Charge

In this case, the regularized charge Q_λ and the chiral charge Q_λ^5 become

$$Q_\lambda = \frac{2}{\lambda} + O(\lambda), \tag{F.2.3a}$$

$$Q_\lambda^5 = 2n_0 + 1 + \frac{Lg}{\pi}A_1 + O(\lambda). \tag{F.2.3b}$$

Now, one should let λ very small, and then the charge Q becomes infinity. But this is of course expected since one counts the number of the particles in the negative energy states which must be infinity. Since this number has no meaning in the vacuum, one can reset the charge Q to zero.

On the other hand, the chiral charge Q^5 is finite when the value of λ is set to zero. This means that the chiral charge of the vacuum is not conserved since A_1 depends on time! This is somewhat strange since the Schwinger model has the chiral symmetry in the classical Lagrangian density. But the regularization induces the non-conservation of the chiral charge and this is called *anomaly*. In four dimensional QED, the same phenomena occur and the presence of the anomaly is indeed confirmed by experiments.

F.2.4. Anomaly Equation

If one makes a time derivative of Q^5, then one obtains

$$\dot{Q}_5 = \frac{Lg}{\pi}\dot{A}_1. \tag{F.2.4}$$

Since Q_5 is described as

$$Q_5 = \int j_5^0(x)\,dx \tag{F.2.5}$$

one can rewrite eq.(F.2.4) as

$$\partial_\mu j_5^\mu = \frac{g}{2\pi}\varepsilon_{\mu\nu}F^{\mu\nu}, \tag{F.2.6}$$

where $\varepsilon_{\mu\nu}$ denotes the anti-symmetric symbol in two dimensions. That is,

$$\varepsilon_{01} = -\varepsilon_{10} = 1, \quad \varepsilon_{00} = \varepsilon_{11} = 0.$$

Eq.(F.2.6) clearly shows that the conservation of the axial vector current does not hold any more due to the anomaly.

Physics of Anomaly

Physics of the anomaly is closely related to the gauge invariant regularization of the energy and charge of the vacuum state. Since we respect most the local gauge invariance in the regularization procedure, the axial vector current conservation is violated. The regularization of the charge is originated from the field quantization, and therefore one may say that the quantization and the local gauge invariance induce the violation of the axial vector current conservation.

F.3. Index of Renormalizability

Field theory models have infinite degrees of freedom, and therefore some of the quantities which are evaluated perturbatively become infinity. When the infinite quantity can be renormalized into some constants such as the coupling constant or fermion mass, this field theory model is called "renormalizable". When one calculates a dimensionless quantity $\hat{O}$ in the perturbation theory, then one can expand the expectation value of $\hat{O}$ in terms of the coupling constant g as

$$\langle \hat{O} \rangle = c_0 + c_1 g + c_2 g^2 + \cdots + c_n g^n + \cdots . \tag{F.3.1}$$

F.3.1. Renormalizable

If the coupling constant g is dimensionless, then one sees that c_1 should depend on the cut-off momentum Λ, at most,

$$c_1 \sim \ln(\Lambda/m). \tag{F.3.2}$$

The logarithmic dependence of the cut-off momentum Λ is not at all serious. This is clear since even if one makes Λ very large, $\ln(\Lambda/m)$ is just some number. In addition, c_n should be described in terms of some function of $\ln(\Lambda/m)$ as

$$c_n \sim f_n(\ln(\Lambda/m)) . \tag{F.3.3}$$

Therefore, one can expect that this field theory model must be renormalizable since the infinity only comes from $\ln(\Lambda/m)$.

F.3.2. Unrenormalizable

On the other hand, if the coupling constant g has a mass inverse dimension, then one sees that

$$c_1 \sim \Lambda. \tag{F.3.4}$$

This is difficult since the $\langle O \rangle$ becomes quickly infinity when Λ becomes very large. In addition, c_n should behave as

$$c_n \sim \Lambda^n. \tag{F.3.5}$$

In this case, it is impossible to adjust the parameters to renormalize the theory since the n-th order perturbative calculations diverge at the different level of the Λ dependence. Namely, to the n-th order perturbation evaluation, one needs a new parameter to adjust a new level of infinity. Therefore, one sees that the field theory model with the coupling constant of the mass inverse dimension cannot be renormalized perturbatively.

If one solves it non-perturbatively, then it should be a different story. However, at the present stage, we cannot say more what should happen to this field theory model if it is solved exactly.

F.3.3. Summary of Renormalizability

Here, we summarize the renormalizability conditions. The coupling constant g is defined in the interaction Lagrangian density in the following shape.

$$\begin{aligned}
\text{QED}\ &:\ \mathcal{L}_I = g\bar{\psi}A_\mu\gamma^\mu\psi, \\
\text{QCD}\ &:\ \mathcal{L}_I = g\bar{\psi}A_\mu^a T^a\gamma^\mu\psi, \\
\text{Thirring}\ &:\ \mathcal{L}_I = g(\bar{\psi}\gamma_\mu\psi)(\bar{\psi}\gamma^\mu\psi), \\
\text{NJL}\ &:\ \mathcal{L}_I = g\left[(\bar{\psi}\psi)^2 + (\bar{\psi}i\gamma_5\psi)^2\right].
\end{aligned}$$

Renormalizability conditions

$g \sim M^1$	super-renormalizable	QED_2, QCD_2
$g \sim M^0$	renormalizable	QED_4, QCD_4, Thirring
$g \sim M^{-1}$	unrenormalizable	
$g \sim M^{-2}$	unrenormalizable	NJL

QED_2: Quantum electrodynamics in two dimensions D.

QCD_2: Quantum chromodynamics in two dimensions D.

QED_4: Quantum electrodynamics in four dimensions D.

QCD_4: Quantum chromodynamics in four dimensions D.

Thirring: current- current interaction model in two dimensions D.

NJL: current- current interaction model in four dimensions D.

F.4. Infinity in Physics

In physics, there appear many kinds of infinities. Mathematically, infinity is simply denoted as ∞. For example, when one has to take the thermodynamic limit, one says that the box length L must be set to infinity $L \to \infty$. However, in physics, the above equation does not make sense since there is no infinity in reality. In other words, if one calculates some quantity in atomic system and has to take the thermodynamic limit, then $L = 6.4 \times 10^6$ m (the earth radius) is of course infinity since the atomic scale is of the order 10^{-6} cm, at most.

In this respect, the physical meaning of $L \to \infty$ is that the box length L must be much larger than any of the relevant scales in the system of the model. If one has to take an infinity in the mathematical sense, then it indicates that the model one considers is too much simplified and should be improved to introduce some relevant scale such that the new scale can play some role in the model.

Also, when one evaluates the momentum integral in the perturbation theory, then the integral ranges from $-\Lambda$ to Λ where Λ is set to infinity. However, the infinity of this cutoff momentum Λ means again that it should be much larger than any of the relevant mass scale m

$$\Lambda \gg m.$$

For example, if the value of Λ is

$$\Lambda \sim 10^8\, m$$

then it is physically sufficiently larger than m. The important point is that one should build a scheme in which physical observables should not depend on the choice of the Λ value if it is sufficiently larger than any of the mass scales.

In this respect, the logarithmic divergence $\ln(\frac{\Lambda}{m})$ is just a number for the infinity of the cutoff momentum Λ. For example, the value of $\Lambda \sim 10^8\, m$ gives

$$\ln\left(\frac{\Lambda}{m}\right) \simeq \ln 10^8 \simeq 18$$

which is not at all a large value. This indicates that one should not worry about the logarithmic divergence $\ln(\frac{\Lambda}{m})$ when one renormalizes the infinity of the self-energy diagrams as discussed in Appendix I.

Appendix G

Path Integral Formulation

The path integral in quantum mechanics can be obtained by rewriting the quantum mechanical amplitude $K(x,x':t)$ into many dimensional integrations over discretized coordinates x_n. We discuss some examples which can be related to physical observables. However, the path integral expression cannot be connected to the dynamics of classical mechanics, even though, superficially, there is some similarity between them. Then, we show Feynman's formulation of the path integral in quantum electrodynamics (QED), which is based on many dimensional integrations over the parameters $q_{\boldsymbol{k},\lambda}$ appearing in the vector potential. This should be indeed connected to the second quantization in field theory models. However, the field theory path integral formulation in most of the textbooks is normally defined in terms of many dimensional integrations over fields. In this case, the path integral does not correspond to the field quantization. Here, we clarify what should be the problems of the path integral formulation over the field variables and why the integrations over fields do not correspond to the field quantization.

G.1. Path Integral in Quantum Mechanics

The path integral formulation in one dimensional quantum mechanics starts from the amplitude $K(x,x':t)$ which is defined by

$$K(x,x':t) = \langle x'|e^{-iHt}|x\rangle, \qquad (G.1.1)$$

where the system is specified by the Hamiltonian H. In the field theory textbooks, one often finds the expression of the amplitude $K(x,x':t)$ in terms of the transition between the state $|x,t\rangle$ and $|x',t'\rangle$ as

$$\langle x',t'|x,t\rangle \longrightarrow \langle x'|e^{-iH(t'-t)}|x\rangle. \qquad (G.1.2)$$

However, the state $|x,t\rangle$ is not an eigenstate of the Hamiltonian and therefore, one cannot prove the rightarrow of eq.(G.1.2). Instead, we should rewrite the amplitude $K(x,x':t)$ so as to understand its physical meaning

$$K(x,x':t) = \langle x'|e^{-iHt}|x\rangle = \sum_n \psi_n(x')\psi_n^\dagger(x)e^{-iE_nt}, \qquad (G.1.3)$$

where $\psi_n(x)$ and E_n should be the eigenstate and the eigenvalue of the Hamiltonian H. We note that eq.(G.1.3) is not yet directly related to physical observables.

G.1.1. Path Integral Expression

We start from the amplitude $K(x',x:t)$

$$K(x',x:t) = \langle x'|e^{-iHt}|x\rangle,$$

where the Hamiltonian in one dimension is given as

$$H = \frac{\hat{p}^2}{2m} + U(x) = -\frac{1}{2m}\frac{\partial^2}{\partial x^2} + U(x). \tag{G.1.4}$$

Here, a particle with its mass m is bound in the potential $U(x)$. Now, one can make n partitions of t and $x'-x$, and therefore we label the discretized coordinate x as

$$x = x_0, x_1, x_2, \ldots, x' = x_n. \tag{G.1.5}$$

In this case, we assume that each x_i and p should satisfy the following completeness relations

$$\int_{-\infty}^{\infty} dx_i |x_i\rangle\langle x_i| = 1, \quad \langle x_i|p\rangle = \frac{1}{\sqrt{2\pi}}e^{ipx_i}, \quad \int_{-\infty}^{\infty} dp|p\rangle\langle p| = 1 \ \ (i = 1,\ldots,n), \tag{G.1.6}$$

Therefore, $K(x',x:t)$ becomes

$$K(x',x:t)$$

$$= \int_{-\infty}^{\infty} dx_1 \cdots \int_{-\infty}^{\infty} dx_{n-1} \langle x'|e^{-iH\Delta t}|x_{n-1}\rangle\langle x_{n-1}|e^{-iH\Delta t}|x_{n-2}\rangle \cdots \langle x_1|e^{-iH\Delta t}|x\rangle, \tag{G.1.7}$$

where Δt is defined as $\Delta t = \frac{t}{n}$. Further, one can calculate the matrix elements, for example, as

$$\langle x_1|e^{-iH\Delta t}|x\rangle = \langle x_1|\exp\left[-i\left(-\frac{\Delta t}{2m}\frac{\partial^2}{\partial x^2} + U(x)\Delta t\right)\right]|x\rangle$$

$$\simeq \exp(-iU(x)\Delta t)\langle x_1|\exp\left(i\frac{\Delta t}{2m}\frac{\partial^2}{\partial x^2}\right)|x\rangle + O((\Delta t)^2). \tag{G.1.8}$$

In addition, $\langle x_1|e^{i\frac{\Delta t}{2m}\frac{\partial^2}{\partial x^2}}|x\rangle$ can be evaluated by inserting a complete set of momentum states

$$\langle x_1|\exp\left(i\frac{\Delta t}{2m}\frac{\partial^2}{\partial x^2}\right)|x\rangle$$

$$= \int_{-\infty}^{\infty}\frac{dp}{2\pi}e^{-i\frac{p^2}{2m}\Delta t}e^{-ip(x-x_1)} = \sqrt{\frac{m}{2i\pi\Delta t}}e^{-i\frac{m(x-x_1)^2}{2\Delta t}}. \tag{G.1.9}$$

Therefore, one finds now the path integral expression for $K(x',x:t)$

$$K(x',x:t) = \lim_{n\to\infty}\left(\frac{m}{2i\pi\Delta t}\right)^{\frac{n}{2}} \times$$

$$\int_{-\infty}^{\infty} dx_1 \cdots \int_{-\infty}^{\infty} dx_{n-1} \exp\left\{ i \sum_{k=1}^{n} \left(\frac{m(x_k - x_{k-1})^2}{2\Delta t} - U(x_k)\Delta t \right) \right\}, \qquad (G.1.10a)$$

where $x_0 = x$ and $x_n = x'$, respectively. Since the classical action S is given as

$$S = \int_0^t dt \left(\frac{1}{2} m\dot{x}^2 - U(x) \right) = \lim_{n\to\infty} \sum_{k=1}^{n} \Delta t \left\{ \frac{m}{2} \left(\frac{x_k - x_{k-1}}{\Delta t} \right)^2 - U(x_k) \right\} \qquad (G.1.11)$$

the amplitude can be symbolically written as

$$K(x', x : t) = \mathcal{N} \int [\mathcal{D}x] \exp\left\{ i \int_0^t \left(\frac{1}{2} m\dot{x}^2 - U(x) \right) dt \right\}, \qquad (G.1.10b)$$

where $\mathcal{N} \int [\mathcal{D}x]$ is defined as

$$\mathcal{N} \int [\mathcal{D}x] \equiv \lim_{n\to\infty} \left(\frac{m}{2i\pi\Delta t} \right)^{\frac{n}{2}} \int_{-\infty}^{\infty} dx_1 \cdots \int_{-\infty}^{\infty} dx_{n-1}.$$

This is indeed amazing in that the quantum mechanical amplitude is connected to the Lagrangian of the classical mechanics for a particle with its mass m in the same potential $U(x)$. Since the procedure of obtaining eq.(G.1.10) is just to rewrite the amplitude by inserting the complete set of the $|x_n\rangle$ states, there is no mathematical problem involved in evaluating eq.(G.1.10).

G.1.2. Physical Mmeaning of Path Integral

However, the physical meaning of the procedure is not at all easy to understand.

It is clear that eq.(G.1.10a) is well defined and there is no problem since it simply involves mathematics. However, there is a big jump from eq.(G.1.10a) to eq.(G.1.10b), even though it looks straightforward. Eq.(G.1.10b) indicates that the first term of eq.(G.1.10b) in the curly bracket is the kinetic energy of the particle in classical mechanics. In this case, however, x_k and x_{k-1} cannot be varied independently as one sees it from classical mechanics since it is related to the time derivative. On the other hand, they must be varied independently in the original version of eq.(G.1.10a) since it has nothing to do with the time derivative in the process of the evaluation. This is clear since, in quantum mechanics, time and coordinate are independent from each other. Therefore, it is difficult to interpret the first term of eq.(G.1.10a) as the kinetic energy term in classical mechanics. Secondly, the procedure of rewriting the amplitude is closely connected to the fact that the kinetic energy of the Hamiltonian is quadratic in p, that is, it is described as $\frac{p^2}{2m}$. In this respect, the fundamental ingredients of the path integral formulation must lie in eq.(G.1.9) which relates the momentum operator p^2 to the time derivative of the coordinate x under the condition that x_k and x_{k-1} are sufficiently close to each other. In this sense, if the kinetic energy operator were linear in p like the Dirac equation, then there is no chance to rewrite the amplitude since the Gaussian integral is crucial in evaluating the integral.

Therefore, it should be difficult to claim that eq.(G.1.10a) can correspond to the dynamics of classical mechanics, even though, superficially, there is some similarity between them. In other words, it is hard to prove that the quantum mechanical expression of $K(x',x:t)$ is related to any dynamics of classical mechanics. One may say that $K(x',x:t)$ happens to have a similar shape to classical Lagrangian, mathematically, but, physically it has nothing to do with the dynamics of classical mechanics.

No Summation of Classical Path

In some of the path integral textbooks, one finds the interpretation that the quantum mechanical dynamics can be obtained by summing up all possible paths in the classical mechanical trajectories. However, if one starts from eq.(G.1.10b) and tries to sum up all the possible paths, then one has to find the functional dependence of the coordinates on time and should integrate over all the possible coordinate configurations as the function of time. This should be quite different from the expression of eq.(G.1.10a). If one wishes to sum up all the possible paths in eq.(G.1.10b), then one may have to first consider the following expression of the coordinate x as the function of time

$$x = x_{cl}(t') + \sum_{n=1}^{\infty} y_n \sin\left(\frac{2\pi n}{t}\right) t',$$

where $x_{cl}(t')$ denotes the classical coordinate that satisfies the Newton equation of motion with the initial conditions of $x_{cl}(0) = x$ and $x_{cl}(t) = x'$. The amplitude y_n is the expansion coefficient. Therefore, the integrations over all the paths should mean that one should integrate over

$$\prod_{n=1}^{\infty} \int_{-\infty}^{\infty} dy_n$$

and, in this case, one can easily check that the calculated result of the integration over all the paths cannot reproduce the proper quantum mechanical result for the harmonic oscillator case. This simply means that the integration over $dx_1 \cdots dx_{n-1}$ in eq.(G.1.10a) and the integration over classical paths are completely different from each other, which is a natural result. This fact is certainly known to some careful physicists, but most of the path integral textbooks are reluctant to putting emphasis on the fact that the classical trajectories should not be summed up in the path integral formulation. Rather, they say that the summation of all the classical paths should correspond to the quantization by the path integral, which is a wrong and misleading statement.

G.1.3. Advantage of Path Integral

What should be any merits of the path integral formulation? Eqs.(G.1.10) indicate that one can carry out the first quantization of the classical system once the Lagrangian is given where the kinetic energy term should have a quadratic shape, that is, $c\dot{x}^2$ with c some constant. In this case, one can obtain the quantized expression just by tracing back from eq.(G.1.10b) to eq.(G.1.10a).

There is an advantage of the path integral formulation. That is, one does not have to solve the differential equation. Instead, one should carry out many dimensional integrations.

It is, of course, not at all clear whether the many dimensional integrations may have some advantage over solving the differential equation or not. However, one can, at least, claim that the procedure of the many dimensional integrations is indeed an alternative method to solving the Schrödinger equation.

G.1.4. Harmonic Oscillator Case

When the potential $U(x)$ is a harmonic oscillator

$$U(x) = \frac{1}{2} m\omega^2 x^2$$

then one can evaluate the amplitude analytically after some lengthy calculations

$$K(x',x:t) = \sqrt{\frac{m\omega}{2i\pi \sin \omega t}} \exp\left\{ i\frac{m\omega}{2}\left[(x'^2 + x^2)\cot \omega t - \frac{2x'x}{\sin \omega t} \right] \right\}. \tag{G.1.12}$$

On the other hand, one finds

$$K(x,x:t) = \langle x|e^{-iHt}|x\rangle = \sum_n e^{-iE_n t}|\psi_n(x)|^2. \tag{G.1.13}$$

Therefore, if one integrates $K(x,x:t)$ over all space, then one obtains

$$\int_{-\infty}^{\infty} dx K(x,x:t) = \sum_{n=0}^{\infty} e^{-iE_n t} = \int_{-\infty}^{\infty} dx \sqrt{\frac{m\omega}{2i\pi \sin \omega t}}\, e^{-im\omega x^2 \tan \frac{\omega t}{2}}$$

$$= \frac{1}{2i \sin \frac{\omega t}{2}}. \tag{G.1.14}$$

Since the last term can be expanded as

$$\frac{1}{2i\sin \frac{\omega t}{2}} = \frac{e^{-\frac{i}{2}\omega t}}{1 - e^{-i\omega t}} = e^{-\frac{i}{2}\omega t}\sum_{n=0}^{\infty} (e^{-i\omega t})^n = \sum_{n=0}^{\infty} e^{-i\omega t(n+\frac{1}{2})} \tag{G.1.15}$$

one obtains by comparing two equations

$$E_n = \omega\left(n + \frac{1}{2}\right) \tag{G.1.16}$$

which is just the right energy eigenvalue of the harmonic oscillator potential in quantum mechanics.

It should be important to note that the evaluation of eq.(G.1.12) is entirely based on the expression of eq.(G.1.10a) which is just the quantum mechanical equation. Therefore, this example of the harmonic oscillator case shows that the rewriting of the amplitude $K(x',x:t)$ is properly done in obtaining eq.(G.1.10a). This does not prove any connection of the $K(x',x:t)$ to the classical mechanics.

G.2. Path Integral in Field Theory

The basic notion of the path integral was introduced by Feynman [56, 57, 58], and the formulation of the path integral in quantum mechanics is given in terms of many dimensional integrations of the discretized coordinates x_n. As one sees from eq.(G.1.10), the amplitude is expressed in terms of many dimensional integrations with the weight factor of e^{iS} where S is the action of the classical mechanics. This was, of course, surprising and interesting. However, as Feynman noted in his original papers, the path integral expression is not more than the ordinary quantum mechanics.

When the classical particle interacts with the electromagnetic field $\boldsymbol{A}$, the amplitude of the particle can be expressed in terms of the many dimensional integrations of the action of the classical particle. However, the electromagnetic field $\boldsymbol{A}$ is already a quantum mechanical object, and therefore, there is no need for the first quantization in the Maxwell equation. However, when one wishes to treat physical processes which involve the absorption or emission of photon, then one has to quantize the electromagnetic field $\boldsymbol{A}$ which is called field quantization or second quantization.

G.2.1. Field Quantization

The field quantization of the electromagnetic field $\boldsymbol{A}$ can be done by expanding the field $\boldsymbol{A}$ in terms of the plane wave solutions

$$\boldsymbol{A}(x) = \sum_{\boldsymbol{k}} \sum_{\lambda=1}^{2} \frac{1}{\sqrt{2V\omega_{\boldsymbol{k}}}} \boldsymbol{\varepsilon}(\boldsymbol{k},\lambda) \left[c_{\boldsymbol{k},\lambda} e^{-ikx} + c_{\boldsymbol{k},\lambda}^{\dagger} e^{ikx} \right]. \tag{G.2.1}$$

The field quantization requires that $c_{\boldsymbol{k},\lambda}$ and $c_{\boldsymbol{k},\lambda}^{\dagger}$ should be operators which satisfy the following commutation relations

$$[c_{\boldsymbol{k},\lambda}, c_{\boldsymbol{k}',\lambda'}^{\dagger}] = \delta_{\boldsymbol{k},\boldsymbol{k}'} \delta_{\lambda,\lambda'} \tag{G.2.2}$$

and all other commutation relations should vanish. This is the standard way of the second quantization procedure even though it is not understood well from the fundamental principle. However, it is obviously required from experiments since electron emits photon when it decays from the $2p_{\frac{1}{2}}$ state to the $1s_{\frac{1}{2}}$ state in hydrogen atom.

G.2.2. Field Quantization in Path Integral (Feynman's Ansatz)

In his original paper, Feynman proposed a new method to quantize the electromagnetic field $\boldsymbol{A}$ in terms of the path integral formulation [56, 57, 58]. Here, we should first describe his formulation of the path integral. For the fermion part, he employed the particle expression, and therefore the path integral is defined in terms of quantum mechanics.

For the gauge field, Feynman started from the Hamiltonian formulation of the electromagnetic field. The Hamiltonian of the electromagnetic field can be expressed in terms of the sum of the harmonic oscillators

$$H_{el} = \frac{1}{2} \sum_{\boldsymbol{k},\lambda} \left(p_{\boldsymbol{k},\lambda}^2 + k^2 q_{\boldsymbol{k},\lambda}^2 \right), \tag{G.2.3}$$

where $p_{k,\lambda}$ is a conjugate momentum to $q_{k,\lambda}$. Here, it should be noted that the $q_{k,\lambda}$ corresponds to the amplitude of the vector potential $\boldsymbol{A}(x)$. The classical $c_{k,\lambda}$ and $c^{\dagger}_{k,\lambda}$ can be expressed in terms of $p_{k,\lambda}$ and $q_{k,\lambda}$ as

$$c_{k,\lambda} = \frac{1}{\sqrt{2\omega_k}}\left(p_{k,\lambda} - i\omega_k q_{k,\lambda}\right), \tag{G.2.4a}$$

$$c^{\dagger}_{k,\lambda} = \frac{1}{\sqrt{2\omega_k}}\left(p_{k,\lambda} + i\omega_k q_{k,\lambda}\right). \tag{G.2.4b}$$

In this case, the Hamiltonian can be written in terms of $c_{k,\lambda}$ and $c^{\dagger}_{k,\lambda}$ as

$$H_{el} = \frac{1}{2}\sum_{k,\lambda}\omega_k\left(c^{\dagger}_{k,\lambda}c_{k,\lambda} + c_{k,\lambda}c^{\dagger}_{k,\lambda}\right). \tag{G.2.5}$$

It should be important to note that the Hamiltonian of eq.(G.2.3) is originated from the Hamiltonian of field theory and it has nothing to do with the classical Hamiltonian of Newton dynamics. For the electromagnetic field, there is no corresponding Hamiltonian of the Newton dynamics.

Feynman's Ansatz

Feynman proposed a unique way of carrying out the field quantization [56, 57, 58]. Since the Hamiltonian of the electromagnetic field can be written as the sum of the harmonic oscillators, he presented the path integral formulation for the electromagnetic field

$$K(q_{k,\lambda}, q'_{k,\lambda}, t) \equiv \mathcal{N}\int[\mathcal{D}q_{k,\lambda}]\exp\left\{\frac{i}{2}\int_0^t\sum_{k,\lambda}\left(\dot{q}^2_{k,\lambda} - k^2 q^2_{k,\lambda}\right)dt\right\} \tag{G.2.6}$$

which should correspond to the quantization of the variables $q_{k,\lambda}$, and this corresponds to the quantization of the $c_{k,\lambda}$ and $c^{\dagger}_{k,\lambda}$. Thus, it is the second quantization of the electromagnetic field.

In this expression, there is an important assumption for the coordinates $q_{k,\lambda}$ which are the parameters appearing in the vector potential. That is, the states $|q_{k,\lambda}\rangle$ should make a complete set. Only under this assumption, one can derive the quantization of the harmonic oscillators. Since this is the parameter space, it may not be easy to prove the completeness of the states $|q_{k,\lambda}\rangle$. Nevertheless, Feynman made use of the path integral expression to obtain the Feynman rules in the perturbation theory for QED. It may also be important to note that the path integral in Feynman's method has nothing to do with the integration of the configuration space. It is clear that one should not integrate out over the configuration space in the path integral since the field quantization should be done for the parameters $c_{k,\lambda}$ and $c^{\dagger}_{k,\lambda}$.

G.2.3. Electrons Interacting through Gauge Fields

When one treats the system in which electrons are interacting through electromagnetic fields, one can write the whole system in terms of the path integral formulation. In this case, however, we treat electrons in the non-relativistic quantum mechanics. The electromagnetic fields are treated just in the same way as the previous section.

$$K(q_{\boldsymbol{k},\lambda}, q'_{\boldsymbol{k},\lambda}, \boldsymbol{r}, \boldsymbol{r}', t)$$

$$\equiv \mathcal{N} \int [\mathcal{D}\boldsymbol{r}][\mathcal{D}q_{\boldsymbol{k},\lambda}] \exp\left\{ i \int_0^t \left(\frac{1}{2} m\dot{\boldsymbol{r}}^2 - g\dot{\boldsymbol{r}} \cdot \boldsymbol{A}(\boldsymbol{r}) + \frac{1}{2} \sum_{\boldsymbol{k},\lambda} \left(\dot{q}_{\boldsymbol{k},\lambda}^2 - k^2 q_{\boldsymbol{k},\lambda}^2 \right) \right) dt \right\}, \qquad (G.2.7)$$

where the vector potential $\boldsymbol{A}$ is given in eq.(G.2.1)

$$\boldsymbol{A}(x) = \sum_{\boldsymbol{k}} \sum_{\lambda=1}^{2} \frac{\boldsymbol{\varepsilon}(\boldsymbol{k},\lambda)}{\sqrt{V\omega_{\boldsymbol{k}}}} \left[\dot{q}_{\boldsymbol{k},\lambda} \cos(\boldsymbol{k}\cdot\boldsymbol{r}) + \omega_{\boldsymbol{k}} q_{\boldsymbol{k},\lambda} \sin(\boldsymbol{k}\cdot\boldsymbol{r}) \right] \qquad (G.2.8)$$

which is now rewritten in terms of the variables $q_{\boldsymbol{k},\lambda}$.

It should be noted that the path integral formulation works only for the electromagnetic field since its field Hamiltonian can be described by the sum of the harmonic oscillators. This is a very special case, and there is only little chance that one can extend his path integral formulation to other field theory models. In particular, it should be hopeless to extend the path integral formulation to the field quantization of quantum chromodynamics (QCD) since QCD includes fourth powers of $q_{\boldsymbol{k},\lambda}$. In this case, one cannot carry out the Gaussian integral in the parameter space of $q_{\boldsymbol{k},\lambda}$.

G.3. Problems in Field Theory Path Integral

In this section, we discuss the problems in the standard treatment of the path integral formulation in field theory models. Normally, one starts from writing the path integral formulation in terms of the many dimensional integrations over field variables [33, 34].

G.3.1. Real Scalar Field as Example

For simplicity, we take a real scalar field in 1+1 dimensions. In most of the field theory textbooks, the amplitude Z is written as

$$Z = \mathcal{N} \int [\mathcal{D}\phi(t,x)] \exp\left[i \int \mathcal{L}(\phi, \partial_\mu\phi)\, dt\, dx \right], \qquad (G.3.1)$$

where the Lagrangian density $\mathcal{L}(\phi, \partial_\mu\phi)$ is given as

$$\mathcal{L}(\phi, \partial_\mu\phi) = \frac{1}{2}\left(\frac{\partial\phi}{\partial t}\right)^2 - \frac{1}{2}\left(\frac{\partial\phi}{\partial x}\right)^2 - \frac{1}{2} m^2\phi^2. \qquad (G.3.2)$$

If we rewrite the path integral definition explicitly in terms of the field variable integrations like eq.(G.1.10), we find

$$Z = \mathcal{N} \lim_{n\to\infty} \int_{-\infty}^{\infty} \cdots \int_{-\infty}^{\infty} \prod_{k,\ell=1}^{n} d\phi_{k,\ell}$$

$$\times \exp\left[i \sum_{k,\ell=1}^{n} \Delta t \Delta x \left(\frac{(\phi_{k,\ell}-\phi_{k-1,\ell})^2}{2(\Delta t)^2} - \frac{(\phi_{k,\ell}-\phi_{k,\ell-1})^2}{2(\Delta x)^2} - \frac{1}{2}m^2\phi_{k,\ell}^2 \right)\right], \quad (G.3.3)$$

where $\phi_{k,\ell}$ is defined as

$$\phi_{k,\ell} = \phi(t_k, x_\ell), \text{ with } t_1 = t, \cdots, t_n = t' \text{ and } x_1 = x, \cdots, x_n = x'. \quad (G.3.4)$$

Also, Δt and Δx are defined as

$$\Delta t = \frac{(t-t')}{n}, \quad \Delta x = \frac{(x-x')}{n}. \quad (G.3.5)$$

Now, we should examine the physical meaning of the expression of the amplitude Z in eq.(G.3.3), and clarify as to what are the problems in eq.(G.3.3) in connection to the field quantization. The first problem is that eq.(G.3.3) does not contain any quantity which is connected to the initial and final states. This is clear since, in eq.(G.3.3), one should integrate over fields as defined and therefore no information of $\phi(t,x)$ and $\phi(t',x')$ is left while, in quantum mechanics version, the amplitude is described by the quantities $\psi_n(x)$ and $\psi_n^\dagger(x')$ which are the eigenstates of the Hamiltonian. The second problem is that the calculated result of the amplitude Z must be only a function of $m, \Delta t, \Delta x$, that is

$$Z = f(m, \Delta t, \Delta x). \quad (G.3.6)$$

This shows that the formulation which is started from many dimensional integrations over the field $\phi(t,x)$ has nothing to do with the second quantization. In addition, Z depends on the artificial parameters Δt and Δx, and this clearly shows that it cannot be related to any physical observables.

This is in contrast with the formulation of eq.(G.2.7) where the amplitude K is specified by the quantum number of the parameter space $q_{\mathbf{k},\lambda}$ which is connected to the state with a proper number of photons, and it must be a function of $m, g, q_{\mathbf{k},\lambda}$

$$K = f(m, g, q_{\mathbf{k},\lambda}, q'_{\mathbf{k},\lambda}). \quad (G.3.7)$$

In addition, the K does not depend on the parameters Δt and Δx, which is a natural result as one can see it from the quantum mechanical path integral formulation.

Finally, we note that the treatment of Feynman is based on the total QED Hamiltonian which is a conserved quantity. On the other hand, eq.(G.3.3) is based on the action which is obtained by integrating the Lagrangian density over space and time. As one knows, the Lagrangian density is not directly related to physical observables. Therefore, unless one can confirm that the path integral is reduced to the quantum mechanical amplitude like eq.(G.1.10), one cannot make use of the field theory path integral formulation. In fact, one cannot rewrite the expression of eq.(G.3.3) in terms of the field theory Hamiltonian density $\mathcal{H}$, contrary to the path integral formulation in quantum mechanics. This shows that the amplitude Z has nothing to do with the amplitude K in eq.(G.2.7). In this respect, the amplitude Z has no physical meaning, and therefore one cannot calculate any physical quantities from the path integral formulation of Z in field theory.

G.3.2. Lattice Field Theory

Most of the numerical calculations in the lattice field theory are based on the path integral formulation of eq.(G.3.3). Unfortunately, the path integral formulation of eq.(G.3.3) has lost its physical meaning, and therefore there is little chance that one can obtain any physics out of numerical simulations of the lattice field theory. In this respect, it is, in a sense, not surprising that the calculated result of Wilson's area law [112] in QED is unphysical with incorrect dimensions as we saw in Chapter 8.

Since Wilson's calculation is presented analytically, it may be worth writing again the result of the Wilson loop calculation in QED as given in Chapter 8

$$W \equiv \mathcal{N} \prod_m \prod_\mu \int_{-\infty}^{\infty} dB_{m\mu} \exp\left(i \sum_P B_{n\mu} + \frac{1}{2g^2} \sum_{n\mu\nu} e^{if_{n\mu\nu}} \right). \tag{8.49}$$

In the strong coupling limit, one can evaluate eq.(8.49) analytically as

$$W \simeq (g^2)^{-\Delta S/a^2}, \tag{G.3.8}$$

where ΔS should be an area encircled by the loop. This has the same behavior as that of eq.(G.3.6) since the lattice constant a is equal to $a = \Delta x = \Delta t$, that is,

$$W = f(g, \Delta S, \Delta x, \Delta t). \tag{G.3.9}$$

This amplitude W has the dependence of the artificial parameters Δt and Δx. This is completely different from eq.(G.3.7), and therefore one sees that the calculation of eq.(8.49) has no physical meaning, contrary to Feynman's treatment which has a right behavior as the function of the field parameters $q_{\boldsymbol{k},\lambda}$.

G.3.3. Physics of Field Quantization

The quantization of fields is required from experiment. Yet, it is theoretically quite difficult to understand the basic physics of the field quantization. The fundamental step of the quantization is that the Hamiltonian one considers becomes an operator after the quantization. The reason why one considers the Hamiltonian is because it is a conserved quantity. In this respect, one cannot quantize the Hamiltonian density since it is not a conserved quantity yet. This is an important reason why one must quantize the field in terms of the creation and annihilation operator $c_{\boldsymbol{k},\lambda}$ and $c^\dagger_{\boldsymbol{k},\lambda}$ in QED. In this respect, it is clear that the field quantization must be done in terms of $c_{\boldsymbol{k},\lambda}$ and $c^\dagger_{\boldsymbol{k},\lambda}$ with which the Hamiltonian in classical QED can be described.

G.3.4. No Connection between Fields and Classical Mechanics

Here, we should make a comment on the discretized coordinates and fields. The discretized space is, of course, artificial, and there is no physics in the discretized fields and equations. In some textbooks, the field equation is derived from the picture that the field is constructed by the sum of springs in which the discretized coordinates of neighboring sites are connected by the spring. This picture can reproduce the field equation for a massless scalar field by adjusting some parameters, even though one started from a non-relativistic classical mechanics. However, this is obviously a wrong picture for a scalar field theory since the field equation has nothing to do with classical mechanics. It is somewhat unfortunate that it may have had some impact on the picture concerning lattice gauge field calculations as an excuse to make use of the discretized classical fields. As we saw in the previous section, the path integral formulation has nothing to do with the dynamics of classical mechanics, and it is, of course, clear that the field theory path integral is never connected to any dynamics of the classical field theory.

G.4. Path Integral Function Z in Field Theory

As we saw in Section G.2, Feynman's path integral formulation in terms of many dimensional integrations in parameter space is indeed a plausible method to quantize the fields in QED. However, people commonly use the expression of the amplitude Z as defined in eq.(G.3.1).

G.4.1. Path Integral Function in QCD

The "new formulation" of the path integral in QCD was introduced by Faddeev and Popov who wrote the S-matrix elements as [32]

$$\langle \text{out}|\text{in}\rangle \equiv Z = \mathcal{N} \int [\mathcal{D}A_\mu^a(x)] \exp\left[i \int \mathcal{L}_{\text{QCD}} d^4x \right], \qquad (G.4.1)$$

where the definition of the path integral volume $[\mathcal{D}A_\mu^a(x)]$ is just the same as the one explained in the previous section

$$\int [\mathcal{D}A_\mu^a(x)] \equiv \lim_{n\to\infty} \int_{-\infty}^{\infty} \cdots \int_{-\infty}^{\infty} \prod_{\mu,a} \prod_{(i,j,k,\ell)}^{n} dA_\mu^a(x_0^i, x_1^j, x_2^k, x_3^\ell). \qquad (G.4.2)$$

The Lagrangian density $\mathcal{L}_{\text{QCD}}$ for QCD is given as

$$\mathcal{L}_{\text{QCD}} = -\frac{1}{2} \text{Tr}\left(G_{\mu\nu} G^{\mu\nu} \right), \qquad (G.4.3)$$

where $G_{\mu\nu}$ denotes the field strength in QCD as given in Chapter 6

$$G_{\mu\nu} = \partial_\mu A_\nu - \partial_\nu A_\mu + ig[A_\mu, A_\nu]. \qquad (G.4.4)$$

However, the expression of the path integral in QCD in eq.(G.4.1) does not correspond to the field quantization. One can also understand why the path integral formulation of QCD cannot be done, in contrast to Feynman's integrations over the parameter space in QED. In QCD, the Hamiltonian for gluons contains the fourth power of $q_{\boldsymbol{k},\lambda}^a$, and therefore one cannot carry out the Gaussian integrations over the parameters $q_{\boldsymbol{k},\lambda}^a$ in QCD. As Feynman stated repeatedly in his original papers, the path integral formulation is closely connected to the Gaussian integration where the kinetic energy term in non-relativistic quantum mechanics is always described in terms of the quadratic term $\frac{p^2}{2m}$. This naturally leads to the conclusion that the path integral formulation cannot be properly constructed in QCD, and this is consistent with the picture that QCD does not have a free Fock space due to its gauge non-invariance.

G.4.2. Fock Space

In quantum field theory, one must prepare Fock space since the Hamiltonian becomes an operator. The second quantized formulation is based on the creation and annihilation operators which act on the Fock space. In Feynman's path integral formulation, he prepared states which determine the number of photons in terms of $|q_{k,\lambda}\rangle$. Therefore, he started from the second quantized expression of the path integral formulation. However, Faddeev and Popov simply employed the same formula of the Lagrangian density, but integrated over the function $\mathcal{D}A_\mu^a(x)$. This cannot specify any quantum numbers of the Fock space, and therefore the integration over the function $\mathcal{D}A_\mu^a(x)$ does not correspond to the field quantization.

Appendix H

New Concept of Quantization

In this textbook, the first quantization of $[x_i, p_j] = i\delta_{ij}$ is employed to derive the Dirac equation. Here, we present a new way of understanding the first quantization. From the local gauge invariance and the Maxwell equation, one can derive the Lagrangian density of the Dirac field without involving the first quantization. This leads to a new concept of the quantization itself.

H.1. Derivation of Lagrangian Density of Dirac Field from Gauge Invariance and Maxwell Equation

Dirac derived the Dirac equation by factorizing eq.(D.1.1) such that the field equation becomes the first order in time derivative. Now, one can derive the Lagrangian density of Dirac field in an alternative way by employing the local gauge invariance and the Maxwell equation as the most fundamental principle [50].

H.1.1. Lagrangian Density for Maxwell Equation

We start from the Lagrangian density that reproduces the Maxwell equation

$$\mathcal{L} = -gj_\mu A^\mu - \frac{1}{4}F_{\mu\nu}F^{\mu\nu}, \qquad (H.1.1)$$

where A^μ is the gauge field, and $F_{\mu\nu}$ is the field strength and is given as

$$F_{\mu\nu} = \partial_\mu A_\nu - \partial_\nu A_\mu.$$

j_μ denotes the current density of matter field which couples to the electromagnetic field. From the Lagrange equation, one obtains

$$\partial_\mu F^{\mu\nu} = gj^\nu \qquad (H.1.2)$$

which is just the Maxwell equation as we also saw in Appendix E.

H.1.2. Four Component Spinor

Now, one can derive the kinetic energy term of the fermion Lagrangian density. First, one assumes that the Dirac fermion should have four components

$$\psi = \begin{pmatrix} \psi_1 \\ \psi_2 \\ \psi_3 \\ \psi_4 \end{pmatrix}.$$

This is based on the observation that electron has spin degree of freedom which is two. In addition, there must be positive and negative energy states since it is a relativistic field, and therefore the fermion field should have 4 components.

16 Independent Components

Now, the matrix elements

$$\psi^\dagger \hat{O} \psi$$

can be classified into 16 independent Lorentz invariant components as

$$\bar{\psi}\psi : \text{ scalar}, \quad \bar{\psi}\gamma_5\psi : \text{ pseudo} - \text{scalar}, \qquad (H.1.3a)$$

$$\bar{\psi}\gamma_\mu\psi : \text{ 4 component vector}, \quad \bar{\psi}\gamma_\mu\gamma_5\psi : \text{ 4 component axial-vector}, \qquad (H.1.3b)$$

$$\bar{\psi}\sigma_{\mu\nu}\psi : \text{ 6 component tensor}, \qquad (H.1.3c)$$

where $\bar{\psi}$ is defined for convenience as

$$\bar{\psi} = \psi^\dagger \gamma_0.$$

These properties are determined by mathematics.

Shape of Vector Current

From the invariance consideration, one finds that the vector current j_μ must be written as

$$j_\mu = C_0 \bar{\psi}\gamma_\mu\psi, \qquad (H.1.4)$$

where C_0 is a constant. Since one can renormalize the constant C_0 into the coupling constant g, one can set without loss of generality

$$C_0 = 1. \qquad (H.1.5)$$

H.2. Shape of Lagrangian Density

Now, one can make use of the local gauge invariance of the Lagrangian density, and one sees that the following shape of the Lagrangian density can keep the local gauge invariance

$$\mathcal{L} = C_1 \bar{\psi} \partial_\mu \gamma^\mu \psi - g \bar{\psi} \gamma_\mu \psi A^\mu - \frac{1}{4} F_{\mu\nu} F^{\mu\nu}, \qquad (H.2.1)$$

where C_1 is a constant. At this point, one requires that the Lagrangian density should be invariant under the local gauge transformation

$$A_\mu \longrightarrow A_\mu + \partial_\mu \chi, \qquad (H.2.2a)$$

$$\psi \longrightarrow e^{-ig\chi} \psi. \qquad (H.2.2b)$$

In this case, it is easy to find that the constant C_1 must be

$$C_1 = i. \qquad (H.2.3)$$

Here, the constant $\hbar$ should be included implicitly into the constant C_1. The determination of $\hbar$ can be done only when one compares calculated results with experiment such as the spectrum of hydrogen atom.

H.2.1. Mass Term

The Lagrangian density of eq.(H.2.1) still lacks the mass term. Since the mass term must be a Lorentz scalar, it should be described as

$$C_2 \bar{\psi} \psi \qquad (H.2.4)$$

which is, of course, gauge invariant as well. This constant C_2 should be determined again by comparing the calculated results of hydrogen atom, for example, with experiment. By denoting C_2 as $(-m)$, one arrives at the Lagrangian density of a relativistic fermion which couples with the electromagnetic fields A^μ

$$\mathcal{L} = i \bar{\psi} \partial_\mu \gamma^\mu \psi - g \bar{\psi} \gamma_\mu \psi A^\mu - m \bar{\psi} \psi - \frac{1}{4} F_{\mu\nu} F^{\mu\nu} \qquad (H.2.5)$$

which is just the Lagrangian density for the Dirac field interacting with electromagnetic fields.

H.2.2. First Quantization

It is important to note that, in the procedure of deriving the Lagrangian density of eq.(H.2.5), one has not made use of the quantization condition of

$$E \to i \frac{\partial}{\partial t}, \quad \boldsymbol{p} \to -i \boldsymbol{\nabla}. \qquad (H.2.6)$$

Instead, the first quantization is automatically done by the gauge condition since the Maxwell equation knows the first quantization in advance. This indicates that there may be some chance to understand the first quantization procedure in depth since this method gives an alternative way of the quantization condition of the energy and momentum.

H.3. Two Component Spinor

The derivation of the Dirac equation in terms of the local gauge invariance shows that the current density that can couple to the gauge field A^μ must be rather limited. Here, we discuss a possibility of finding field equation for the two component spinor. When the field has only two components,

$$\phi = \begin{pmatrix} \phi_1 \\ \phi_2 \end{pmatrix}$$

then one can prove that one cannot make the current j_μ that couples with the gauge field A_μ. This can be easily seen since the matrix elements

$$\phi^\dagger \hat{O} \phi$$

can be classified into 4 independent variables as

$$\phi^\dagger\phi : \text{ scalar}, \quad \phi^\dagger\sigma_k\phi : \text{ 3 component vector.} \tag{H.3.1}$$

Therefore, there is no chance to make four vector currents which may couple to the gauge field A_μ. This way of making the Lagrangian density indicates that it should be difficult to find a Lagrangian density of relativistic bosons.

H.4. Klein–Gordon Equation

In the new picture, the correspondence between $(E, \boldsymbol{p})$ and the differential operators $(i\frac{\partial}{\partial t}, -i\boldsymbol{\nabla})$ can be obtained as a consequence from the Dirac equation

$$\text{Dirac equation} \Longrightarrow E \to i\frac{\partial}{\partial t}, \quad \boldsymbol{p} \to -i\boldsymbol{\nabla}.$$

Therefore, it should not be considered as the fundamental principle any more. This equation can be applied to the non-relativistic motion since the Dirac equation can be reduced to the Schrödinger equation in the non-relativistic limit. However, one cannot apply the above relations to the relativistic kinematics since they are not the fundamental principle any more. In this respect, the derivation of the Schrödinger equation by the replacement of the energy and momentum by eq.(H.2.6)

$$E = \frac{\boldsymbol{p}^2}{2m} \Longrightarrow i\frac{\partial\psi}{\partial t} = -\frac{1}{2m}\boldsymbol{\nabla}^2\psi$$

can be justified, but the Klein–Gordon equation

$$E^2 = \boldsymbol{p}^2 + m^2 \Longrightarrow -\frac{\partial^2\psi}{\partial t^2} = \left(-\boldsymbol{\nabla}^2 + m^2\right)\psi$$

may not be justified any more. Therefore, this picture shows that the Klein–Gordon equation should not be taken as the fundamental equation for the elementary particles.

At the present stage, there is no way to derive the Klein–Gordon equation within the new picture. Therefore, it is most likely that there should not exist any elementary Klein–Gordon field of scalar bosons in nature, and in fact there is no elementary scalar boson observed up to now. However, the study of scalar field is mathematically an interesting subject, and one may learn some new aspects in boson fields in a future study.

In nature, there exist bosons, and they are all composite objects which have integer spins. For the composite boson which is composed of fermion and anti-fermion, a Klein–Gordon type equation should be obtained from the two particle Dirac equation. However, it may not be very easy to derive the equation for the center of mass motion for the two particle system in the Dirac equation since it involves the relativistic kinematics. But still theoretically it must be a doable calculation.

H.5. Incorrect Quantization in Polar Coordinates

In the standard quantum mechanical treatment, the particle Hamiltonian in polar coordinates in classical mechanics cannot be quantized in the canonical formalism. In order to see this situation more explicitly, one can write a free particle Hamiltonian with its mass m in polar coordinates

$$H = \frac{{p_r}^2}{2m} + \frac{{p_\theta}^2}{2mr^2} + \frac{{p_\varphi}^2}{2mr^2 \sin^2\theta}, \tag{H.5.1}$$

where p_r, p_θ, p_φ are the generalized momenta in polar coordinates in classical mechanics. If one quantizes the Hamiltonian with the canonical quantization conditions

$$[r, p_r] = i, \quad [\theta, p_\theta] = i, \quad [\varphi, p_\varphi] = i \tag{H.5.2}$$

then one obtains

$$H = -\frac{1}{2m}\left(\frac{\partial^2}{\partial r^2} + \frac{1}{r^2}\frac{\partial^2}{\partial \theta^2} + \frac{1}{r^2 \sin^2\theta}\frac{\partial^2}{\partial \varphi^2}\right) \tag{H.5.3}$$

which is not a correct Hamiltonian for a free particle in polar coordinates. The correct Hamiltonian for a free particle is, of course, given as

$$H = -\frac{1}{2m}\left(\frac{1}{r^2}\frac{\partial}{\partial r}r^2\frac{\partial}{\partial r}\right) - \frac{1}{2mr^2}\left(\frac{1}{\sin\theta}\frac{\partial}{\partial \theta}\sin\theta\frac{\partial}{\partial \theta} + \frac{1}{\sin^2\theta}\frac{\partial^2}{\partial \varphi^2}\right) \tag{H.5.4}$$

which is obtained by transforming the Schrödinger equation in Cartesian coordinates into polar coordinates.

Now, we can explain the reason why the quantization condition is valid for the Cartesian coordinates, but not for the polar coordinates. In the new picture, the differential operators first appear in the Dirac equation and then the momentum is identified as the corresponding differential operator. Therefore, the Schrödinger equation is obtained from the Dirac equation, but not from the Hamiltonian of classical mechanics in the canonical formalism by replacing the generalized momenta by the corresponding differential operators.

H.6. Interaction with Gravity

The chain of the fundamental equations from classical mechanics to Dirac equation is now reversed

$$\text{Dirac equation} \Longrightarrow \text{Schrödinger equation} \Longrightarrow \text{Newton equation.}$$

At this point, however, one realizes that there is no way to include the gravitational force into the new system. This may be a serious defect of this new concept of the first quantization. However, if the gravity is included in this formalism, then it means that the quantum theory of gravity can be constructed.

Appendix I

Renormalization in QED

This textbook treats mainly the non-perturbative aspects of quantum field theory of fermions and therefore we have only briefly discussed the perturbation theory in QED in Chapter 5. In this Appendix, we present some of the basic ingredients in the renormalization scheme in the perturbative evaluation in QED. However, we only describe the basic concept of the renormalization scheme since calculations in detail are found in the textbook of Bjorken-Drell [14]. The essential difference between the perturbative evaluation and the non-perturbative treatment should be found in the Fock space which can be prepared in advance in the perturbative treatment while one does not know the eigenstates of the Hamiltonian in the non-perturbative calculation where one has to solve the eigenvalue equation of eq.(3.1).

I.1. Hilbert Space of Unperturbed Hamiltonian

First, we start from the QED Lagrangian density $\mathcal{L}$ which is composed of the unperturbed Lagrangian density $\mathcal{L}_0$ and the interaction term $\mathcal{L}_I$

$$\mathcal{L}_0 = \bar{\psi}(\not{p} - m)\psi - \frac{1}{4}F_{\mu\nu}F^{\mu\nu}, \tag{I.1.1}$$

$$\mathcal{L}_I = -eA^{\mu}\bar{\psi}\gamma_{\mu}\psi. \tag{I.1.2}$$

In this case, the unperturbed Hamiltonian H_0 can be constructed from the Lagrangian density $\mathcal{L}_0$ as given in eq.(5.28a) in Chapter 5. The Hilbert space of the quantized Hamiltonian $\hat{H}_0$ can be well constructed since one finds the exact eigenvalues and eigenstates of eq.(3.1) for the $\hat{H}_0$. In this case, the Fock space can be specified by the box length L and the cutoff momentum Λ as well as by the energies and momenta of the free fermion and free gauge field states

$$(L,\Lambda): \begin{pmatrix} E_p = \sqrt{\boldsymbol{p}_n^2 + m^2}, & \boldsymbol{p}_n = \dfrac{2\pi \boldsymbol{n}}{L} \\ \omega_k = |\boldsymbol{k}_n|, & \boldsymbol{k}_n = \dfrac{2\pi \boldsymbol{n}}{L} \end{pmatrix}, \tag{I.1.3}$$

$$n_i = 0, \pm 1, \ldots, \pm N \ \text{ with } \ \Lambda = \frac{2\pi N}{L},$$

where the maximum number of freedom N is taken to be the same between the fermion and the gauge fields. The perturbative evaluation can be made within this Hilbert space and one can calculate physical quantities in terms of the S-matrix formulation.

In the evaluation of the second order perturbation energy, there are some diagrams which are divergent. However, the momentum integral has the cutoff due to the Λ, and therefore, the second order energies are described as the function of $\ln(\Lambda/m)$. Among the divergent diagrams, the photon self-energy contribution (vacuum polarization diagram) is special, and therefore we will teat it separately in Appendix J. Here, we discuss the fermion self-energy diagram and the vertex correction contribution.

I.2. Necessity of Renormalization

Before explaining the divergent contribution of the self-energy diagrams, we should understand reasons why we should make a renormalization procedure [59, 103, 107]. The fermion self-energy diagram itself is obviously unphysical since the fermion with its momentum p goes to the same fermion with the same momentum p after emitting a photon and absorbing the same photon. This process itself cannot be observed and examined experimentally. Why is it then necessary to consider the fermion self-energy diagram into the renormalization procedure?

I.2.1. Intuitive Picture of Fermion Self-energy

The answer to the above question is simple. We should consider the following situation in which the fermion is found in the different quantum state from the free state. For example, if we consider electron in hydrogen atom, then its quantum state should be very different from the free state. In this case, the fermion self-energy diagram gives a physical contribution which is observed as the Lamb shift energy. Basically, in the renormalization procedure, we consider a counter term which can exactly cancel the divergent contribution of the second order self-energy diagrams in the free state of fermions. However, in atom, this cancellation does not have to take place in an exact fashion, and this small effect of difference gives rise to the Lamb shift energy in $1s_{\frac{1}{2}}$ and $2s_{\frac{1}{2}}$ states in hydrogen atom.

I.2.2. Intuitive Picture of Photon Self-energy

On the other hand, the photon self-energy is quite different in that photon can never become a bound state. It is always found as a free state. Therefore, there is no physical process in which the contribution of the photon self-energy is detectable in some way or the other. In fact, there has been no physical processes which show the effect of the simple self-energy diagrams.

I.3. Fermion Self-energy

Now, the fermion self-energy $\Sigma(p)$ is obtained from the corresponding Feynman diagram as

$$\Sigma(p) = -ie^2 \int \frac{d^4k}{(2\pi)^4} \gamma_\mu \frac{1}{\not{p} - \not{k} - m} \gamma^\mu \frac{1}{k^2}. \tag{I.3.1}$$

This can be calculated to be

$$\Sigma(p) = \frac{e^2}{8\pi^2} \ln\left(\frac{\Lambda}{m}\right)(-\not{p} + 4m) + \text{finite terms}. \tag{I.3.2}$$

Therefore, the Lagrangian density of the free fermion part

$$\mathcal{L}_\mathrm{F} = \bar{\psi}(\not{p} - m)\psi \tag{I.3.3}$$

should be modified, up to one loop contributions, by the counter term $\delta\mathcal{L}_\mathrm{F}$

$$\delta\mathcal{L}_\mathrm{F} = \bar{\psi}\left[\frac{e^2}{8\pi^2} \ln\left(\frac{\Lambda}{m}\right)(-\not{p} + 4m)\right]\psi.$$

In this case, the total Lagrangian density of fermion becomes

$$\mathcal{L}'_\mathrm{F} = \mathcal{L}_\mathrm{F} + \delta\mathcal{L}_\mathrm{F} = \bar{\psi}[(1+B)\not{p} - (1+A)m]\psi + \text{finite terms}, \tag{I.3.4}$$

where

$$A = -\frac{e^2}{2\pi^2} \ln\left(\frac{\Lambda}{m}\right), \quad B = -\frac{e^2}{8\pi^2} \ln\left(\frac{\Lambda}{m}\right). \tag{I.3.5}$$

Now, one defines Z_2 and δm as

$$Z_2 \equiv 1 + B = 1 - \frac{e^2}{8\pi^2} \ln\left(\frac{\Lambda}{m}\right), \tag{I.3.6a}$$

$$\delta m \equiv \frac{3e^2 m}{8\pi^2} \ln\left(\frac{\Lambda}{m}\right). \tag{I.3.6b}$$

Here, one can introduce the wave function renormalization and the bare mass m_0

$$\psi_\mathrm{b} \equiv \sqrt{Z_2}\psi = \sqrt{1+B}\psi, \tag{I.3.7a}$$

$$\begin{aligned} m_0 &\equiv Z_2^{-1} m(1+A) \\ &= m\left(1 + \frac{e^2}{8\pi^2} \ln\left(\frac{\Lambda}{m}\right)\right)\left(1 - \frac{e^2}{2\pi^2} \ln\left(\frac{\Lambda}{m}\right)\right) \simeq m - \delta m, \end{aligned} \tag{I.3.7b}$$

where one should always keep up to order of e^2. In this case, one can rewrite $\mathcal{L}'_F$ as

$$\mathcal{L}'_F = \bar{\psi}_b(\not{p} - m_0)\psi_b + \text{finite terms} \tag{I.3.8}$$

which has just the same shape as the original one, and thus it is renormalizable.

I.4. Vertex Corrections

The vertex corrections can be evaluated as

$$\Lambda^\mu(p', p) = -ie^2 \int \frac{d^4k}{(2\pi)^4} \gamma^\nu \frac{1}{\not{p}' - \not{k} - m} \gamma^\mu \frac{1}{\not{p} - \not{k} - m} \gamma_\nu \frac{1}{k^2}$$

$$= \frac{e^2}{8\pi^2} \ln\left(\frac{\Lambda}{m}\right) \gamma^\mu + \text{finite terms.} \tag{I.4.1}$$

Therefore, the counter term of the interaction Lagrangian density $\delta\mathcal{L}_I$ becomes

$$\delta\mathcal{L}_I = \frac{e^3}{8\pi^2} \ln\left(\frac{\Lambda}{m}\right) A^\mu \bar{\psi}\gamma_\mu\psi.$$

In this case, the total interaction Lagrangian density can be written as

$$\mathcal{L}'_I = -eA^\mu\bar{\psi}\gamma_\mu\psi + \delta\mathcal{L}_I = -Z_1 eA^\mu\bar{\psi}\gamma_\mu\psi + \text{finite terms,} \tag{I.4.2}$$

where Z_1 is defined as

$$Z_1 \equiv 1 - \frac{e^2}{8\pi^2} \ln\left(\frac{\Lambda}{m}\right). \tag{I.4.3}$$

From eqs.(I.3.6) and (I.4.3), one finds

$$Z_1 = Z_2. \tag{I.4.4}$$

This can be also understood from the following identity

$$\frac{\partial}{\partial p_\mu}\left(\frac{1}{\not{p} - m}\right) = -\frac{1}{\not{p} - m}\gamma^\mu \frac{1}{\not{p} - m}. \tag{I.4.5}$$

Therefore, one finds

$$\Lambda^\mu(p, p) = -\frac{\partial\Sigma(p)}{\partial p_\mu} \tag{I.4.6}$$

which leads to eq.(I.4.4). The interaction Lagrangian density can be rewritten in terms of the bare quantities

$$\psi_b \equiv \sqrt{Z_2}\psi \tag{I.4.7}$$

as

$$\mathcal{L}'_I = -Z_1 eA^\mu\bar{\psi}\gamma_\mu\psi = -Z_1 e\frac{1}{Z_2} A^\mu\bar{\psi}_b\gamma_\mu\psi_b = -e_b A^\mu\bar{\psi}_b\gamma_\mu\psi_b + \text{finite terms,} \tag{I.4.8}$$

where the bare charge e_b turns out to be

$$e_b = e. \tag{I.4.9}$$

Therefore, all the infinite quantities are renormalized into the physical constants as well as the wave function. It should be important that the charge e is not affected from the renormalization.

I.5. New Aspects of Renormalization in QED

The renormalization scheme in QED is best studied and most reliable. Nevertheless there is one important modification in the renormalization procedure, and it is connected to the treatment of the photon self-energy diagram. It turns out that the photon self-energy contribution should not be considered for the renormalization procedure since the contribution itself is unphysical. The detailed discussion of the photon self-energy in QED is treated in Appendix J. This gives rise to some important consequences. The most important of all is that the charge is not affected by the renormalization procedure. This means that there is no renormalization group equation. Here, we should first like to discuss the old version of the renormalization group equation in QED and then compare the new simple picture which has no renormalization group equation since the charge has no renormalization effect.

I.5.1. Renormalization Group Equation in QED

(1) Old Version

For the renormalization group equation, people start from the following equation for the renormalized coupling constant e_b

$$e_b = eZ_1\lambda^{\frac{\varepsilon}{2}} \frac{1}{Z_2\sqrt{Z_3}} = e\lambda^{\frac{\varepsilon}{2}} \left(1+\frac{e^2}{12\pi^2\varepsilon}\right), \tag{I.5.1}$$

where λ is a parameter which is introduced in the dimensional regularization and ε is an infinitesimally small number. Here, Z_3 is defined as

$$Z_3 = 1-\frac{e^2}{6\pi^2\varepsilon}$$

which comes from the wave function renormalization of the photon self-energy contribution. Even though eq.(I.5.1) is obtained from renormalizing all of the self-energy diagrams together with the vertex corrections, this renormalization of the coupling constant e_b is basically originated from the photon self-energy contribution since the renormalization constant Z_1 of the vertex correction and the fermion wave function renormalization Z_2 cancel with each other because of $Z_1 = Z_2$. Since the e_b should not depend on the scale λ, one can obtain the renormalization group equation

$$\lambda\frac{\partial e}{\partial\lambda} = -\frac{e}{2}\left(1-\frac{e^2}{4\pi^2\varepsilon}\right)\left(\varepsilon+\frac{e^2}{12\pi^2}\right) = -\frac{\varepsilon}{2}e+\frac{1}{12\pi^2}e^3+O(e^5). \tag{I.5.2}$$

When one makes $\varepsilon \to 0$, then one finds

$$\lambda\frac{\partial e}{\partial\lambda} = \frac{1}{12\pi^2}e^3 \tag{I.5.3}$$

which is the renormalization group equation for QED. This equation can be easily solved for e. The expression for the running coupling constant $\alpha(\lambda) \equiv \frac{e^2}{4\pi}$ is given as

$$\alpha(\lambda) = \frac{\alpha(\lambda_0)}{1-\frac{2\alpha(\lambda_0)}{3\pi}\ln(\frac{\lambda}{\lambda_0})}, \tag{I.5.4}$$

where λ_0 denotes the renormalization point for the coupling constant. This is the standard procedure to obtain the behavior of the coupling constant as the function of the λ.

However, the λ does not appear in the Hilbert space of the unperturbed Hamiltonian H_0 and this means that one cannot interpret the result of eq.(I.5.4) in terms of the physical observables which should be found in the unperturbed Fock space. Therefore $\alpha(\lambda)$ does not have any physical meaning at all.

(2) New Version

The fact that the photon self-energy contribution should not be considered for the renormalization procedure has many important consequences. The most important is that the charge is not influenced from the renormalization as one can see it from eq.(I.4.9) and, therefore, there is no renormalization group equation in QED. The idea of the renormalization group equation itself is gone.

I.6. Renormalization in QCD

When one wishes to carry out the perturbation calculation, then one should first construct the free Fock space since all of the perturbative evaluations aim at describing physical observables in terms of the free state physical quantities. However, QCD has intrinsic difficulties to construct the free Fock space.

I.6.1. Fock Space of Free Fields

The normal perturbation theory always starts from the free field Hilbert space which is made from the unperturbed Hamiltonian H_0. This free field Fock space must be constructed from the physical states as long as one stays in the perturbation theory. However, in QCD, there is no free quark state in nature due to the gauge non-invariance of the quark color charge. In fact, if one defines the unperturbed Lagrangian density $\mathcal{L}_0$ for QCD as

$$\mathcal{L}_0 = \bar{\psi}(\not{p} - m)\psi - \frac{1}{4}(\partial_\mu A_\nu^a - \partial_\nu A_\mu^a)(\partial^\mu A^{\nu,a} - \partial^\nu A^{\mu,a})$$

then, one finds that it is not invariant under the local gauge transformation of eqs.(6.7) and (6.8). Even for the free fermion part, it transforms under eq.(6.7) as

$$\bar{\psi}'(\not{p} - m)\psi' = \bar{\psi}(\not{p} - m)\psi - g(\partial^\mu \chi^a) j_\mu^a = \bar{\psi}(\not{p} - m)\psi - g\partial^\mu(\chi^a j_\mu^a) + g(\partial^\mu j_\mu^a)\chi^a$$

which is not invariant since the vector current is not conserved ($\partial^\mu j_\mu^a \neq 0$) for QCD, contrary to QED.

I.6.2. Renormalization Group Equation in QCD

Therefore, one cannot find any perturbation scheme in QCD, and there is no renormalization. In this respect, the gluon coupling constant is not affected from the renormalization. This is always true even if one could define some type of the perturbation theory by hand. Therefore, there is no renormalization group equation for QCD, and thus the concept of the asymptotic freedom has lost its physical meaning.

I.6.3. Serious Problems in QCD

However, OCD has much more serious problems than the lack of the QCD renormalization group equation. As we stress above, we cannot carry out the perturbative calculations since there is no free Fock state in QCD. This means that one cannot describe any physical quantities in terms of the free Fock state terminology since the free states of quarks and gluons do not exist as the physical states. In this respect, the evaluation of QCD cannot be based on the perturbation scheme and therefore one has to invent a new scheme to solve QCD. This is a very important problem which should be solved in some way or the other in future [35].

I.7. Renormalization of Massive Vector Fields

Since the self-energy of boson fields should not be considered for the renormalization procedure, there are many important consequences concerning the renormalizability of the model field theory. Here, we should briefly discuss the renormalization procedure of the massive vector fields which interact with fermion fields. The simplest Lagrangian density for two flavor leptons which couple to the SU(2) vector fields W_μ^a can be written as

$$\mathcal{L} = \bar{\Psi}(i\partial_\mu \gamma^\mu + m)\Psi - gJ_\mu^a W^{\mu,a} + \frac{1}{2}M^2 W_\mu^a W^{\mu,a} - \frac{1}{4}G_{\mu\nu}^a G^{\mu\nu,a}, \qquad (I.7.1)$$

where M denotes the mass of the vector boson. The fermion wave function Ψ has two components.

$$\Psi = \begin{pmatrix} \psi_1 \\ \psi_2 \end{pmatrix}. \qquad (I.7.2)$$

Correspondingly, the mass matrix can be written as

$$m = \begin{pmatrix} m_1 & 0 \\ 0 & m_2 \end{pmatrix}. \qquad (I.7.3)$$

The fermion current J_μ^a and the field strength $G_{\mu\nu}^a$ are defined as

$$J_\mu^a = \bar{\Psi}\gamma_\mu \tau^a \Psi, \quad G_{\mu\nu}^a = \partial_\mu W_\nu^a - \partial_\nu W_\mu^a. \qquad (I.7.4)$$

This Lagrangian density is almost the same as the standard model Lagrangian density apart from the Higgs fields, as far as the basic structure of the field theory is concerned.

I.7.1. Renormalizability

Now, one can easily examine the renormalizability of the model field theory, and one can prove that the fermion self-energy has the logarithmic divergence which is just the same as the QED case. Also, the vertex correction in this model field theory has the logarithmic divergence and is the same as the QED case as long as the renormalizability is concerned. Since one should not consider the self-energy of the massive vector fields, one can safely renormalize the divergent contributions of the fermion self-energy as well as the vertex correction.

In this sense, this is just similar to checking the renormalizability of the *final* Lagrangian density of the standard model, and it is indeed renormalizable. This is quite important since the final Lagrangian density of the Weinberg-Salam model is quite successful for describing the experiment even though the Higgs mechanism itself is not a correct procedure in terms of the symmetry breaking physics as discussed in Chapter 4.

Appendix J

Photon Self-energy Contribution in QED

In the renormalization procedure of QED, one considers the vacuum polarization which is the contribution of the self-energy diagram of photon

$$\Pi^{\mu\nu}(k) = ie^2 \int \frac{d^4 p}{(2\pi)^4} \mathrm{Tr}\left[\gamma^\mu \frac{1}{\not{p}-m}\gamma^\nu \frac{1}{\not{p}-\not{k}-m}\right]. \qquad (J.0.1)$$

This integral obviously gives rise to the quadratic divergence (Λ^2 term). However, when one considers the counter term of the Lagrangian density which should cancel this quadratic divergence term, then the counter Lagrangian density violates the gauge invariance since it should correspond to the mass term in the gauge field Lagrangian density. Therefore, one has to normally erase it by hand, and in the cutoff procedure of the renormalization scheme, one subtracts the quadratic divergence term such that one can keep the gauge invariance of the Lagrangian density. Here, we should notice that the largest part of the vacuum polarization contributions is discarded, and this indicates that there must be something which is not fully understandable in the renormalization procedure. Physically, it should be acceptable to throw away the Λ^2 term since this infinite term should not be connected to any physical observables. Nevertheless we should think it over why the unphysical infinity appears in the self-energy diagram of photon.

On the other hand, the quadratic divergence term disappears in the treatment of the dimensional regularization scheme. Here, we clarify why the quadratic divergence term does not appear in the dimensional regularization treatment. That is, the treatment of the dimensional regularization employs the mathematical formula which is not valid for the evaluation of the momentum integral. Therefore, the fact that there is no quadratic divergence term in the dimensional regularization is simply because one makes a mistake by applying the invalid mathematical formula to the momentum integral. This is somewhat surprising, but now one sees that the quadratic divergence is still there in the dimensional regularization, and this strongly indicates that we should reexamine the effect of the photon self-energy diagram itself.

J.1. Momentum Integral with Cutoff Λ

This procedure of the photon self-energy is well explained in the textbook of Bjorken and Drell, and therefore we describe here the simplest way of calculating the momentum integral. The type of integral one has to calculate can be summarized as

$$\int d^4p \frac{1}{(p^2-s+i\varepsilon)^n} = i\pi^2 \int_0^{\Lambda^2} w\,dw \frac{1}{(w-s+i\varepsilon)^n}, \quad \text{with } w=p^2, \tag{J.1.1}$$

where i appears because the integral is rotated into the Euclidean space and this corresponds to $D=4$ in the dimensional regularization as we will see it below.

J.1.1. Photon Self-energy Contribution

The photon self-energy contribution $\Pi^{\mu\nu}(k)$ in eq.(J.0.1) can be easily evaluated as

$$\Pi_{\mu\nu}(k) = \frac{4ie^2}{(2\pi)^4}\int_0^1 dz \int d^4p \left[\frac{2p_\mu p_\nu - g_{\mu\nu}p^2 + sg_{\mu\nu} - 2z(1-z)(k_\mu k_\nu - k^2 g_{\mu\nu})}{(p^2-s+i\varepsilon)^2}\right]$$

$$= \frac{\alpha}{2\pi}\int_0^1 dz \int_0^{\Lambda^2} dw \left[\frac{w(w-2s)g_{\mu\nu} + 4z(1-z)w(k_\mu k_\nu - k^2 g_{\mu\nu})}{(w-s+i\varepsilon)^2}\right], \tag{J.1.2}$$

where s is defined as $s = m^2 - z(1-z)k^2$. This can be calculated to be

$$\Pi_{\mu\nu}(k) = \Pi^{(1)}_{\mu\nu}(k) + \Pi^{(2)}_{\mu\nu}(k),$$

where

$$\Pi^{(1)}_{\mu\nu}(k) = \frac{\alpha}{2\pi}\left(\Lambda^2 + m^2 - \frac{k^2}{6}\right) g_{\mu\nu} \tag{J.1.3a}$$

$$\Pi^{(2)}_{\mu\nu}(k) = \frac{\alpha}{3\pi}(k_\mu k_\nu - k^2 g_{\mu\nu})\left[\ln\left(\frac{\Lambda^2}{m^2 e}\right) - 6\int_0^1 dz z(1-z)\ln\left(1 - \frac{k^2}{m^2}z(1-z)\right)\right]. \tag{J.1.3b}$$

Here, the $\Pi^{(1)}_{\mu\nu}(k)$ term corresponds to the quadratic divergence term and this should be discarded since it violates the gauge invariance. The $\Pi^{(2)}_{\mu\nu}(k)$ term can keep the gauge invariance, and therefore one can renormalize it into the new Lagrangian density.

J.1.2. Finite Term in Photon Self-energy Diagram

After the renormalization, one finds a finite term which should affect the propagator change in the process involving the exchange of the transverse photon $\boldsymbol{A}$. The propagator $\frac{1}{q^2}$ should be replaced by

$$\frac{1}{q^2} \Longrightarrow \frac{1}{q^2}\left[1 + \frac{2\alpha}{\pi}\int_0^1 dz z(1-z)\ln\left(1 - \frac{q^2 z(1-z)}{m^2}\right)\right], \tag{J.1.4}$$

where q^2 should become $q^2 \approx -\boldsymbol{q}^2$ for small $\boldsymbol{q}^2$. It should be important to note that the correction term arising from the finite contribution of the photon self-energy should affect only for the renormalization of the vector field $\boldsymbol{A}$. Since the Coulomb propagator is not affected by the renormalization procedure of the transverse photon (vector field $\boldsymbol{A}$), one should not calculate its effect on the Lamb shift.

J.2. Dimensional Regularization

In the evaluation of the momentum integral, one can employ the dimensional regularization [72, 73] where the integral is replaced as

$$\int \frac{d^4p}{(2\pi)^4} \longrightarrow \lambda^{4-D} \int \frac{d^Dp}{(2\pi)^D}, \tag{J.2.1}$$

where λ is introduced as a parameter which has a mass dimension in order to compensate the unbalance of the momentum integral dimension. This is the integral in the Euclidean space, but D is taken to be $D = 4 - \varepsilon$ where ε is an infinitesimally small number.

J.2.1. Photon Self-energy Diagram with $D = 4 - \varepsilon$

In this case, the photon self-energy $\Pi_{\mu\nu}(k)$ can be calculated to be

$$\Pi_{\mu\nu}(k) = i\lambda^{4-D}e^2 \int \frac{d^Dp}{(2\pi)^D} \mathrm{Tr}\left[\gamma_\mu \frac{1}{\not{p} - m} \gamma_\nu \frac{1}{\not{p} - \not{k} - m}\right]$$

$$= \frac{\alpha}{3\pi}\left(k_\mu k_\nu - g_{\mu\nu}k^2\right)\left[\frac{2}{\varepsilon} + \text{finite term}\right], \tag{J.2.2}$$

where the finite term is just the same as eq.(J.1.3). In eq.(J.2.2), one sees that the quadratic divergence term ($\Pi_{\mu\nu}^{(1)}(k)$) is missing. This is surprising since the quadratic divergence term is the leading order contribution in the momentum integral, and whatever one invents in the integral, there is no way to erase it unless one makes a mistake.

J.2.2. Mathematical Formula of Integral

Indeed, in the treatment of the dimensional regularization, people employ the mathematical formula which is invalid for the integral in eq.(J.2.2). That is, the integral formula for $D = 4 - \varepsilon$

$$\int d^Dp \frac{p_\mu p_\nu}{(p^2 - s + i\varepsilon)^n} = i\pi^{\frac{D}{2}}(-1)^{n+1}\frac{\Gamma(n-\frac{1}{2}D-1)}{2\Gamma(n)} \frac{g_{\mu\nu}}{s^{n-\frac{1}{2}D-1}} \quad (\text{for } n \geq 3) \tag{J.2.3}$$

is only valid for $n \geq 3$ in eq.(J.2.3). For $n = 3$, the integral should have the logarithmic divergence, and this is nicely avoided by the replacement of $D = 4 - \varepsilon$. However, the $n = 2$ case must have the quadratic divergence and the mathematical formula of eq.(J.2.3) is meaningless. In fact, one should recover the result of the photon self-energy contribution $\Pi_{\mu\nu}(k)$ of eqs.(J.1.3) at the limit of $D = 4$, apart from the $(2/\varepsilon)$ term which corresponds to the logarithmic divergence term.

J.2.3. Reconsideration of Photon Self-energy Diagram

In mathematics, one may define the gamma function in terms of the algebraic equations with complex variables. However, the integral in the renormalization procedure is defined only in real space integral, and the infinity of the integral is originated from the infinite degrees of freedom which appear in the free Fock space as discussed in Appendix I.

Therefore, one sees that the disappearance of the quadratic divergence term in the evaluation of $\Pi_{\mu\nu}(k)$ in the dimensional regularization is not due to the mathematical trick, but simply due to a simple-minded mistake. In this respect, it is just accidental that the $\Pi_{\mu\nu}^{(1)}(k)$ term in the dimensional regularization vanishes to zero. Indeed, one should obtain the same expression of the $\Pi_{\mu\nu}^{(1)}(k)$ term as eq.(J.1.3a) when one makes $\varepsilon \to 0$ in the calculation of the dimensional regularization. This strongly suggests that we should reconsider the photon self-energy diagram itself in the renormalization procedure [45].

J.3. Propagator Correction of Photon Self-energy

In order to examine whether the inclusion of the photon self-energy contribution is necessary for the renormalization procedure or not, we should consider the effect of the finite contribution from the photon self-energy diagram. As we see, there is a finite contribution of the transverse photon propagator to physical observables after the renormalization of the photon self-energy. The best application of the propagator correction must be the magnetic hyperfine splitting of the ground state ($1s_{\frac{1}{2}}$ state) in the hydrogen atom since this interaction is originated from the vector field $\boldsymbol{A}$ which gives rise to the magnetic hyperfine interaction between electrons and nucleus.

J.3.1. Lamb Shift Energy

In some of the textbooks [14], the correction term arising from the finite contribution of the photon self-energy diagram is applied to the evaluation of the Lamb shift energy in hydrogen atom. However, this is not a proper application since only the renormalization of the vector field $\boldsymbol{A}$ should be considered. This is closely connected to the understanding of the field quantization itself. One sees that the second quantization of the electromagnetic field should be made only for the vector field $\boldsymbol{A}$, and this is required from the experimental observation that photon is created from the vacuum of the electromagnetic field in the atomic transitions. Therefore, it is clear that the renormalization becomes necessary only for the vector field $\boldsymbol{A}$.

Classical Picture of Polarization

This classical picture of the fermion pair creations (Uehling potential) should come from the misunderstanding of the structure of the vacuum state [65, 66, 110]. In the medium of solid state physics, the polarization can take place when there is an electric field present. In this case, the electric field can indeed induce the electric dipole moments in the medium, and this corresponds to the change of the charge density. However, this is a physical process which can happen in the real space (configuration space). On the other hand, the fermion pair creation in the vacuum in field theory is completely different in that the negative energy states are all filled in momentum space, and the time independent field of A^0 which is only a function of coordinates cannot induce any changes on the vacuum state. Therefore, unless some time dependent field is present in the reaction process, the pair creation of fermions cannot take place in physical processes. Therefore, in contrast to the common belief, there is, unfortunately, no change of the charge distribution in QED vacuum, even at the presence of two charges, and this is basically because the Coulomb field is not time dependent.

J.3.2. Magnetic Hyperfine Interaction

The magnetic hyperfine interaction between electron and proton in hydrogen atom can be written with the static approximation in the classical field theory as

$$H' = -\int \boldsymbol{j}_e(\boldsymbol{r}) \cdot \boldsymbol{A}(\boldsymbol{r})\, d^3r, \tag{J.3.1}$$

where $\boldsymbol{j}_e(\boldsymbol{r})$ denotes the current density of electron, and $\boldsymbol{A}(\boldsymbol{r})$ is the vector potential generated by proton and is given as

$$\boldsymbol{A}(\boldsymbol{r}) = \frac{1}{4\pi}\int \frac{\boldsymbol{J}_p(\boldsymbol{r}')}{|\boldsymbol{r}-\boldsymbol{r}'|}\, d^3r', \tag{J.3.2}$$

where $\boldsymbol{J}_p(\boldsymbol{r})$ denotes the current density of proton. The hyperfine splitting of the ground state in the hydrogen atom can be calculated as

$$\Delta E_{hfs} = \langle 1s_{\frac{1}{2}}, I : F | H' | 1s_{\frac{1}{2}}, I : F\rangle, \tag{J.3.3}$$

where I and F denote the spins of proton and atomic system, respectively. This can be explicitly calculated as

$$\Delta E_{hfs} = (2F(F+1)-3)\frac{\alpha g_p}{3M_p}\int_0^\infty F^{(1s)}(r)G^{(1s)}(r)\, dr, \tag{J.3.4}$$

where g_p and M_p denote the g-factor and the mass of proton, respectively. $F^{(1s)}(r)$ and $G^{(1s)}(r)$ are the small and large components of the radial parts of the Dirac wave function of electron in the atom. In the nonrelativistic approximation, the integral can be expressed as

$$\int_0^\infty F^{(1s)}(r)G^{(1s)}(r)\, dr \simeq \frac{(m_r\alpha)^3}{m_e}, \tag{J.3.5a}$$

where m_e is the mass of electron and m_r denotes the reduced mass defined as

$$m_r = \frac{m_e}{1+\frac{m_e}{M_p}}. \tag{J.3.6}$$

It should be noted that, in eq.(J.3.5), m_e appears in the denominator because it is originated from the current density of electron. Therefore, the energy splitting between $F=1$ and $F=0$ atomic states in the nonrelativistic limit with a point nucleus can be calculated from Eq.(J.3.4) as

$$\Delta E_{hfs}^{(0)} = \frac{8\alpha^4 m_r^3}{3m_e M_p}. \tag{J.3.7}$$

J.3.3. QED Corrections for Hyperfine Splitting

There are several corrections which arise from the various QED effects such as the anomalous magnetic moments of electron and proton, nuclear recoil effects and relativistic effects. We write the result

$$\Delta E_{hfs}^{(QED)} = \frac{4g_p\alpha^4 m_r^3}{3m_e M_p}(1+a_e)\left(1+\frac{3}{2}\alpha^2\right)(1+\delta_R), \qquad (J.3.8)$$

where a_e denotes the anomalous magnetic moment of electron. The term $(1+\frac{3}{2}\alpha^2)$ appears because of the relativistic correction of the electron wave function

$$\int_0^\infty F^{(1s)}(r)G^{(1s)}(r)\,dr = \frac{(m_r\alpha)^3}{m_e}\left(1+\frac{3}{2}\alpha^2+\cdots\right). \qquad (J.3.5b)$$

The term δ_R corresponds to the recoil corrections and can be written as [16]

$$\delta_R = \alpha^2\left(\ln 2-\frac{5}{2}\right)-\frac{8\alpha^3}{3\pi}\ln\alpha\left(\ln\alpha-\ln 4+\frac{281}{480}\right)+\frac{15.4\alpha^3}{\pi}. \qquad (J.3.9)$$

Now the observed value of $\Delta E_{hfs}^{(exp)}$ is found to be [30]

$$\Delta E_{hfs}^{(exp)} = 1420.405751767 \text{ MHz}.$$

Also, we can calculate $\Delta E_{hfs}^{(0)}$ and $\Delta E_{hfs}^{(QED)}$ numerically and their values become

$$\Delta E_{hfs}^{(0)} = 1418.83712 \text{ MHz}, \quad \Delta E_{hfs}^{(QED)} = 1420.448815 \text{ MHz}.$$

Therefore, we find the deviation from the experimental value as

$$\frac{\Delta E_{hfs}^{(exp)}-\Delta E_{hfs}^{(QED)}}{\Delta E_{hfs}^{(0)}} \simeq -30 \text{ ppm}. \qquad (J.3.10)$$

J.3.4. Finite Size Corrections for Hyperfine Splitting

In addition to the QED corrections, there is a finite size correction of proton and its effect can be written as

$$\Delta E_{hfs}^{(FS)} = \Delta E_{hfs}^{(0)}(1+\varepsilon), \qquad (J.3.11)$$

where the ε term corresponds to the Bohr-Weisskopf effect [17, 37, 114]

$$\varepsilon \simeq -m_e\alpha R_p, \qquad (J.3.12)$$

where R_p denotes the radius of proton. It should be noted that the perturbative treatment of the finite proton size effect on the hyperfine splitting overestimates the correction by a factor of two. Now, the calculated value of ε becomes

$$\varepsilon \simeq -17 \text{ ppm}.$$

Therefore, the agreement between theory and experiment is quite good.

J.3.5. Finite Propagator Correction from Photon Self-energy

The hyperfine splitting of the $1s_{\frac{1}{2}}$ state energy including the propagator correction can be written in terms of the momentum representation in the nonrelativistic limit as

$$\Delta E_{hfs}^{(VP)} = (2F(F+1)-3)\frac{16}{3\pi}\frac{\alpha^5 m_r^4}{m_e M_p}\int_0^\infty \frac{q^2 dq}{(q^2+4(m_r\alpha)^2)^2}\left(1+M^R(q)\right)$$

$$\equiv \Delta E_{hfs}^{(0)}(1+\delta_{vp}). \qquad (J.3.13)$$

$M^R(q)$ denotes the propagator correction and can be written as

$$M^R(q) = \frac{2\alpha}{\pi}\int_0^1 dz z(1-z)\ln\left(1+\frac{q^2}{m_e^2}z(1-z)\right). \qquad (J.3.14)$$

We can carry out numerical calculations of the finite term of the renormalization in the photon self-energy diagram, and we find

$$\delta_{vp} \simeq 18 \text{ ppm} \qquad (J.3.15)$$

which tends to make a deviation larger between theory and experiment of hyperfine splitting in hydrogen atom. This suggests that the finite correction from the photon self-energy contribution should not be considered for the renormalization procedure.

J.3.6. Magnetic Moment of Electron

The finite terms of the vacuum polarization contribute to the magnetic moment of electron. The evaluation of the magnetic moment of electron which is often described as g factor can be carried out within the renormalization scheme of QED. Here, we only quote the calculated results of one and two loop diagrams which are up to the order of α^2.

One Loop Calculation

The calculation of the electron g factor is just the result of the vertex correction in the renormalization procedure, and it is given as

$$g = 2\left(1+\frac{\alpha}{2\pi}+\cdots\right). \qquad (J.3.16)$$

It should be important to note that the diagram that contains the vacuum polarization in the one loop calculation vanishes to zero due to the condition that the magnetic moment of electron is measured at the zero momentum transfer.

Two Loop Calculation

The g factor of electron is calculated up to the two loop order, and it is given as

$$g = 2\left(1+\frac{\alpha}{2\pi}+W_0\left(\frac{\alpha}{\pi}\right)^2+W_{vp}\left(\frac{\alpha}{\pi}\right)^2+\cdots\right), \qquad (J.3.17)$$

where W_0 denotes the contributions of the two loop diagrams without the vacuum polarization term while W_{vp} is the contribution of the two loop order from the vacuum polarization term. They are calculated to be [3]

$$W_0 = \frac{3}{4}\zeta(3) - \frac{\pi^2}{2}\ln 2 + \frac{5}{12}\pi^2 - \frac{279}{144} = -0.34416639, \tag{J.3.18a}$$

$$W_{vp} = -\frac{\pi^2}{3} + \frac{119}{36} = 0.015687, \tag{J.3.18b}$$

where $\zeta(3)$ is given as $\zeta(3) = 1.202056903$. It should be noted that the contribution of the finite term of the vacuum polarization is a few percents of the whole two loop order calculations, and in this sense it cannot be a very important contribution to the electron magnetic moment. In addition, the vacuum polarization in two loop order should have a $(\ln\Lambda)^2$ term which should give rise to a conceptual difficulty in the renormalization procedure.

Now, the observed g factor of electron is given [111]

$$g_{\text{exp}} = 2\ (1 + 0.001\ 159\ 652\ ). \tag{J.3.19a}$$

The calculated g factor of electron without the vacuum polarization diagram becomes

$$g_{\text{no-vp}} = 2\ (1 + 0.001\ 159\ 553\ ) \tag{J.3.19b}$$

while the total value of the calculated g factor of electron becomes

$$g_{\text{total}} = 2\ (1 + 0.001\ 159\ 637\ ) \tag{J.3.19c}$$

where we take the value of the fine structure constant of α as $\alpha^{-1} = 137.036$.

Three Loop Calculation

The three loop calculations are found to be around [77]

$$g_{\text{threeloopl}} = 2\left(1 + \cdots + c_3\left(\frac{\alpha}{\pi}\right)^3\right) + \cdots \tag{J.3.19d}$$

where $c_3 \simeq 1.18$. In this sense, one sees that the total value of the two loop calculations seems to reproduce the experimental number better than the one without the vacuum polarization diagram.

Ambiguity in Fine Structure Constant α

From the observed value of the Zeeman splitting energy in electron

$$\Delta E_{exp} = \frac{e\hbar g}{2m_e c}B = \frac{\sqrt{4\pi\alpha}\hbar g}{2m_e c}B \tag{J.3.20}$$

one can extract the g factor value. In this case, we should examine the accuracy of the constants which appear in eq.(J.3.20). Here, B denotes the external magnetic field. The experimental accuracies of$\hbar, m_e, c, B$ are all reliable up to 8th digit. However, we should be careful for the value of the coupling constant α since the experimental value of g_{exp} is very sensitive to the value of α. In fact, the difference between the α values at the 8-th digit may well be responsible for the g factor value at the corresponding accuracy. In this respect, it is most important that the experimental determination of the α value should be done at the accuracy of the 8-th digit independently from the $g-2$ experiment.

J.4. Spurious Gauge Conditions

The evaluation of the vacuum polarization contribution gives rise to the quadratic divergence and, if this should exist, there is no way to renormalize it into the standard renormalization scheme [14]. In fact, Pauli and Villars proposed [101] that the quadratic divergence term should be evaded by the requirement that the calculated result should be gauge invariant when renormalizing it into the Lagrangian density. This requirement of the gauge invariance should be based on the following relation for the vacuum polarization tensor $\Pi^{\mu\nu}(k)$ as

$$k_\mu \Pi^{\mu\nu}(k) = 0. \qquad (J.4.1)$$

However, this equation does not hold and it is indeed a spurious equation even though it has been employed as the *gauge condition*. The proof of eq.(J.4.1) in most of the field theory textbooks is simply a mathematical mistake basically due to the wrong replacement of the integration variable in the infinite integral case. This problem of mathematics was, of course, realized and stated in the text book of Bjorken and Drell [14], but they accepted this relation up to the last step of their calculation of the vacuum polarization contributions.

J.4.1. Gauge Condition of $\Pi^{\mu\nu}(k)$

If one carries out the self-energy diagram of photon, then one obtains the vacuum polarization tensor which is given in eqs.(J.1.3), and this is obviously inconsistent with eq.(J.4.1). For a long time, people believe that the $\Pi^{\mu\nu}(k)$ should satisfy the relation of eq.(J.4.1)

$$k_\mu \Pi^{\mu\nu}(k) = 0$$

and this equation is called "gauge condition of $\Pi^{\mu\nu}(k)$". Now, we present the proof of the above relation as discussed in the text book of Bjorken and Drell [14], and show that the proof of eq.(J.4.1) is a simple mathematical mistake. Therefore, this gauge condition is spurious, and the relation has no physical foundation at all. The standard method of the proof starts by rewriting the $k_\mu \Pi^{\mu\nu}(k)$ as

$$k_\mu \Pi^{\mu\nu}(k) = ie^2 \int \frac{d^4p}{(2\pi)^4} \mathrm{Tr}\left[\left(\frac{1}{\not{p}-\not{k}-m+i\varepsilon} - \frac{1}{\not{p}-m+i\varepsilon}\right)\gamma^\nu\right]. \qquad (J.4.2)$$

In the first term, the integration variable should be replaced as

$$q = p - k$$

and thus one can prove that

$$k_\mu \Pi^{\mu\nu}(k) = ie^2 \left\{\int \frac{d^4q}{(2\pi)^4} \mathrm{Tr}\left[\frac{1}{\not{q}-m+i\varepsilon}\gamma^\nu\right] - \int \frac{d^4p}{(2\pi)^4} \mathrm{Tr}\left[\frac{1}{\not{p}-m+i\varepsilon}\gamma^\nu\right]\right\} = 0. \quad (J.4.3)$$

At a glance, this proof looks plausible. However, one can easily notice that the replacement of the integration variable is only meaningful when the integral is finite.

In order to clarify the mathematical mistake in eq.(J.4.3), we present a typical example which shows that one cannot make a replacement of the integration variable when the integral is infinity. Let us now evaluate the following integral

$$Q = \int_{-\infty}^{\infty} \left((x-a)^2 - x^2\right) dx. \qquad (J.4.4)$$

If we replace the integration variable in the first term as $x' = x - a$, then we can rewrite eq.(J.4.4) as

$$Q = \int_{-\infty}^{\infty} \left(x'^2 dx' - x^2 dx\right) = 0. \qquad (J.4.5)$$

However, if we calculate it properly, then we find

$$Q = \int_{-\infty}^{\infty} \left((x-a)^2 - x^2\right) dx = \int_{-\infty}^{\infty} \left(a^2 - 2ax\right) dx = a^2 \times \infty \qquad (J.4.6)$$

which disagrees with eq.(J.4.5). If one wishes to carefully calculate eq.(J.4.4) by replacing the integration variable, then one should do as follows

$$Q = \lim_{\Lambda \to \infty} \left[\int_{-\Lambda-a}^{\Lambda-a} x'^2 dx' - \int_{-\Lambda}^{\Lambda} x^2 dx \right] = \lim_{\Lambda \to \infty} 2a^2 \Lambda. \qquad (J.4.7)$$

It is clear by now that the replacement of the integration variable in the infinite integral should not be made, and this is just the mistake which has been accepted as the gauge condition of the $\Pi^{\mu\nu}(k)$ in terms of eq.(J.4.1). Therefore, one sees that the requirement of the gauge condition of the vacuum polarization is unphysical.

J.4.2. Physical Processes Involving Vacuum Polarizations

In nature, there are a number of Feynman diagrams which involve the vacuum polarization. The best known physical process must be the π^0 decay into two photons, $\pi^0 \to \gamma + \gamma$. This process of the Feynman diagrams can be well calculated in terms of the nucleon and anti-nucleon pair creation where these fermions couple to photons [94]. In this calculation, one knows that the loop integral gives a finite result since the apparent logarithmic divergence vanishes to zero due to the kinematical cancellation. Also, the physical process of photon-photon scattering involves the box diagrams where electrons and positrons are created from the vacuum state. As is well known, the apparent logarithmic divergence of this box diagrams vanishes again due to the kinematical cancellation, and the evaluation of the Feynman diagrams gives a finite number. This is clear since all of the perturbative calculations employ the free fermion basis states which always satisfy the current conservation of $\partial_\mu j^\mu = 0$. In these processes, one does not have any additional "gauge conditions" in the evaluation of the Feynman diagrams. In this respect, if the process is physical, then the corresponding Feynman diagram should become finite without any further constraints of the gauge invariance.

J.5. Renormalization Scheme

Here, it is shown that there occurs no wave function renormalization of photon in the exact Lippmann-Schwinger equation for the vector potential [87]. The Lippmann-Schwinger equation for the fermion field ψ becomes

$$\psi(x) = \psi_0(x) + g\int G_F(x,x')A_\mu(x')\gamma^\mu\psi(x')d^4x', \qquad (J.5.1)$$

where $G_F(x,x')$ denotes the Green function which satisfies the following equation

$$(i\partial_\mu\gamma^\mu - m)G_F(x,x') = \delta^4(x-x'). \qquad (J.5.2)$$

$\psi_0(x)$ is the free fermion field solution. The Green function $G_F(x,x')$ can bc cxplicitly written as

$$G_F(x,x') = \int \frac{1}{p_\mu\gamma^\mu - m + i\varepsilon} e^{ip(x-x')} \frac{d^4p}{(2\pi)^4}. \qquad (J.5.3)$$

On the other hand, the Lippmann-Schwinger equation for the vector field $\boldsymbol{A}$ becomes

$$\boldsymbol{A} = \boldsymbol{A}_0 + g\int G_0(x,x')\boldsymbol{j}(x')d^4x', \qquad (J.5.4)$$

where $\boldsymbol{A}_0$ denotes the free field solution of the vector field. Here, the Green function $G_0(x,x')$ satisfies

$$\left(\frac{\partial^2}{\partial t^2} - \boldsymbol{\nabla}^2\right) G_0(x,x') = \delta^4(x-x'). \qquad (J.5.5)$$

This can be explicitly written as

$$G_0(x,x') = \int \frac{1}{-p_0^2 + \boldsymbol{p}^2 + i\varepsilon} e^{ip(x-x')} \frac{d^4p}{(2\pi)^4}. \qquad (J.5.6)$$

Since we employ the Coulomb gauge fixing, the equation of motion for the A_0 field becomes a constraint equation and thus can be solved exactly as

$$A_0(\boldsymbol{r}) = \frac{g}{4\pi}\int \frac{j_0(\boldsymbol{r}')}{|\boldsymbol{r}-\boldsymbol{r}'|} d^3r'. \qquad (J.5.7)$$

J.5.1. Wave Function Renormalization−Fermion Field

When we carry out the perturbation expansion, we can obtain the integral equations in powers of the coupling constant g as

$$\psi(x) = \psi_0(x) + g\int G_F(x,x')A_\mu\gamma^\mu\psi_0(x')d^4x' +$$

$$g^2\int G_F(x,x')A_\mu(x')\gamma^\mu G_F(x',x'')A_\nu(x'')\gamma^\nu\psi_0(x'')d^4x'd^4x'' + \cdots. \qquad (J.5.8)$$

This equation clearly shows that the fermion field should be affected by the perturbation expansion, and if it diverges, then we have to renormalize the wave function so as to absorb the infinity. Indeed, the infinity is logarithmic divergence and can be well renormalized into the wave function ψ_0.

J.5.2. Wave Function Renormalization – Vector Field

The vector field $\boldsymbol{A}$ can be determined from eq.(J.5.4) only when the fermion numbers are conserved. The best example can be found when the annihilation of the fermion pair takes place. In this case, we can write eq. (J.5.4) as

$$\langle 0|\boldsymbol{A}|f\bar{f}\rangle = g\int G_0(x,x')\langle 0|\boldsymbol{j}(x')|f\bar{f}\rangle d^4x', \tag{J.5.9}$$

where $|f\bar{f}\rangle$ denotes the fermion and anti-fermion state. Now, we can consider the following physical process of $e^+e^- \to e^+e^-$ which can be described in terms of the T-matrix as

$$T = -g\langle e\bar{e}|\int \boldsymbol{j}\cdot\boldsymbol{A}|e\bar{e}\rangle = -g^2\int\langle e\bar{e}|\boldsymbol{j}(x)|0\rangle G_0(x,x')\langle 0|\boldsymbol{j}(x')|e\bar{e}\rangle d^3x d^4x', \tag{J.5.10}$$

where one can see that there appears no self-energy of photon term whatever one evaluates any physical processes in the Lippmann-Schwinger equation. From this equation, one finds that the vector field $\boldsymbol{A}$ cannot be affected by the renormalization procedure, and it always stays as a free state of photon. Since this is the exact equation of motion, there is no other possibility for the vector field. In this respect, it is just simple that the gauge field $\boldsymbol{A}$ always behaves as a free photon state in the evaluation of any Feynman diagrams.

J.5.3. Mass Renormalization – Fermion Self-energy

The evaluation of the self-energy of fermions can be carried out in a straight forward way, and one can obtain the self-energy which has a logarithmic divergence of the momentum cut-off Λ. Since electron has a mass, one can renormalize this logarithmic divergence term into the new mass term. In this procedure, there is no conceptual difficulty and indeed one can relate this renormalized effect to the observed value of the Lamb shift in hydrogen atom, which is indeed a great success of the QED renormalization scheme.

J.5.4. Mass Renormalization – Photon Self-energy

The evaluation of the self-energy of photon gives rise to the energy which has a quadratic divergence. There is no way to renormalize it into the renormalization scheme of QED since photon has no mass term. This clearly indicates that one should not take the contributions of the photon self-energy diagrams since they violate the Lorentz invariance. In this respect, one can now realize that the quadratic divergence term should be discarded because it is not consistent with the Lorentz invariance, and it has nothing to do with the gauge invariance.

The energy of photon is calculated in the system where fermion is at rest. The energy of photon with its momentum $\boldsymbol{k}$ must be described as $E_k = |\boldsymbol{k}|$, and there is no other expression. Therefore, the Lagrangian density of the vector field A_μ should be always written as

$$\mathcal{L}_0 = -\frac{1}{4}F_{\mu\nu}F^{\mu\nu} \tag{J.5.11}$$

and there should not be any modifications possible.

It may be interesting to note that Tomonaga stated a half century ago that the self-energy of photon should vanish to zero, even though he did not present any concrete proof at that time [108]. This claim must come from the understanding that the vacuum polarization energy of electromagnetic fields calculated by Heisenberg and Euler [65, 66] has no chance to be renormalized into the original Lagrangian density of electromagnetic fields. This is indeed the essence of the renormalization scheme.

References

[1] [Textbooks]:
There are many field theory textbooks available for graduate students. Among them, the following two books may be suitable for the first year of graduate students,

(i) J.D. Bjorken and S.D. Drell, "Relativistic Quantum Mechanics", (McGraw-Hill Book Company,1964).

(ii) F. Mandl and G. Shaw, "Quantum field theory", (John Wiley & Sons, 1993).

When one wishes to learn physics in a concrete fashion,

(iii) F. Gross, "Relativistic quantum mechanics and field theory", (John Wiley & Sons, 1993)
may be suitable for graduate students.

If one wishes to understand realistic calculations in some old type of field theory,

(iv) K. Nishijima, " Fields and Particles ", (W.A. Benjamin,INC)
may be the best for all.

[2] [References]: Below are some lists of references which may be helpful for further readings in the context of symmetry breaking physics.

[3] G.S. Adkins, *Phys. Rev.* **D39**, 3798 (1989)

[4] M. Amoretti, C. Amsler, G. Bonomi, A. Bouchta, P. Bowe, C. Carraro, C. L. Cesar, M. Charlton, M. Collier, M. Doser, V. Filippini, K. Fine, A. Fontana, M. Fujiwara, R. Funakoshi, P. Genova, J. Hangst, R. Hayano, M. Holzscheiter, L. Jorgensen, V. Lagomarsino, R. Landua, D. Lindelof, E. Rizzini, M. Macri, N. Madsen, G. Manuzio, M. Marchesotti, P. Montagna, H. Pruys, C. Regenfus, P. Riedler, J. Rochet, A. Rotondi, G. Rouleau, G. Testera, A. Variola, T. Watson, and D. van der Werf, *Nature* **419**, 456 (2002)

[5] N. Andrei and J. H. Lowenstein, *Phys. Rev. Lett.* **43**, 1698 (1979)

[6] T. Asaga and T. Fujita, "No area law in QCD", hep-th/0511172

[7] A. Badertscher, S. Dhawan, P.O. Egan, V. W. Hughes, D.C. Lu, M.W. Ritter, K.A. Woodle, M. Gladisch, H. Orth, G. zu Putlitz, M. Eckhause and J. Kane, *Phys. Rev. Lett.* **52**, 914 (1984)

[8] T. B. Bahder, *Phys. Rev.* **D68**, 063005 (2003)

[9] I. Bars and M.B. Green, *Phys. Rev.* **D17**, 537 (1978)

[10] H. Bergknoff, *Nucl. Phys.* **B122**, 215 (1977)

[11] H.A. Bethe, *Zeits f. Physik* **71**, 205 (1931)

[12] H.A. Bethe, *Phys. Rev.* **72**, 339 (1947)

[13] H. Bergknoff and H.B. Thacker, *Phys. Rev. Lett.* **42**, 135 (1979)

[14] J.D. Bjorken and S.D. Drell, "Relativistic Quantum Mechanics", (McGraw-Hill Book Company,1964)

[15] N.N. Bogoliubov, *J. Phys. (USSR)* **11**, 23 (1947)

[16] G.T. Bodwin and D.R. Yennie, *Phys. Rev.* **D37**, 498 (1988)

[17] A. Bohr and V.F. Weisskopf, *Phys. Rev.* **77**, 94 (1950)

[18] S.J. Brodsky, H.C. Pauli, and S.S. Pinsky, *Phys. Rep.* **301**, 299 (1998)

[19] M. Burkardt, *Phys. Rev.* **D53**, 933 (1996)

[20] M. Burkardt, F. Lenz, and M. Thies, *Phys. Rev.* **D65**, 125002 (2002)

[21] A. Chodos, R.L. Jaffe, K. Johnson, C.B. Thorn and V.F. Weisskopf, *Phys. Rev.* **D9**, 3471 (1974)

[22] S. Coleman, *Comm. Math. Phys.* **31**, 259 (1973)

[23] S. Coleman, *Phys. Rev.* **D11**, 2088 (1975)

[24] R. F. Dashen, B. Hasslacher and A. Neveu, *Phys. Rev.* **D11**, 3432 (1975)

[25] P.A.M. Dirac, *Proc. Roy. Soc.* **A117**, 610 (1928)

[26] P.A.M. Dirac, *Proc. Roy. Soc.* **A118**, 351 (1928)

[27] P.A.M. Dirac, *Rev. Mod. Phys.* **21**, 392 (1949)

[28] A. Einstein, *Annalen Phys.* **49**, 769 (1916)

[29] T. Eller, H.C. Pauli and S. Brodsky, *Phys. Rev.* **D35**, 1493 (1987)

[30] L. Essen, R.W. Donaldson, M.J. Bangham and E.G. Hope, *Nature* **229**, 110 (1971)

[31] M. Faber and A.N. Ivanov, *Eur. Phys. J.* **C20**, 723 (2001)

[32] L.D. Fadeyev and Y.N. Popov, *Phys. Lett.* **25B**, 29 (1967)

[33] T. Fujita, "Critical Review of Path Integral Formulation", arXiv:0801.1933

[34] T. Fujita, "Physical observables in path integral formulation", in *New Fundamentals in Fields and Particles*, Transworld Research Network (2009), p.31−p.45

[35] T. Fujita, "Physical observables in gauge field theory" in *New Fundamentals in Fields and Particles*, Transworld Research Network (2009), p.1−p.20

[36] T. Fujita, "Quantum Gravity without General Relativity", arXiv:0804.2518

[37] T. Fujita and A. Arima, *Nucl. Phys.* **A254**, 513 (1975)

[38] T. Fujita and M. Hiramoto, *Phys. Rev.* **D58**, 125019 (1998)

[39] T. Fujita, M. Hiramoto and T. Homma, "New spectrum and condensate in two dimensional QCD", hep-th/0306085.

[40] T. Fujita, M. Hiramoto, T. Homma and H. Takahashi, *J. Phys. Soc. Japan* **74**, 1143 (2005)

[41] T. Fujita, M. Hiramoto and H. Takahashi, "No Goldstone boson in NJL and Thirring models", hep-th/0306110.

[42] T. Fujita, M. Hiramoto and H. Takahashi, "Historical Mistake in Spontaneous Symmetry Breaking", hep-th/0410220.

[43] T. Fujita, M. Hiramoto and H. Takahashi, "Boson after symmetry breaking in quantum field theory", in *Focus on Boson Research* (Nova Science Publisher, 2005)

[44] T. Fujita, T. Kake and H. Takahashi, *Ann. Phys.* **282**, 100 (2000)

[45] T. Fujita, N. Kanda, H. Kato, H. Kubo, Y. Munakata, S. Oshima and K. Tsuda, "New Renormalization Scheme of Vacuum Polarization in QED", arXiv:0901.3421

[46] T. Fujita and N. Kanda, "Novel Solution of Mercury Perihelion Shift", physics.gen-ph/0911.2086

[47] T. Fujita and N. Kanda, "Physics of Leap Second", physics.gen-ph/0911.2087

[48] T. Fujita, S. Kanemaki, A. Kusaka and S. Oshima, "Mystery of Real Scalar Klein-Gordon Field", physics/0610268

[49] T. Fujita, S. Kanemaki and S. Oshima, "Gauge Non-invariance of Quark-quark Interactions", hep-ph/0511326

[50] T. Fujita, S. Kanemaki and S. Oshima, " New Concept of First Quantization", hep-th/0601102.

[51] T. Fujita, T. Kobayashi, and H. Takahashi, *Phys. Rev.* **D68**, 068701 (2003)

[52] T. Fujita, T. Kobayashi, M. Hiramoto, H. Takahashi, *Eur. Phys. J.* **C39**, 511 (2005)

[53] T. Fujita, A. Kusaka, K. Tsuda and S. Oshima, "Unphysical Gauge Fixing in Higgs Mechanism", arXiv:0806.2957

[54] T. Fujita and A. Ogura, *Prog. Theor. Phys.* **89**, 23 (1993)

[55] T. Fujita, Y. Sekiguchi and K. Yamamoto, *Ann. Phys.* **255**, 204 (1997)

[56] R.P. Feynman, *Rev. Mod. Phys.* **20**, 367 (1948)

[57] R.P. Feynman, *Phys. Rev.* **76**, 749 (1949)

[58] R.P. Feynman, *Phys. Rev.* **80**, 440 (1950)

[59] R.P. Feynman, *Phys. Rev.* **76**, 769 (1949)

[60] J. Goldstone, *Nuovo Cimento* **19**, 154 (1961)

[61] J. Goldstone, A. Salam and S. Weinberg, *Phys. Rev.* **127**, 965 (1962)

[62] W. Gordon, *Z. Physik* **40**, 117 (1926)

[63] F. Gross, "Relativistic quantum mechanics and field theory", (John Wiley & Sons, 1993)

[64] P. G. Harris, C. A. Baker, K. Green, P. Iaydjiev, S. Ivanov, D. J. R. May, J. M. Pendlebury, D. Shiers, K. F. Smith, M. van der Grinten and P. Geltenbort, *Phys. Rev. Lett.* **82**, 904 (1999)

[65] W. Heisenberg, *Zeits. f. Physik* **90**, 209 (1934)

[66] W. Heisenberg and H. Euler, *Zeits. f. Physik* **98**, 714 (1936)

[67] P. W. Higgs, *Phys. Lett.* **12**,132 (1964)

[68] M. Hiramoto and T. Fujita, *Phys. Rev.* **D66**, 045007 (2002)

[69] M. Hiramoto and T. Fujita, "No massless boson in chiral symmetry breaking in NJL and Thirring models", hep-th/0306083.

[70] G. 't Hooft, *Nucl. Phys.* **B75**, 461 (1974)

[71] G. 't Hooft, in 'New Phenomena in Subnuclear Physics', *Proceedings of the 14th Course of the International School of Subnuclear Physics*, Erice, 1975, edited by A. Zichichi (Plenum, New Pork, 1977)

[72] G. 't Hooft and M. Veltman, *Nucl. Phys.* **B44**,189 (1972)

[73] G. 't Hooft and M. Veltman, *Nucl. Phys.* **B50**,318 (1972)

[74] K. Hornbostel, S.J. Brodsky, and H.C. Pauli, *Phys. Rev.* **D41**, 3814 (1990)

[75] Y. Hosotani and R. Rodriguez, *J. Phys.* **A31**, 9925 (1998)

[76] S. Huang, J.W. Negele and J. Polonyi, *Nucl. Phys.* **B307**, 669 (1988)

[77] V.W. Hughes and T. Kinoshita, *Rev. Mod. Phys.* **71**, S133 (1999)

[78] P. Jordan and E. Wigner, *Z. Phys.* **47**, 631 (1928)

[79] K.P. Jungmann, *Lect. Notes Phys.* **570**, 81 (2001)

[80] S. Kanemaki, A. Kusaka, S. Oshima and T. Fujita, "Problems of scalar bosons", in *New Fundamentals in Fields and Particles*, ed. by T. Fujita, Transworld Research Network (2009), p.47−p.60

[81] B. Klaiber, in *Lectures in Theoretical Physics,* 1967, edited by A. Barut and W. Britten (Gordon and Breach, NY, 1968)

[82] O. Klein, *Zeits. f. Physik* **41**, 407 (1927)

[83] H. Kleinert and B. Van den Bossche, *Phys. Lett.* **B 474**, 336 (2000)

[84] S. Klevansky, *Rev. Mod. Phys.* **64**, 649 (1992)

[85] H. Kubo, T. Fujita, N. Kanda, H. Kato, Y. Munakata, S. Oshima and K. Tsuda, "A New Relation between Lamb Shift Energies", quant-ph/1003.5050

[86] F. Lenz, M. Thies, S. Levit and K. Yazaki, *Ann. Phys.* **208**, 1 (1991)

[87] B.A. Lippmann and J. Schwinger, *Phys. Rev.* **79**, 469 (1950)

[88] A. Luther, *Phys. Rev.* **B14**, 2153 (1976)

[89] F. Mandl and G. Shaw, "Quantum Field Theory", (John Wiley & Sons, 1993)

[90] N. S. Manton, *Ann. Phys.* **159**, 220 (1985)

[91] N.D. Mermin and H. Wagner, *Phys. Rev. Lett.* **17**, 1133 (1966)

[92] C.W. Misner, K.S. Thorne and J.A. Wheeler, "Gravitation" (Freeman, 1973)

[93] Y. Nambu and G. Jona-Lasinio, *Phys. Rev.* **122**, 345 (1961)

[94] K. Nishijima, "Fields and Particles", (W.A. Benjamin, INC, 1969)

[95] K. Odaka and S. Tokitake, *J. Phys. Soc. Japan* **56**, 3062 (1987)

[96] A. Ogura, T. Tomachi and T. Fujita, *Ann. Phys.* **237**, 12 (1995)

[97] C.J. Oram, J.M. Bailey, P.W. Schmor, C.A. Ery, R.F. Kiefl, J.B. Warren, G.M. Marshall and A. Olin, *Phys. Rev. Lett.* **52**, 910 (1984)

[98] R. Orbach, *Phys. Rev.* **112**, 309 (1958)

[99] S. Oshima, S. Kanemaki and T. Fujita, "Problem of Real Scalar Klein-Gordon Field", hep-th/0512156.

[100] B.W. Parkinson and J.J. Spilker, eds., *Global Positioning System, Progress in Astronautics and Aeronautics,* **163, 164** (1996)

[101] W. Pauli and F. Villars, *Rev. Mod. Phys.* **21**, 434 (1949)

[102] E. Schrödinger, *Ann. Physik* **81**, 109 (1926)

[103] J. Schwinger, *Phys. Rev.* **74**, 1439 (1948)

[104] J. Schwinger, *Phys. Rev.* **128**, 2425 (1962)

[105] H.B. Thacker, *Rev. Mod. Phys.* **53**, 253 (1981)

[106] T. Tomachi and T. Fujita, *Ann. Phys.* **223**, 197 (1993)

[107] S. Tomonaga, *Prog. Theor. Phys.* **1**, 27 (1946)

[108] S. Tomonaga, "Complete Works of Tomonaga Shinichiro", Vol. 10, (Misuzu Shobo, 2001, in Japanese)

[109] W. Thirring, *Ann. Phys. (N.Y)* **3**, 91 (1958)

[110] E.A. Uehling, *Phys. Rev.* **48**, 55 (1935)

[111] R.S. Van Dyck, P.B Schwinberg and H.G. Dehmelt, *Phys. Rev. Lett.* **59**, 26 (1987)

[112] K.G. Wilson, *Phys. Rev.* **10**, 2445 (1974)

[113] K.A. Woodle, A. Badertscher, V. W. Hughes, D.C. Lu, M.W. Ritter, M. Gladisch, H. Orth, G. zu Putlitz, M. Eckhause and J. Kane, *Phys. Rev.* **A41**, 93 (1990)

[114] A.C. Zemach, *Phys. Rev.* **104**, 1771 (1956)

[115] A.B. Zhitnitsky, *Phys. Lett.* **B165**, 405 (1985)

Index